Green Communication Technologies for Future Networks

This book explores all the energy-efficient communication technologies used for various communication systems and every aspect of these systems, such as green electronics, network protocols, handover, codes, antenna, and the role of artificial intelligence and IoT, including the energy management strategies. It identifies the development of sustainable plans and programs at the communication level within the current legislative framework.

Features:

- Gives a fundamental description of the green communications including granularities of green wired and wireless systems.
- Describes a comprehensive review of innovations, challenges, and opportunities for green communication.
- Provides guiding principles on how to build the green communication network.
- Includes a holistic view of both wireless and wired green communication systems with an emphasis on applications and challenges in each area.
- Suggests various ways of benchmarking and measuring the performance of green communication systems.

This book will be of great interest to graduate students and researchers in green technologies, communications, wireless communication, optical communication, underwater communication, microwave and satellite communication, networking, the internet of things, and energy management.

Green Communication Technologies for Future Networks

An Energy-Efficient Perspective

Edited by

Gurjit Kaur and Akanksha Srivastava

CRC Press
Taylor & Francis Group
Boca Raton London New York

CRC Press is an imprint of the
Taylor & Francis Group, an **informa** business

First edition published 2023
by CRC Press
6000 Broken Sound Parkway NW, Suite 300, Boca Raton, FL 33487-2742

and by CRC Press
4 Park Square, Milton Park, Abingdon, Oxon, OX14 4RN

CRC Press is an imprint of Taylor & Francis Group, LLC

ISBN: 9781032206301 (hbk)
ISBN: 9781032206318 (pbk)
ISBN: 9781003264477 (ebk)

DOI: 10.1201/9781003264477

Typeset in Times
by Newgen Publishing UK

Contents

Editor Bio

Dr. Gurjit Kaur is Professor in the Department of Electronics & Communication Engineering at the Delhi Technological University (DTU), Delhi, India. She has been a topper throughout her academic education. As a testimonial to the same, she has been awarded by Chief Minister S. Prakash Singh Badal for being the topper in Punjab state. After that, she was awarded a 'Gold Medal' by former President of India Dr. A P J Abdul Kalam for being the overall topper of the Punjab Technical University, Jalandhar in B. Tech program. She also received an Honour by Guru Harkrishan Education Society for being the topper among all the colleges and all the disciplines of PTU, Jalandhar. She then proceeded to PEC University of Technology, Chandigarh to complete her M. Tech in 2003 and also earned her Ph.D. degree from Punjab University, Chandigarh in 2010 with distinction.

She has spent over 19 years of her academic career towards research and teaching in the field of electronics and communication in well-reputed institutes like PEC University of Technology, Punjab University, Jaypee Institute of Information and Technology, Gautam Buddha University, and Delhi Technological University, Delhi. During this academic tenure, she has guided 4 Ph.D. students and more than 55 M.Tech students. Her research interests mainly include optical communication systems, optical networks, optical CDMA, wireless communication system, high-speed interconnect, and IOT. She has also authored five books at the international and national level. Her book, "Green and Smart Technologies for Smart Cities" was published by CRC Press, Taylor & Francis Group in December 2019 and another book entitled, "Integrating Artificial Intelligence to Solve Pervasive IoT Issues" was published by Elsevier, Cambridge, MA, USA. Her three books, i.e., "Handbook of Research on Big Data and the IoT", "Examining Cloud Computing Technologies through the Internet of Things", and "Artificial Intelligence and IoT-Based Technologies for Sustainable Farming and Smart Agriculture" were published by IGI-Global, International Publisher of Progressive Information Science and Technology Research. She also authored a national level book entitled "Optical Communication," which was published by Galgotia Publications in 2005. She has presented her research work as short courses/tutorials in many national and international conferences.

In recognition of her outstanding contribution as an active researcher, Dr. Kaur also received the Bharat Vikas Award by the Institute of Self Reliance in National Seminar on Diversity of Cultural and Social Environment at Bhubaneswar, Odisha, India & Prof. Indira Parikh 50 Women in Education Leader Award in World Education Congress, Mumbai, India in 2017. Her career accomplishments over the past sixteen years include over 90 peer-reviewed publications, including 55 journal publications, 35 conference publications, and over ten invited talks/seminars. Due to her pioneer research, her name is also listed in Who's Who in Science and Engineering Directory, USA and her biography was also published by International Biographic Center, Great Britain as 2000 Outstanding Scientists.

She also worked as a convener for two international conferences, i.e., International ICIAICT 2012, which was organized by CSI, Noida Chapter and International conference EPPICTM 2012, which was held in collaboration with MTMI, USA, University of Maryland Eastern Shore, USA and Frostburg State University, USA at School of Information and Communication Technology, Gautam Buddha University, Greater Noida, India. She organized various FDPs and workshops. To date, she has delivered many expert talks on the emerging optical trends for all domain applications at various universities and served as a reviewer of multiple journals like IEEE Transactions on Communication etc.

Ms. Akanksha Srivastava is currently working as a Principal Investigator under Women Scientists Scheme-A (WOS-A) of the Department of Science and Technology (DST) project in the Electronics and Communication Engineering Department of Delhi Technological University Delhi, India. Before joining Delhi Technological University, she worked as a Women Scientist at the National Institute of Technology, Hamirpur, H.P., India. She is pursuing her Ph.D. degree in the field of green communication systems from Delhi Technological University. She has received her B. Tech. degree in 2010 and M. Tech. degree in 2014 in the field of Electronics and Communication Engineering from Dr. A.P.J. Abdul Kalam Technical University Lucknow, India.

Apart from this, she has also served as an Assistant Professor in the Department of Electronics and Communication Engineering at PSIT-College of Engineering, Kanpur, U.P., India.

Her current research includes green communication, energy efficient wireless networks, cognitive radio networks, next-generation networks, and heterogeneous networking. She has published several papers in various reputed International Journals and Conferences. She has contributed numerous book chapters with many publishers like Taylor & Francis Group, Elsevier, Springer, etc. She served as a reviewer for numerous journals and national/international conferences. She worked as an organizing member for various conferences and workshops.

She has been awarded the Commendable Research Award from Delhi Technological University for her research work. She was selected as a Student Ambassador for the IEEEXtreme 14.0 coding competition from the IEEE Delhi section. She has been awarded for "India Innovation Challenge Design Contest-2018" by Department of Science and Technology (DST) and Texas Instruments.

She is a member of the Institute of Electrical and Electronics Engineers (IEEE), IEEE Communications Society (ComSoc), IEEE Women in Engineering (WIE), IEEE Young Professionals, IEEE Microwave Theory and Techniques Society, IEEE Internet of Things Community, IEEE Signal Processing Society and the International Association of Engineers (IAENG).

Contributors

Bahuguna Yash
Department of Electronics and Communication Engineering, Delhi Technological University, Delhi, India

Chauhan Dushyant Singh
Department of Electronics and Communication Engineering, Delhi Technological University, Delhi, India

Gupta Mani Shekhar
Indian Institute of Technology, Delhi, India

Jindal Shilpa
Department of Electronics and Communication Engineering, CCET, Chandigarh, India

Kaur Gurjit
Department of Electronics and Communication Engineering, Delhi Technological University, Delhi, India

Kumar Ashok
Department of Electronics and Communication Engineering, National Institute of Technology, Hamirpur, India

Kumar Dinesh
Department of Electronics and Communication Engineering, Delhi Technological University, Delhi, India

Kumar Manjeet
Department of Electronics and Communication Engineering, Delhi Technological University, Delhi, India

Masih Asbah
Department of Electronics and Communication Engineering, Delhi Technological University, Delhi, India

Mittal Sakshi
Department of Electronics and Communication Engineering, Delhi Technological University, Delhi, India

Parashar Yaman
Aston University, United Kingdom

Priya Puja
Department of Electronics and Communication Engineering, Delhi Technological University, Delhi, India

Rani Rekha
Department of Electronics and Communication Engineering, Delhi Technological University, Delhi, India

Singh Arshpreet
Shiv Nadar School, Noida, India

Sisodiya Divya
Department of Electronics and Communication Engineering, Delhi Technological University, Delhi, India

Srivastava Akanksha
Department of Electronics and Communication Engineering, Delhi Technological University, Delhi, India

Preface

In the present scenario, telecommunication has become the most crucial part of the communication system. It is upgrading day by day, and with the growth of new technologies, numerous environmental issues have arisen because they consume a large amount of power. During the last few years, CO_2 emission has increased radically. CO_2 emission leads to the greenhouse effect, which further leads to global warming. Green technology has drawn a huge amount of attention with the development of the modern world. Similarly, with the development in communication technology the industries and researchers are focusing to make this communication as green as possible. Therefore, energy-efficient green communications technologies, architecture, and networking have attracted significant attention from industry and academia. Energy-efficient technologies need to be designed, which consume less power and produce a better quality of service (QoS).

The future communication system will be highly complicated and composite networks due to the combination of several wireless and wired networks, where each network has its different protocols and properties. This combined network has multiple challenges for network scheduling, operation, troubleshooting, and managing, but the most critical problem will be energy management. In the ongoing scenario, the communication system is shifting from a user-centric approach to a device-oriented system. Therefore, for designing the energy-efficient communication system, several technologies, such as renewable energy, cognitive radio, heterogeneous architecture, Internet of Things, artificial intelligence, MIMO, device to device, base station on/off, carrier aggregation, millimeter wavelength, beamforming etc., are playing an important role.

To make an energy-efficient communication system, each and every component of the system should be concentrated on lower energy consumption. This term "green communication" is composed of two terms, i.e., green and communication. Here green means energy efficient and the communication refers to exchanging information for practical purposes. In contrast, the term "green communications" are referred to as the usage of energy-efficient techniques for communications and networking purposes. This newly invented green communication technologies not only can save energy and reduce the operational expenditure but also can achieve sustainable advancement due to less dependence on fossil fuel and the exploitation of renewable energy resources. The optimal usage of energy resources is the main requirement of the telecommunication industry. To find the optimal use of these resources, researchers need to explore optimization techniques. This book will be a source of motivation for the researcher/authors and an opportunity to exchange knowledge of energy efficiency and the optimization tools for developing a green communication system. This book will provide a comprehensive guide on the fundamental concepts, applications, algorithms, protocols, new trends, challenges solutions at each component level of the communication system.

The different scientists, researchers, and practitioners wrote chapters on the realistic solutions for the current energy crisis, i.e., how to reduce carbon emissions and greenhouse gas emissions. After an overview of the role of ICT in climate change and environmental pollution worldwide, this book will provide an up-to-date definition of the green communication system and its main expected features, focusing on technology innovation, smart energy managing techniques, energy efficient technologies, and their implementation. This book will explore the energy management strategies to minimize the need of energy in various communication systems. A new range of renewable energy technologies modified for installation will be described in this book.

The objective of this book is to explore green communication technologies, green electronics, and green computing technologies for designing future networks. This book will learn how IoT solutions and energy-efficient technologies will disrupt business models and bring new approaches to green communication systems. The fourteen chapters of this book explore the different communication technologies in an energy-efficient way.

Chapter 1, 'Green Communication Systems for Future Networks' provides the essential background and the need for green communication technologies. Various techniques for a green communication system are discussed. This chapter shows how the evolution in the field of communication systems has led academicians and scientists to do research to design communication systems as energy efficient as possible. The environmental and energy efficiency concerns that are further increasing the demand for green communication have been mentioned in this chapter along with the various techniques for a green communication system.

Chapter 2, 'Green Wireless Communication Systems', defines how green network energy efficiency (EE) strategies can be applied with an existing wireless network to fulfil requirements such as enhanced data rate, less energy consumption, interference-free transmission, low radiation effect, extended capacity and promised quality of services (QoS). In this chapter, various techniques for a green wireless communication system are described. This chapter discusses the various energy-efficient techniques, i.e., ultra-dense network, device-to-device communication, beamforming, massive multiple-input multiple-output, hardware improvement, etc., and their pros and cons.

Chapter 3, 'Green Transmission Technologies and Network Protocols', gives the state of the art for current green communication networking technologies addressing low-power design and EE within all layers of the communication system protocol stack. First, an artificial intelligence (AI)-based multi-tier green network architecture for next-generation networks (5G and beyond) has been proposed. The EE metrics used in various green technologies such as cognitive radio networks, ultra-dense networks, femtocells, massive MIMO etc. for performance evaluation is summarized and major challenges dedicated to this topic have been discussed.

Chapter 4, 'Green Ad-Hoc Communication System Based on NOMA Technique', describes non-orthogonal multiple access (NOMA) with an energy-harvesting (EH) relaying system for futuristic green communication-based ad-hoc networks. Relays can harvest energy from the received radio frequency (RF) signal and choose the best relay for further communication through the decode-and-forward (DF) technique to users. In this chapter, the authors have proposed a system, which operates in two phases. During the 1st phase, all relays harvest energy from the received NOMA RF signal aired by the base station (BS). The 2nd phase splits into two equal intervals as one used for signal reception at all EH relays from BS and others for further aired signal to users through the best EH relay. The proposed system performance can be determined through closed forms of both users' outage probabilities. Simulations out beats the existing orthogonal multiple access (OMA)-based relaying green ad-hoc networks. It also shows the duration of EH optimally impacts over proposed system performance. Results are validated through Monte-Carlo operations.

Chapter 5, 'Green Optical Communication Systems', gives a green optical communication network in detail like its features, different technologies to design it, characteristics of its elements, various optical switching techniques, energy-aware optical network, applications and challenges of a green optical communication network. This chapter explores the optical communication system on the land, undersea optical communication system, deep space optical communication system, and free space optical communication system. This chapter also explores the energy-efficient technologies to design all these networks. The last chapter also discusses the research opportunities in an optical communication network.

Chapter 6, 'Green Hybrid Wireless-Optical Broadband Networks', introduces the concept of the hybridizing of both wireless and wired techniques. Various energy-efficient aspects are enlightened by using energy-efficient topologies, optimal placement of ONU, energy-efficient protocols to minimize the energy wastage in the network. Furthermore, it also provides the future directions to increase the further energy efficiency of hybrid wireless optical broadband along with its applications.

Chapter 7, 'Green Underwater Communication Systems', concentrates on the energy-efficient underwater optical communication system. In this chapter, the authors have discussed the attributes of underwater systems, features of their components, various technologies to develop them, different underwater communication methods, energy-efficient underwater systems, challenges, and applications of underwater communication systems. At the end authors also discussed some research opportunities in this communication network.

Chapter 8, 'Green Optical Networks Based on Machine Learning Techniques', provides various types of information about optical networks based on machine learning techniques. In this chapter the author discussed about resource efficiency, capacity of network, and routing and resource allocation. In this chapter, mature mathematical tools of machine learning algorithms are explored for optical communication systems and networks.

Chapter 9, 'Green Electronics for Future Communication Systems', defines the role of green manufacturing, the responsibilities of an individual and manufacturer. This chapter explores the various electronics component and the techniques through which these components can become energy efficient. Finally, the challenges and opportunities in this area are highlighted.

Chapter 10, 'Green Internet of Things (IoT) Technologies for Communication Systems', focuses on the present work and possible technologies for future green IoT. The Internet of Things (IoT) has shown tremendous growth in present-day technology. There are various benefits of using IoT technology still they suffer from many drawbacks like IoT network consumes energy, produces harmful, pollution, and generates E-waste. This chapter explores the energy-efficient way for IoT.

Chapter 11, 'Green Cloud Computing Based Future Communication Systems', gives several features, algorithms, and challenges of green cloud computing. As the need for cloud computing is enhancing with days, demand for more power consumption has also increased. And this is one of the major challenges in future cloud computing infrastructure. High power consumption enhances more operation costs and this leads to an enhancement in carbon footprint. So, it becomes very essential to lower the power usage in cloud computing for environmentally friendly solutions. Hence, power coherent remedies are essential to lower the effects of cloud computing on the environment. To solve this problem a new concept called 'green cloud computing' has been coming into the picture and this chapter describes it.

Chapter 12, 'Green Artificial-Intelligence-Based Communication Systems', provides the essential background for green AI-based communication systems. Green AI-based communication systems are being implanted to solve the crisis of big data. Green AI-based communication systems are used for efficient energy utilization thus reducing the carbon footprint. These systems provide a user-friendly environment and reduce operational costs. Energy harvesting technology for green AI-based communication systems is discussed in this chapter. Later the chapter covers features and techniques of green AI-based communication systems. Architecture, drawbacks, and applications of green AI-based communication systems are also discussed.

Chapter 13, 'Green Microwave and Satellite Communication Systems', provides various types of information about microwaves and satellites in a communication system. Microwaves and satellites play a crucial role for long distance transmission. In this chapter the authors discussed the microwave and satellite communication along with the energy-efficient techniques. This chapter also explains various energy-efficient devices, which can be used for the design of microwave and satellite communication system. Further, the applications and research opportunities of microwave and satellite communications are discussed.

Chapter 14, 'Green Energy-Efficient Wired and Wireless Charging Techniques for IoT Enabled Healthcare Systems', provides various types of information about wired and wireless charging techniques for IoT applications. In this chapter the authors have discussed that wireless power transfer charging can provide a promising way to harness the modern IoT based modern biomedical applications with adequate required energy and lesser sophistication and complexity. This chapter also explains a brief introduction to the technology, its types, applications, system modeling, challenges, safety standards, and prospective future work in a concise and informative manner.

Overall, this book describes how the technology provider can meet the tech-savvy customer's needs and prepare for the changing face of society by discussing innovations of green technologies for future network applications in a detailed way. It will act as a broad reference book for the identification of development sustainable plans and programs at the communication system within the current legislative framework.

Dr Gurjit Kaur
Ms Akanksha Srivastava

Abbreviations

ACO	Ant Colony Optimization Algorithm
ADC	Analog to Digital Converter
AI	Artificial Intelligence
ANN	Artificial Neural Networks
AOUVs	Autonomous Operated Underwater Vehicles
AP	Access Point
BS	Base Station
CADARA	Capacity and Delay Aware Routing Algorithm
CDOM	Colored Dissolved Organic Materials
CIPA	Channel Inversion Power Allocation
CNF	Cellulose Nanofibers
CO_2	Carbon Dioxide
COOFDM	Coherent Optical Orthogonal Frequency Division Multiplexing
CPA	Constant Power Allocation Technique
CPE	Compute Power Competence
CRNs	Cognitive Radio Networks
CU	Cognitive User
CUE	Carbon Usage Effectiveness
D2D	Device to Device
DARA	Delay Aware Routing Algorithm
DCeP	Datacenter Energy Productivity
DCiE	Datacenter Infrastructure Competence
DCP	Data Center Productivity
DF	Decode-and-Forward
DNN	Deep Neural Network
DSP	Digital Signal Processing
DVFS	Dynamic Voltage and Frequency Scaling
EE	Energy Efficiency
EH	Energy-Harvesting
EM	Electro-magnetic
ERE	Energy Reuse Effectiveness
ERF	Energy Reuse Factor
FCC	Federal Communications Commission
FIWI	Fiber Wireless
FSO	Free Space Optics
GA	Greedy Approach
GEC	Green Energy Coefficient
GPS	Global Position System
GROWNet	Grid Based Reconfigurable Optical Wireless Network
H2H	Human to Human
HetNets	Heterogeneous Networks

HVAC	Heating, Ventilation, and Air Conditioning
ICTs	Information and Communication Technologies
IOP	Intrinsic Optical Property
IoT	Internet of Things
ISI	Inter Symbol Interference
LOS	Line of Sight
LSA	Link State Advertisement
M2M	Machine to Machine
MANET	Mobile Ad Hoc Network
MARIN	Metro Access Ring Integrated Network
MHRA	Minimum Hop Routing Algorithm
MILP	Mixed Integer Linear Programming
MIMO	Multiple Input Multiple Output
MIP	Mixed Integer Programming Approach
MISO	Multiple Input Single Output
ML	Machine Learning
MMIMO	Massive MIMO
MMwave	Millimetre wave
MN	Mobile Nodes
NL	Network Layer
NOMA	Non-Orthogonal Multiple Access
OC	Optical Communication
OLED	Organic LED
OLT	Optical Line Terminals
OMA	Orthogonal Multiple Access
ONU	Optical Network Unit
OOK	Modulation Techniques On Off Keying
PAN	Personal Area Network
PFPA	Proportional Fair Power Allocation
PL	Physical Layer
PON	Passive Optical Network
PRS	Partial Relay Selection
PTRA	Predictive Throughput Touting Algorithm
PU	Primary User
PUE	Power Usage Effectiveness
QAM	Quadrature Amplitude Multiplexing
QoS	Quality of Service
RADAR	Risk and Delay Aware Routing Algorithm
RF	Radio Frequency
RFID	Radio Frequency Integrated Circuit
RL	Reinforced Learning
ROUVs	Remotely Operated Underwater Vehicles
SAA	Simulated Annealing Approach
SIMO	Single Input Multiple Output
SISO	Single Input Single Output

SL	Supervised Learning
SOA	Service Oriented Architecture
SoC	System on Chip
SPRA	Shortest Path Routing Algorithm
SSL	Semi Supervised Learning
SVM	Support Vector Machine
SWaP	Space, Wattage, and Performance
SWIPT	Simultaneous Wireless Information and Power Transfer
TCP	Transmission Control Protocol
TDMA	Time Division Multiple Access
UDN	Ultra Dense Network
UOWC	Underwater Optical Wireless Communication
USL	Unsupervised Learning
VLC	Visible Light Communication
VM	Virtual Machine
VSF	Volume Scattering Function
WDM	Wavelength Division Multiplexing
WEEE	Waste Electrical and Electronic Equipment
WFPA	Water Filling Power Allocation
WLANs	Wide Local Area Networks
WOBAN Hybrid	Wireless Optical Broadband Access Network Hybrid
WOC	Wireless Optical Channel
WSN	Wireless Sensor Network
WUE	Water Usage Effectiveness

1 Green Communication Systems for Future Networks

Akanksha Srivastava and Gurjit Kaur

CONTENTS

1.1 INTRODUCTION OF A GREEN COMMUNICATION SYSTEM

Communication technology is a tool that transmits and receives information among devices and people with the help of technology. This transfer of information among people assists them in taking decisions, solving problems, and managing machines. A communication network is defined as a network that exchanges information between two nodes known as the source and destination. The medium through which the information moves from one node to another node is known as a channel. Figure 1.1 shows the block diagram of the communication system.

- **Information Source**: The information source originates from the message/information signal. Generally, the information can be in any form, voice, text, video, image, etc. The output of the information source is processed by the input transducer.
- **Input Transducer**: An input transducer is placed after the information source, which converts the message signal into the electrical form because a transducer

DOI: 10.1201/9781003264477-1

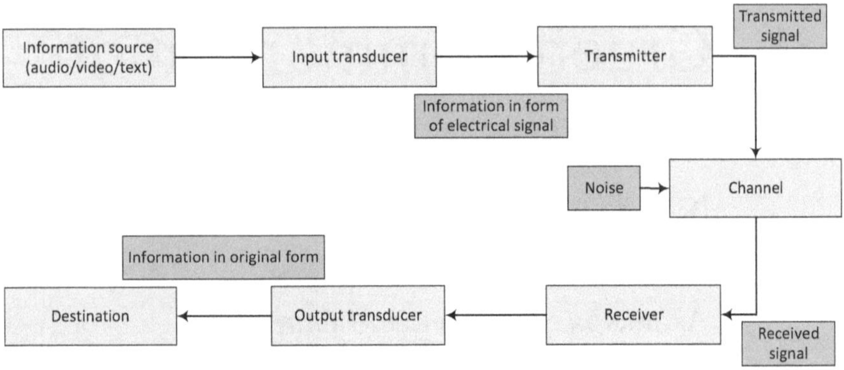

FIGURE 1.1 Communication system.

is a device that transforms one form of energy into another form. For example, in the case of cellular communication, a microphone transforms the message signal from sound waves to an electrical signal.

- **Transmitter:** A transmitter is a device that processes the message signal into an appropriate form to transmit over the communication channel. This process is known as modulation. For example, in the case of a radio communication system, the collected electrical signal is processed to the audio frequencies range and usually amplified. While in the case of wire communication, no processing is required. In the case of long-distance communication, before the modulation process signal amplification is essential. The key function performed by the transmitter is modulation. In the modulation process, the message signal is superimposed upon the high-frequency carrier signal.
- **The Channel:** The communication channel is a medium through which the message signal travels between the source and destination. It provides a pathway for exchanging information. The communication channels are classified into two types:
  1. **Guided channel:** In the guided media the source and destination are physically connected via a cable or a wire. These types of channels are also known as wire channels. Examples of guided media are optical fiber, coaxial cable, and twisted pair.
  2. **Unguided channel:** In the unguided media, there is no direct connection between the source and destination. These types of channels are also known as wireless channels. Examples of unguided media are satellite channels, microwave channels, radio waves' communication, infrared communication.

     The signal receives distortion at the channel due to the presence of noise. Noise is a random signal, and it interferes with the desired signal.
- **Receiver:** The receiver recovers the message from the transmitted signal and reproduces it in the form of an electrical signal. To extract the message signal from the carrier wave is known as demodulation or detection.
- **Output Transducer:** Output transducer converts the electrical signal into the original message signal and this signal is received at the destination side.

TABLE 1.1
Number of Mobile Phone Subscribers and Smartphone Users Worldwide

Year	Mobile Phones Subscribers (Billions)	Smartphone Users (Billions)
2020	6.95	6.05
2021	7.10	6.37
2022	7.26	6.64
2023	7.33	6.92
2024	7.41	7.13
2025	7.49	7.33

Source: (Bankmycell 2021).

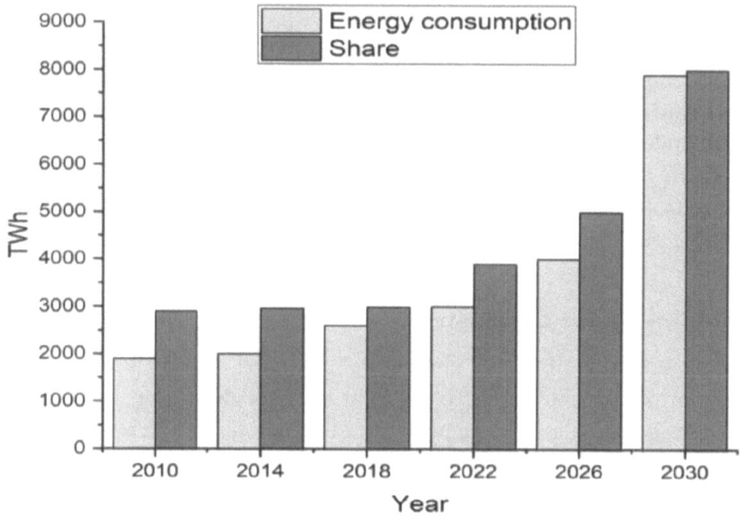

FIGURE 1.2 Energy consumption of ICT and its share.

These are the elements of a communication system in all the communication technologies. Now the question is how to make the communication system green? The design of a green communication system is based on utilizing energy-efficient products and technologies. From the last decade, the world experienced a prodigious extension in the number of mobile subscribers. Table 1.1 shows the number of mobile phone subscribers and smartphone users worldwide.

The increased number of mobile devices require a huge amount of power for communication purposes. It will result in a large amount of energy wastage and greenhouse gas emissions. Figure 1.2 reveals that information and communication technology (ICT) will consume a large amount of electricity in 2030, and it will hold an expected annual growth rate between 6% and 9%. Figure 1.3 shows the percentage of energy consumption of different parts of the ICT system (Srivastava et al. 2020).

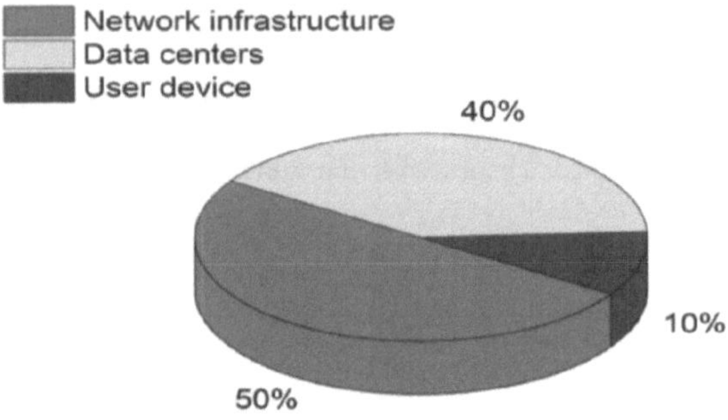

FIGURE 1.3 Energy consumption of different parts for ICT.

1.2 NEED OF GREEN COMMUNICATION TECHNOLOGIES

A lot of important aspects come under the discussion of the need for green communication technologies. An optimal energy-saving strategy is required for the development of the future green communication system. For this strategy, it is essential to understand the services of the current and future communication network, technology roadmap, and various additional data for analysis. Figure 1.4 shows various techniques used for designing of green communication system (Parasher et al. 2019).

1.2.1 Impacts on the Environment

Currently, the emission of CO_2 is getting high attention due to climate changes and the global warming problem. The equipment used for designing the telecommunication systems contains a large number of rare materials and abundant metals. The extraction of these metals through mining and processing of the waste material is the biggest challenge environment. There are various phases in the processing of any component and different amount of CO_2 releases at different phases. Figure 1.5 shows the recycling process of cell phone with percentage of CO_2 emission.

1.2.2 Effects of Electromagnetic Radiation

Currently, the communication system plays an important role in human life. Nowadays, tremendous efforts are performed by the industries for the development of the communication system. But they are not thinking about their unfavorable outcomes on animals and humans. The installation of communication towers (base stations, antennas, etc) near residential areas is always a matter of discussion. These towers radiate electromagnetic radiation (EMR), which creates long-term impacts on humans. The base-station (BS) towers are recognized as the key source of EMR, but mobile phone devices also cause many health risks because these are directly in association connected with the human body generally near the ear and head. Figure 1.6 shows the harmful effects of electromagnetic radiation (Gong et al. 2014).

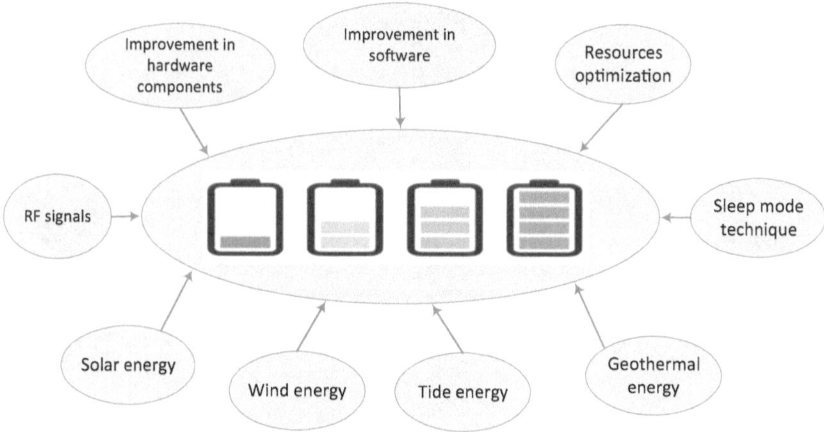

FIGURE 1.4 Various techniques used for the design of the green communication system.

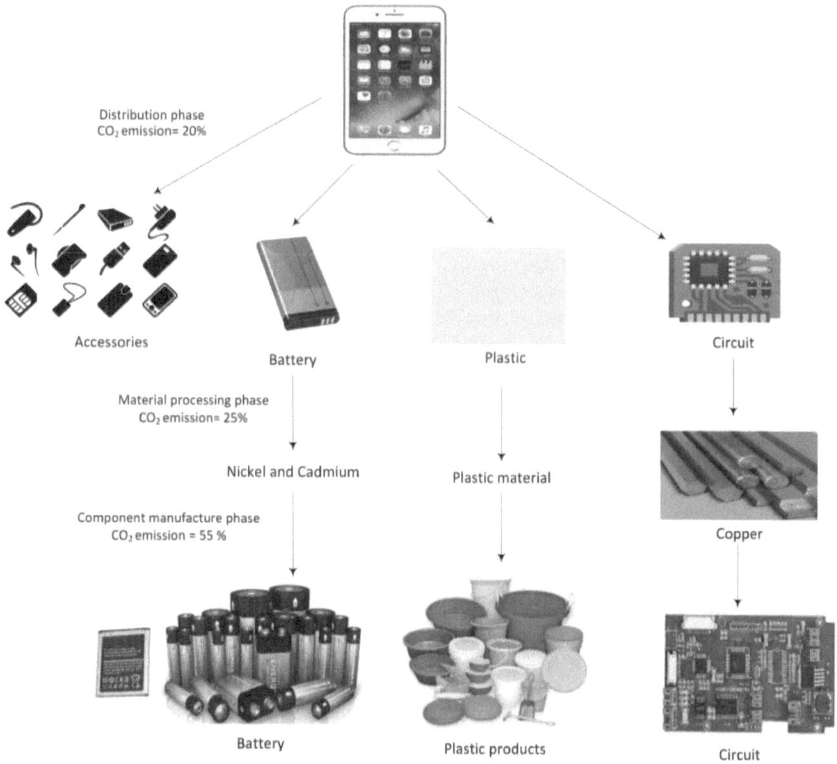

FIGURE 1.5 Recycling of cell phone and percentage of CO_2 emission in each stage.

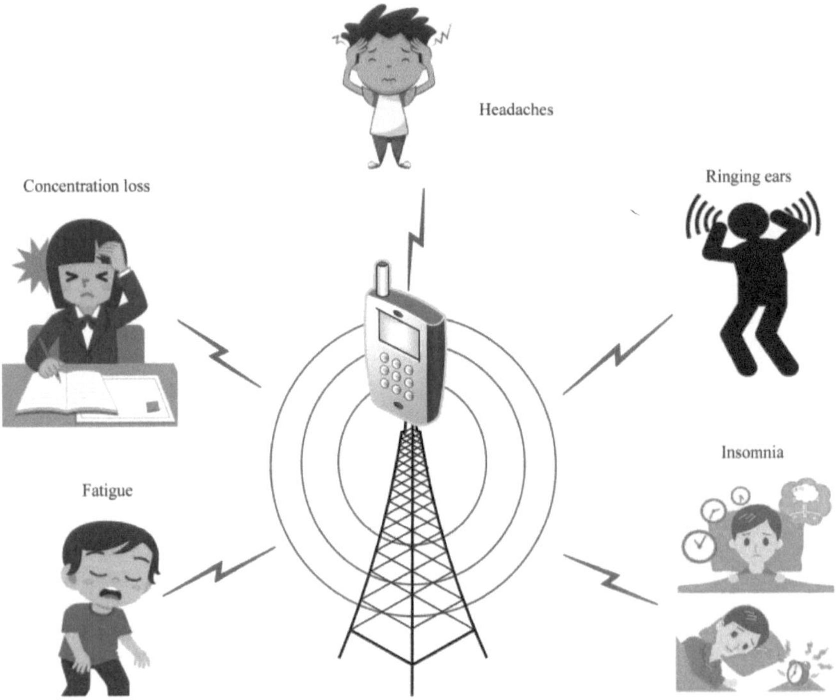

FIGURE 1.6 Harmful effects of electromagnetic radiation.

1.2.3 Carbon Footprint Analysis

The evaluation of carbon footprint is also considered as an estimation of greenhouse gas (GHG) emissions. Manufacturing or recycling of any product causes greenhouse gas emissions, and it results in global warming. It starts with an evaluation of carbon and methane emissions (Srivastava et al. 2019). There are three core standards around carbon footprint analysis the GHG Protocol, PAS 2050, and ISO/TS 14067 (LeBlanc 2019).

1.3 TYPES OF GREEN COMMUNICATION SYSTEMS

The progression of the communication system needs the evolution of green technologies. Some of the green communication systems are discussed below.

1.3.1 Green Wireless Communication System

The conventional design of wireless communication systems initially concentrates on omnipresent access and high capacity. Nowadays the demand for a wireless communication system is increasing due to the astronomical growth of real-time services, multimedia applications, and data requirements (Gupta et al. 2019). In the wireless communication system, multimedia applications consume different amounts of energy and this energy is calculated by some specific tools. There are various wireless technologies used for transferring data. These technologies are Wi-Fi, Wi-Max, Bluetooth, Zigbee, Infrared, etc. The data transmission based on Bluetooth technology does not demand large bandwidth and power (Wang et al. 2018). During wireless data transmission, privacy and security are the major problems of networks. In the literature, several energy-efficient approaches are introduced by researchers for designing the green wireless communication system. These approaches are improvement in hardware components, cell zooming, base station ON/OFF, integration of renewable energy resources, heterogeneous architecture, beamforming, cognitive radio networks, SAR shielding, etc. Figure 1.7 represents the combination of various energy efficient approaches in one frame.

1.3.2 Green Cloud-Computing-Based Future Communication Systems

Currently, cloud computing is an emerging technology that supports web applications and high-performance computing (HPC). It provides cost-effective and highly scalable infrastructure. The demand for cloud infrastructure has been increasing day by day, which results in high energy consumption from data centers (Gavaskar et al. 2016). This increased energy consumption leads to high operational costs and huge carbon emissions. Therefore, green cloud computing is required to reduce the adverse effects of cloud computing on the environment. Figure 1.8 shows the architecture of green cloud computing (Gupta et al. 2021). This architecture does not require prior information of the system, and it improves the quality of service (QoS) while reducing energy consumption by physical resources. In this architecture, multi-layer user applications are designed on the cloud infrastructure. Figure 1.8 shows that each layer of application is designed on a different layer of the physical machines. The main objective of this architecture is to attain high energy efficiency and utilize the resources properly.

1.3.3 Green Optical Communication System

Optical communication system performs a vital role in ICT. This technology works as a base for long-haul networks, and it is also an essential component of switches and interconnects (Căilean et al. 2017). Due to the usage of passive elements and minimum optical fiber propagation losses, optical communication systems do not depend on amplifiers. It leads to small energy consumption. When passive optical network (PON)-based fiber to the home (FTTH) is installed, the energy consumption increases exponentially, and it results in a large burden in OPEX (Lu 2018). After analyzing the environmental consequences and cost-related issues, it is very critical to develop the technologies to enhance the energy efficiency of optical communication systems. There are mainly two methods through which the energy efficiency of the optical systems can be improved.

1. Device-level method: Design the small-size transistors because they demand low power for the system operation. It decreases the power dissipation of

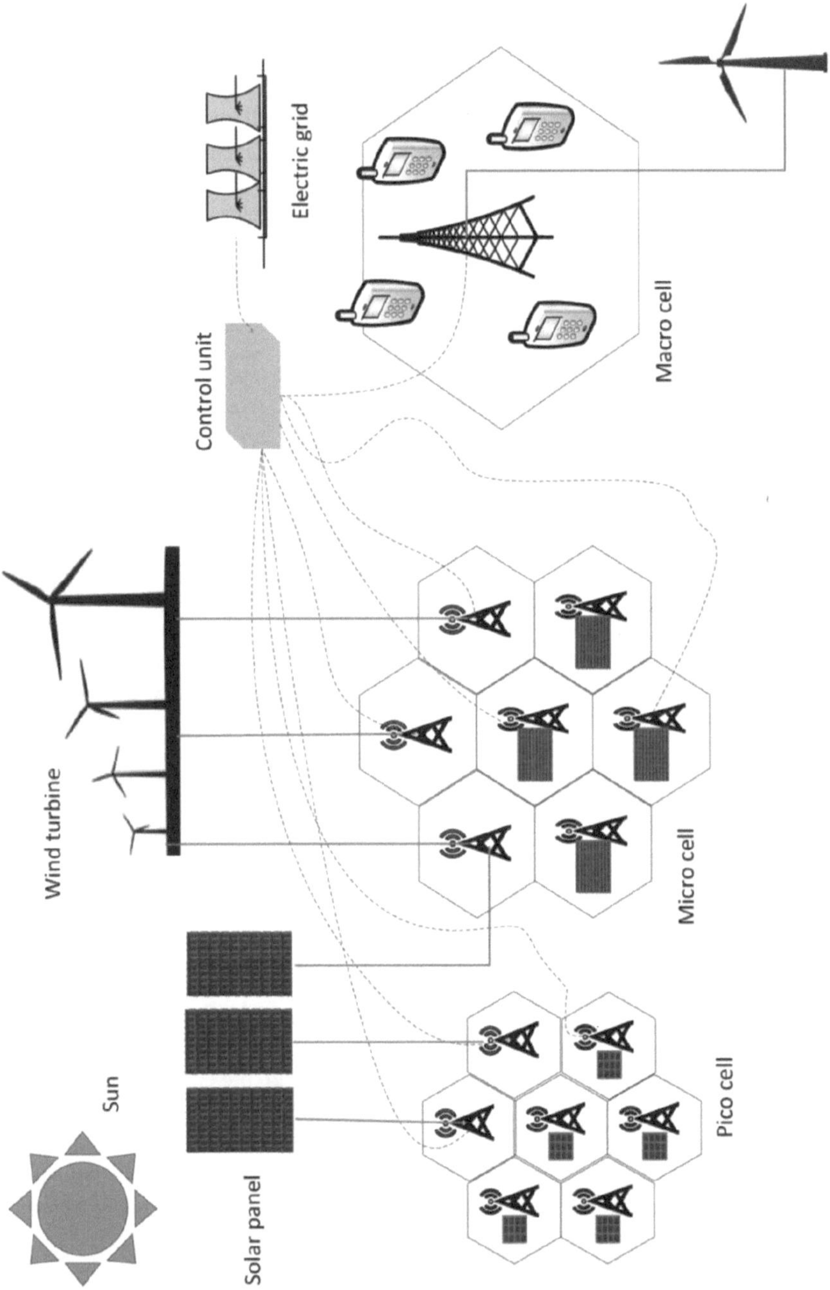

FIGURE 1.7 Combination of various energy efficient approaches in one frame.

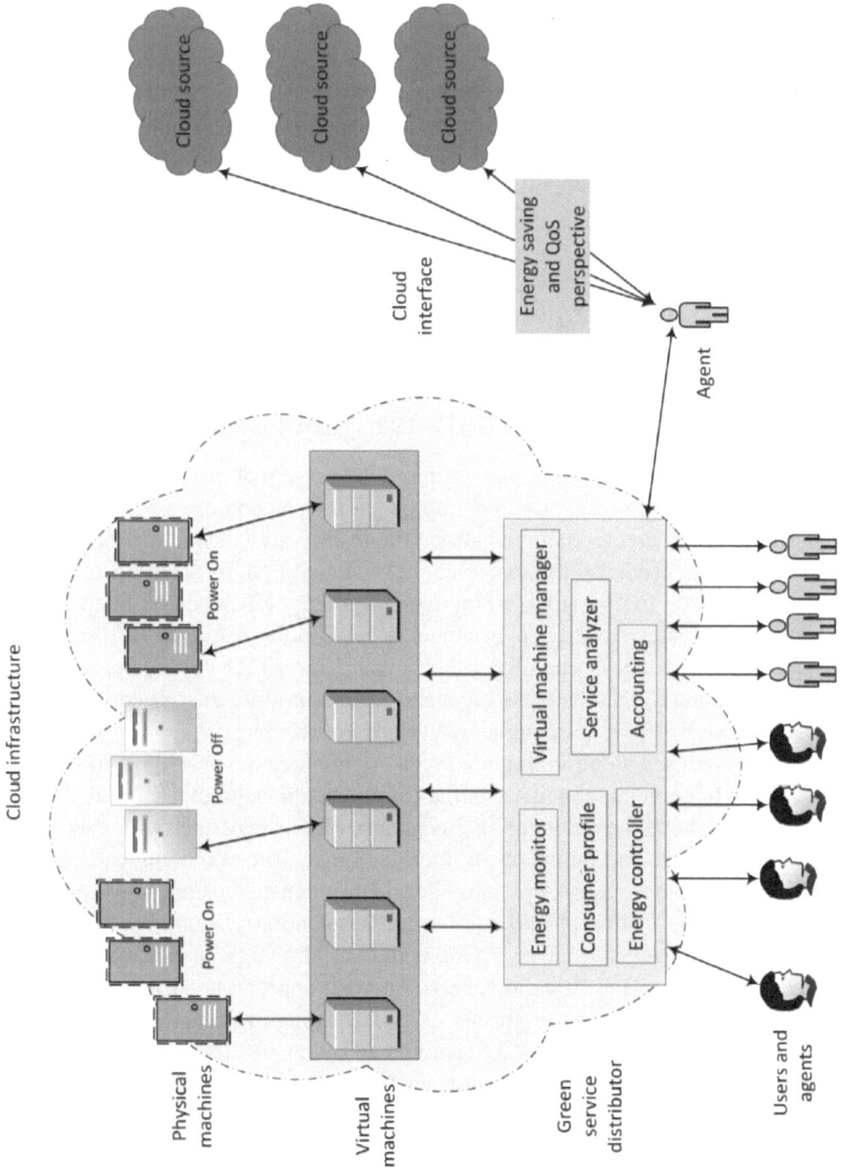

FIGURE 1.8 Architecture of green cloud computing.

the chip. Usage of low power produces less heat and therefore also mitigates energy utilized by the cooling system.

2. Proper network optimization: Several types of research have been performed for network optimization. In the field of coherent optics, digital signal processors (DSPs), application-specific integrated circuits (ASICs), analog to digital converters, digital to analog converters are the highest power-consuming components (Gupta et al. 2020). Various methods have been suggested to enhance energy efficiency in this area. A large amount of energy can be saved by using the sleep mode method. The energy efficiency of these systems can be improved by collectively managing the software part such as coding techniques, modulation techniques, and hardware parts such as DSP, driver, DAC/ADC of the system.

Figure 1.9 shows the proposed framework for attaining the goal of a green optical communication system. The figure indicates the necessary techniques and step-by-step requirements of the green optical communication system.

1.3.4 Green Internet of Things (IoT)-Based Communication System

The Internet of Things (IoT) is an emerging technology that connects millions of devices with the help of sensors. It is the most developing technology for constructing the framework of the green communication system (Srivastava et al. 2021). A huge amount of energy is needed for transferring the data among millions of devices. The objective of the green IoT system is to improve the energy efficiency of communication systems and create a healthy environment. IoT technology utilizes RFID sensors for connecting the devices. Therefore, the demand for RFID will increase exponentially in the future. This increased demand results in more energy consumption because the active RFIDs require higher battery power (Kedia et al. 2017). The green IoT technology focuses on designing the green communication system from the initial phase to the implementation phase of the communication system (Xu et al. 2018). The green IoT supports the reduction in greenhouse effects generated by the existing communication system and optimizes its operation to achieve maximum energy efficiency. There are some major steps through which green IoT technology can convert the communication system into a green communication system (Foteinos et al. 2013). A small size network can save a large amount of energy and increase energy efficiency. The combination of various networks with appropriate strategy management leads towards low carbon footprints. By transmitting only desired information (extract noise from information), a large amount of energy can be saved. Figure 1.10 shows the green IoT-based communication system.

1.3.5 Green Artificial-Intelligence-Based Communication System

The expansion in data traffic with increased user demand has directed the evolution of artificial-intelligence (AI)-based communication systems (Srivastava et al. 2021). However, in the present scenario, a specific portion of the total available energy is

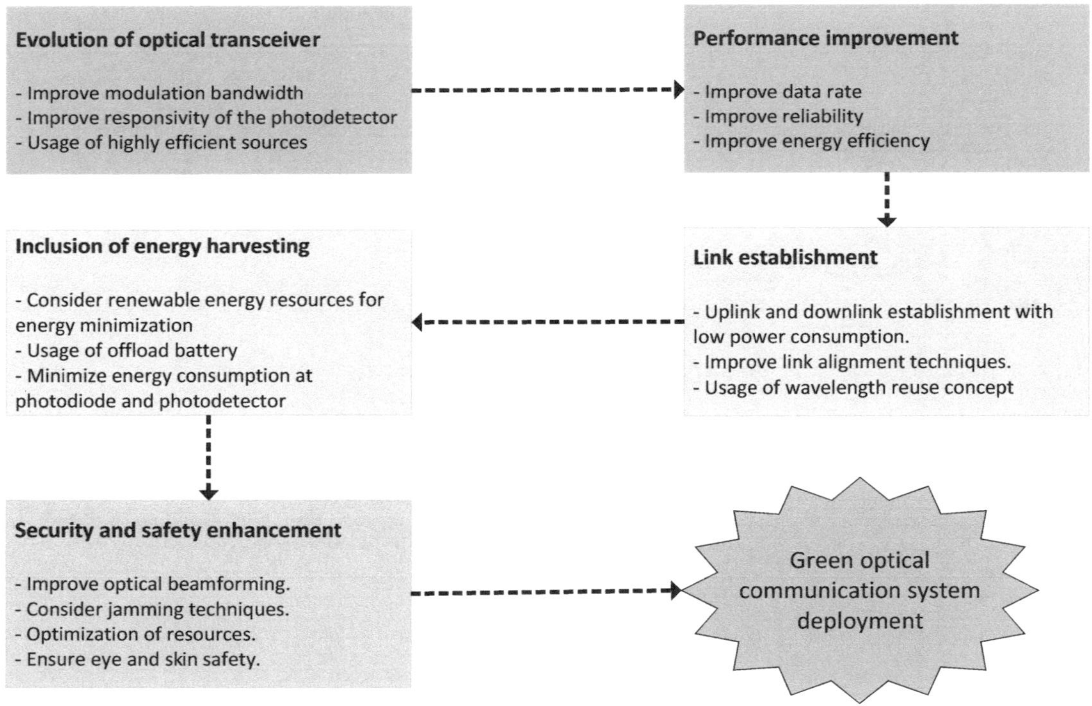

FIGURE 1.9 Proposed framework for attaining the goal of a green optical communication system.

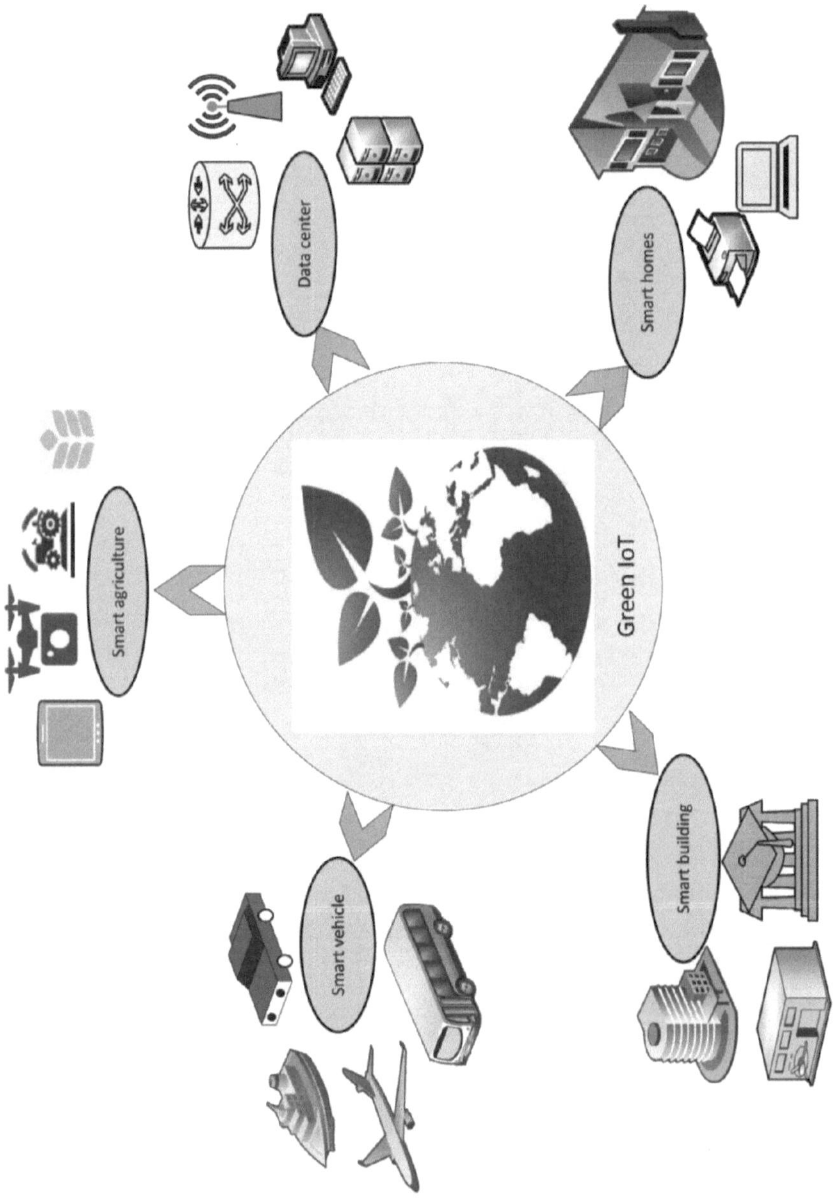

FIGURE 1.10 Green IoT-based communication system.

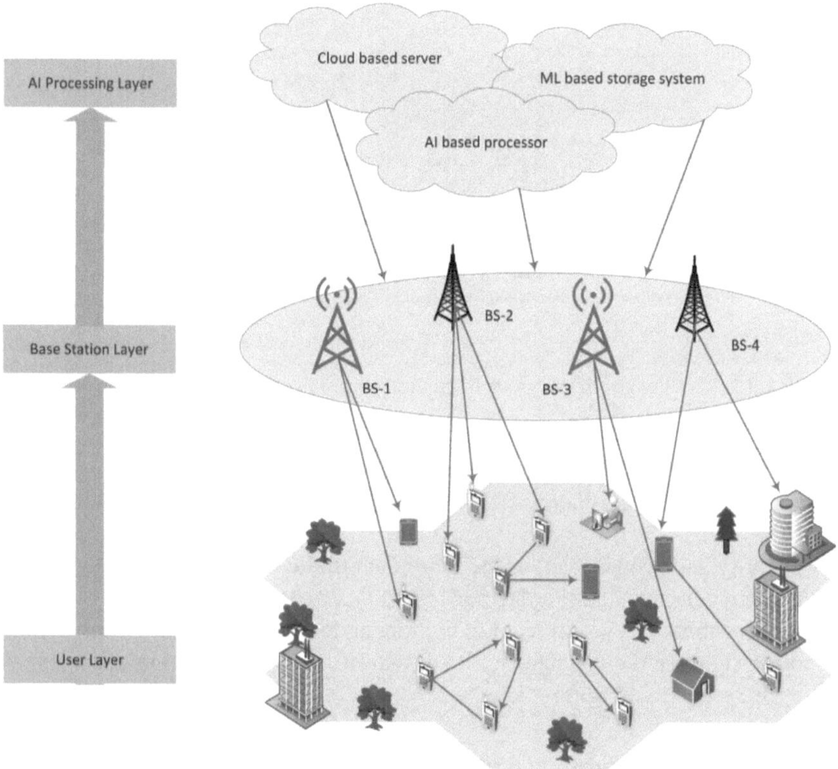

FIGURE 1.11 Green AI-based communication system.

utilized by information communication technology. Therefore, this situation leads to the requirement of green AI-based communication systems (Chen et al. 2016). These systems provide efficient usage of energy. Multiple energy-efficient techniques employed in communication systems are discussed in this book. AI-based systems provide communication system management in an advanced manner, as they can envision various channel conditions, user position, traffic load situation, mobility patterns, etc (Xin et al. 2016). Since many access points, base stations, and central control units also perform the function of storage and computation, AI is utilized to optimize the computation offloading and content caching policies. This optimization decreases the consumption of energy and hence supplies the energy-efficient solution. Figure 1.11 shows the green AI-based communication system.

1.4 CHALLENGES OF GREEN COMMUNICATION SYSTEMS

Several challenges associated with green communication systems are listed below:

- There must be a proper trade-off between flexibility and energy efficiency of the communication system (Sachan et al. 2018).

FIGURE 1.12 Challenges of the green communication system.

- The proper channel modeling plays an important role in system designing and it also enhances the energy efficiency of the communication system (Srivastava et al. 2021).
- Security and proper load balancing of data traffic are important challenges for modern communication system (Piran et al. 2020).
- Besides this, the system reconfiguration, and multi-domain scheduling are the other significant challenges in the design of the green communication system as represented in Figure 1.12.

1.5 RESEARCH OPPORTUNITIES OF GREEN COMMUNICATION SYSTEMS

Based on the literature survey the research opportunities are discussed below:

- In the supermassive MIMO technique, the array of multiple antennas demands a large amount of overhead. Therefore, with the help of the spatial diversity concept, this problem can be solved (Mach et al. 2015).
- The interference-effect of terahertz communication in multiple cells needs to be reduced (Gupta et al. 2021).
- The cost factor associated with communication technologies is a notable research opportunity for the future.
- To obtain a green wireless communication system, advanced and intelligent algorithms are required to be developed (Li et al. 2017).

CONCLUSION

This chapter focuses on the significant role of green technologies in the future communication system. The chapter illustrates the importance of green techniques, utilization of green technologies in the communication system, research opportunities, suitable materials for green technologies, energy-efficient metrics, and research challenges in this domain. The environmental and energy efficiency concerns are further increasing the demand for green communication. Therefore, in this chapter several important initiatives and up-gradation performed for the design of green communication systems are explored in detail. This may prove to be an important step towards a green future. Different types of communication methods with the integration of green technologies are also discussed.

ACKNOWLEDGEMENT

This work is supported by Women Scientist Scheme-A, Department of Science and Technology, New Delhi, Government of India under Grant File No: SR/WOS-A/ET-154/2017.

REFERENCES

Căilean A., Dimian M., (2017) "Current challenges for visible light communications usage in vehicle applications: a survey Commun. Surv. Tutorials, IEEE, 19 (4), pp. 2681–2703.

Chen L., Yu F. R., Ji H., Rong B., Li X., and Leung V. C., (2016) "Green fullduplex self-backhaul and energy harvesting small cell networks with massive mimo," IEEE Journal on Selected Areas in Communications, vol. 34, no. 12, pp. 3709–3724.

Gavaskar S., Anisha A., Renit C., Shiney T.S., (2016) Mobile apps for Green Cloud Computing performance measure. In Proceedings of the International Conference on Energy Efficient Technologies for Sustainability (ICEETS), Nagercoil, India, 7–8 April 2016; pp. 865–869.

Gong, J., Thompson, J.S., Zhou, S., Niu, Z., (2014) Base station sleeping and resource allocation in renewable energy powered cellular networks. IEEE Trans. Commun. 62 (11), 3801–3813.

Gupta M. S., Srivastava A., Kumar K., (2019) "Seamless Vertical Handover for Efficient Mobility Management in Cooperative Heterogeneous Networks," 2nd International Conference on Data Engineering and Communication Technology (ICDECT-2017), Springer, pp.145–153, Pune, India.

Gupta M.S., Kumar K., (2020) "Application Aware Networks' Resource Selection Decision Making Technique using Group Mobility in Vehicular Cognitive Radio Networks", Vehicular Communications, Elsevier, vol.26, pp. 100263.

Gupta M.S., Kumar K., (2021) "Group Mobility Assisted Network Selection Framework in 5G Vehicular Cognitive Radio Networks," Physical Communication, Elsevier, vol. 51, pp. 101578.

Gupta, M.S., Kumar, K., (2021) "AFSOS: An auction framework and Stackelberg game oriented optimal network's resource selection technique in cognitive radio networks," IEEE Transactions on Network and Service Management, 2021.

Kedia D., and Kaur G., (2017) "Examining Different Applications of Cloud-Based IoT". In "Examining Cloud Computing Technologies through the Internet of Things, IGI-Global, International Publisher of Progressive Information Science and Technology Research Books and Journals, ISBN13: 978152253445, pp. 1125–1146.

LeBlanc R., (2019) www.thebalancesmb.com/carbon-footprint-vs-life-cycle-2878059

Li, J., Liu, Y., Zhang, Z., Ren, J., Zhao, N., (2017) Towards green IoT networking: performance optimization of network coding-based communication and reliable storage. IEEE Access 5, 8780–8791.

Lu W.-C. (2018) "The impacts of information and communication technology, energy consumption, financial development, and economic growth on carbon dioxide emissions in 12 Asian countries Mitig. Adapt. Strategies Glob. Change, 23 (8), pp. 1351–1365.

Mach, P., Becvar, Z., Vanek, T., (2015) In-band device-to-device communication in OFDMA cellular networks: a survey and challenges. IEEE Commun. Surv. Tutor. 17 (4), 1885–1922.

Parasher, Y., Singh, P., and Kaur, G. (2019) Green Smart Town Planning. Green and Smart Technologies for Smart Cities, 19–41.

Piran M. J., Pham Q.-V., Islam S.M. R., Cho S., Bae B., Suh D. Y., and Han Z., (2020) "Multimedia communication over cognitive radio networks from QoS/QoE perspective: A comprehensive survey" in 2020 "Journal of Network and Computer Applications" Elsevier.

Sachan A., Nigam S., and Bajpai A., (2018) "An Energy Efficient Virtual-MIMO Communication for Cluster Based Cooperative Wireless Sensor Network," 9th International Conference on Computing, Communication and Networking Technologies (ICCCNT), pp. 1–6, IEEE.

Srivastava A., Gupta M. S., Kaur G., (2019) "Green Smart Cities", Green and Smart Technologies for Smart Cities, CRC Press, ISBN: 978-0-42945483-7, pp. 18.

Srivastava A., Gupta M. S., Kaur G., (2020) "Energy efficient transmission trends towards future green cognitive radio networks (5G): Progress, taxonomy and open challenges", Journal of Network and Computer Applications, Elsevier, vol. 168, pp. 102760.

Srivastava A., and Kaur G, (2021) "Resource management for traffic imbalance problem in green cognitive radio networks", Physical Communication. Elsevier, vol. 48, pp. 101437.

Srivastava A., Gupta M. S., Kaur G, (2021) "Impact of Artificial Intelligence to Solve Pervasive Issues of Sensor Networks of Internet of Things", Artificial Intelligence to Solve Pervasive Internet of Things Issues, Academic Press, ISBN: 978-0-12-818576-6, pp. 367–376.

Srivastava A., Gupta M. S., Kaur G., (2021) "Impact of Intelligence on Future Green Communication", Artificial Intelligence to Solve Pervasive Internet of Things Issues, Academic Press, ISBN: 978-0-12-818576-6, pp. 1–12.

Wang S., Wang W., and Yang H. (2018) "Comparison of product carbon footprint protocols: case study on medium-density fiber board in China", International Journal of Environmental Research and Public Health, MDPI, vol. 15, pp. 2060.

Xin Y., Wang D., Li J., Zhu H., Wang J., and You X., (2016) "Area spectral efficiency and area energy efficiency of massive mimo cellular systems," IEEE Transactions on Vehicular Technology, vol. 65, no. 5, pp. 3243–3254.

Xu G. X., Yuan C., Chen W., Tao X., and Sun Y., (2018) "Adaptive cell zooming and sleeping for green heterogeneous ultradense networks," IEEE Transactions on Vehicular Technology, vol. 67, no. 2, pp. 1612–1621.

2 Green Wireless Communication Systems

Akanksha Srivastava and Gurjit Kaur

CONTENTS

2.1 INTRODUCTION OF GREEN WIRELESS COMMUNICATION SYSTEM

The objective of reducing energy consumption in the area of telecommunication systems inspires researchers to search for novel technologies to achieve green communication. With the fast progression of information and communication technology (ICT), energy consumption is also growing very fast. Now the mobile operators have become the top energy consumers (Jens et al. 2015). After the deployment of the 4G communication system, the consumption of energy has increased in a huge amount. Therefore, in the present scenario, the requirement has shifted from a high data rate system to an energy-efficient system. The energy efficiency (EE) of the wireless communication network can be enhanced by reducing the power consumption involved in

DOI: 10.1201/9781003264477-2

TABLE 2.1
Number of Connected Devices to IoT Fom 2015–2025

Year	Total Number of Connected Devices
2015	15.41 billion
2016	17.68 billion
2017	20.35 billion
2018	23.14 billion
2019	26.66 billion
2020	30.73 billion
2021	35.82 billion
2022	42.62 billion
2023	51.11 billion
2024	62.12 billion
2025	75.44 billion

Source: Statista 2019.

the working operation of the network. There has been a large extension in mobile and wireless networks in the last fifteen years. Nowadays, an intelligent system is adopted to combine various networks, and this intelligent system has decreased the size of the world. This intelligent system is associated with the Internet of Things (IoT), which connects a large number of devices, as represented in Table 2.1.

Several wireless innovations at regular intervals must be able to incorporate notable characteristics from numerous advanced technologies. Some wireless technologies work only for shorter distances like Bluetooth, and some cover a large range like cellular networks. As the wireless systems are developing, the data rate, portability, EE, and spectral efficiency (SE) should also increase. A survey about the growing wireless innovations is given below. In the 1975s, the first generation (1G) of the wireless communication system was created and consisted of wired communication devices like landline telephones (Meraj et al. 2015). In these systems, the transmission signals were carried by the established wired communication setup. In the 1990s, the second-generation communication systems were invented, which were supporting voice and restricted data services. The technology was updated, and the 2G system extended its services. It started to support extra services like text, paging, and telephone message. In 2000, the third-generation (3G) communication systems came and these systems offered improved sound, video, and multimedia services. It can support video streaming and online gaming. In 2010, the fourth-generation (4G) communication system was developed, which improved video conferencing and real-time services. 4G systems allow more information transmission to devices and vehicles moving at high speeds. With the increasing demand of users, 4G can as of now be just replaced with 5G with an innovation named Beam Division Multiple Access (BDMA) and Filter Bank Multicarrier (FBMC) (Srivastava et al. 2019). Figure 2.1 shows the upgradation in the wireless communication generations.

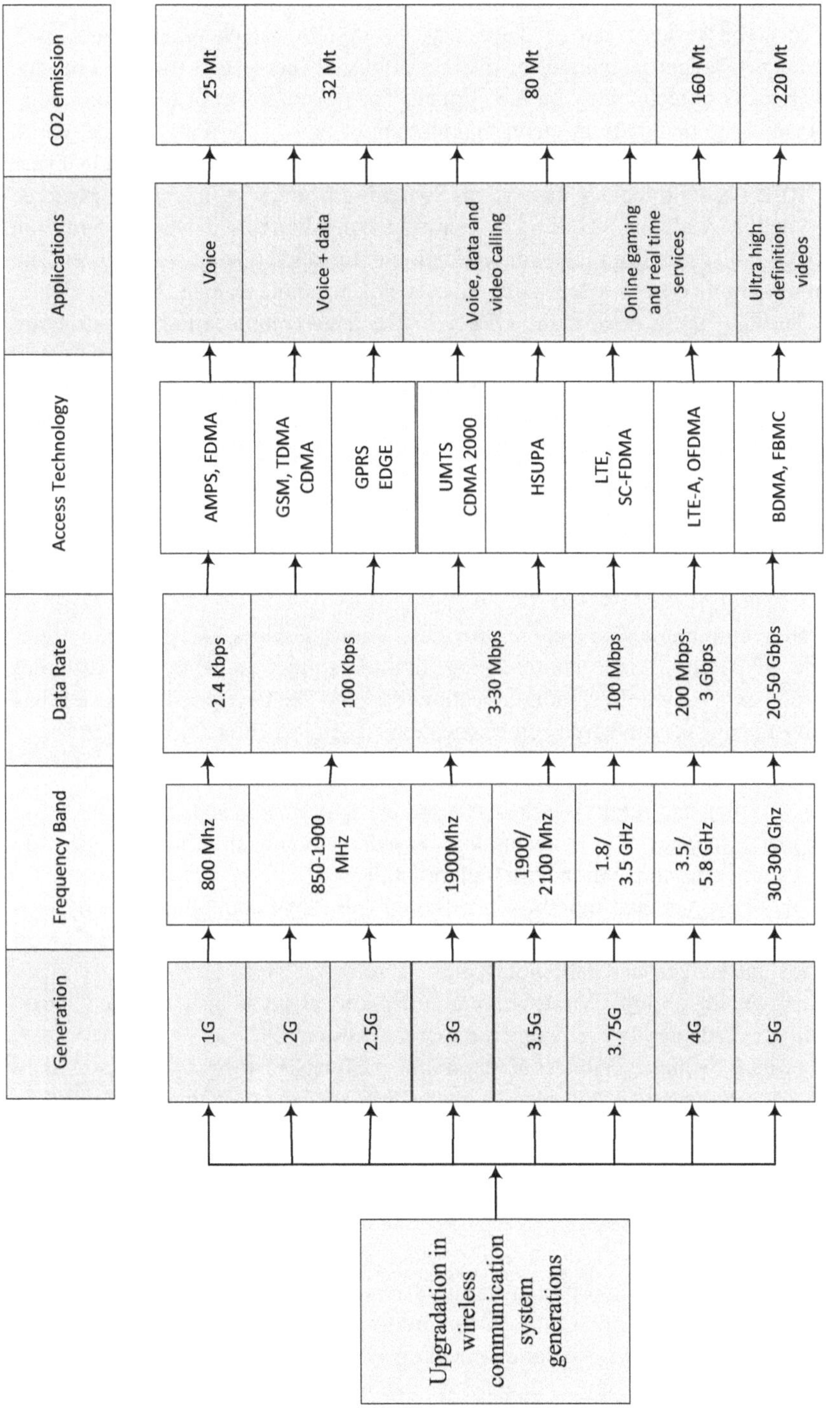

FIGURE 2.1 Evolution of wireless communication generations.

2.2 NEED OF GREEN WIRELESS COMMUNICATION SYSTEMS

The word green systems can be defined as the system, which practices the merging of energy-saving techniques at distinct points of the system to overcome the adverse impacts of technology on the climate. The development of telecommunication infrastructure demands an increasing amount of power. Currently, in the world, the number of phone towers, which spend more than 5 billion liters of diesel in a year (Ghosh 2012), is continuously increasing. Switching from diesel to another resource of energy will save almost $2 billion in operating expenditures and decrease 5 million tons of CO_2 emissions from telecommunication companies (Gupta et al. 2019). The usage of renewable energy sources like solar, wind, etc., can create millions of carbon credits. Through these credits, the OPEX of the telecommunication infrastructure could be balanced, and also, save the energy bill. The green wireless communication system has several aspects. It can be categorized broadly in terms of designing green telecommunication equipment, green telecommunication networks, and appropriate telecommunication waste disposal (Cayamcela et al. 2018). These factors are illustrated briefly in this chapter.

2.3 TYPES OF WIRELESS COMMUNICATION SYSTEMS

The wireless communication system provides various services like paging, cellular telephone, TV, Radio, video conferencing, online gaming, etc. Based on different applications, several wireless communications systems are developed because it has many advantages over the wired communication.

- Easy to plan and setup: Wireless systems are easy to plan and expand.
- Higher efficiency: Latest technology-based communication systems provide fast speed communication with high reliability.
- Global coverage and mobility: It offers global range communication by providing improved networking in areas such as rural areas, suburban areas, urban areas, hilly areas, and battlefields, etc.
- Improved flexibility: Wireless communications systems give excellent flexibility and adaptability as compared to wired systems.
- Cost-effectiveness: Wireless systems are inexpensive to install because they do not require cables and wires. Besides this, the cost of maintaining wireless systems is low.

Wireless communication systems are divided into three types:

- Simplex: The simplex communication system supports only one-way data transmission. An example of a simplex communication system is radio broadcasting.
- Half Duplex: This type of communication system supports bi-directional communication but not simultaneously. An example of a half-duplex communication system is a walkie-talkie.

- Full Duplex: This type of communication system supports a bi-directional communication system simultaneously. An example of a full-duplex communication system is mobile communication.

In this chapter, we discuss some major wireless communication systems such as IR wireless communication, satellite communication, microwave radio, Wi-Fi, infrared communication, radio broadcast.

2.3.1 Satellite Communications

In the field of telecommunications, satellite communications perform an important role. It is broadly expanded on the earth and connects users globally. In satellite communication, there are two major elements like the ground-based element and the space-based element. The fixed and mobile transceiver types of equipment are placed at the ground-based element, and the satellite itself is placed at the space-based element (Kodheli et al. 2020). Figure 2.2 shows the scenario of satellite communication.

2.3.2 Microwave Communication

Receiving and forwarding information by using microwaves is known as microwave communication. This information can be in any form such as text, voice, video, television. A large portion of the EM spectrum is covered by microwaves; therefore, it can be applied in various applications. Some of the EM spectrum applications are shown in Table 2.2. This table shows that the different applications work on different frequency bands.

TABLE 2.2
EM Spectrum Applications, at Different Wavelengths

Wavelength	Frequency	Application
714–400 nm	420–750 THz	Visible light
1550–850 nm	352–193 THz	Fiber optic connection
0.005–0.001 m	60–300 GHz	Millimeter-wave connection
0.3–0.005 m	1–60 GHz	Microwave links
0.380–0.135 m	800–2200 MHz	Mobile base station
1–0.1 m	300–3000 MHz	UHF point to point
3.5–2.8 m	88–108 MHz	FM transmission
165–10 m	1.8–30 MHz	Shortwave radio
1.8 km–188 m	180–1600 kHz	AM radio transmission
10–1 km	30–300 kHz	Long-wave radio transmission
100–10 km	3–30 kHz	Sub-marine data transmission
6000–5000 km	50–60 Hz	Electricity transmission

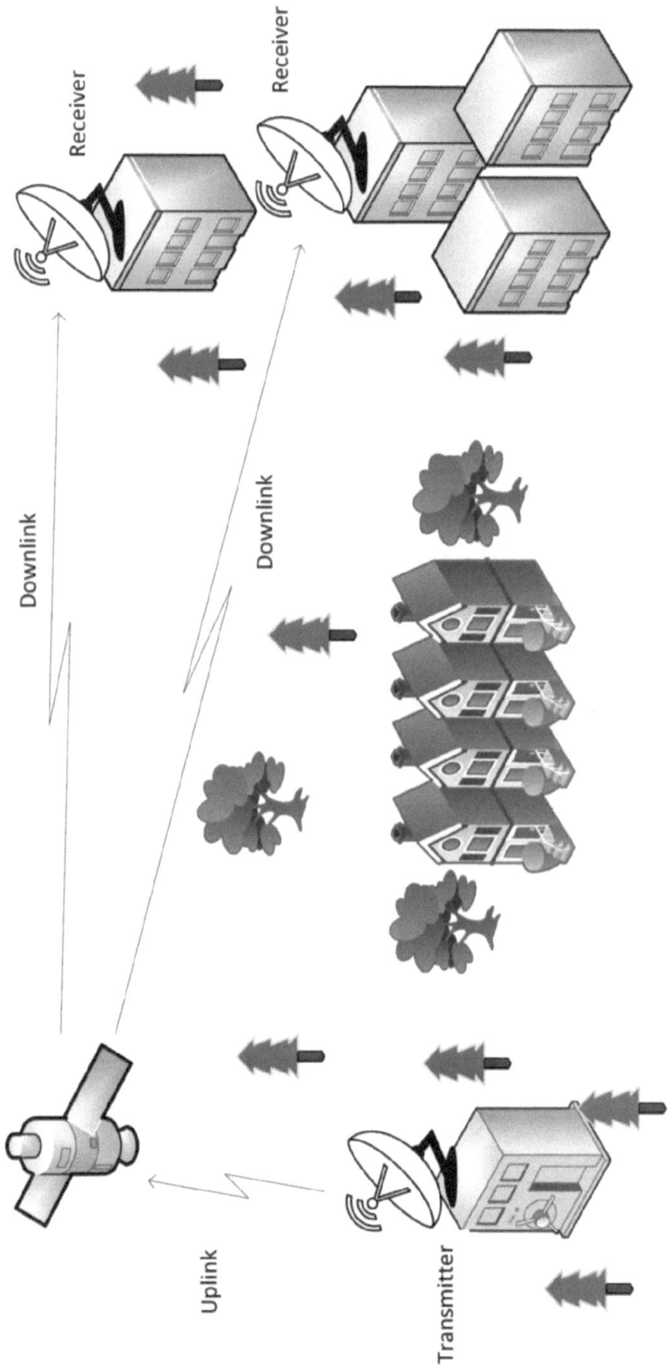

FIGURE 2.2 Satellite communication.

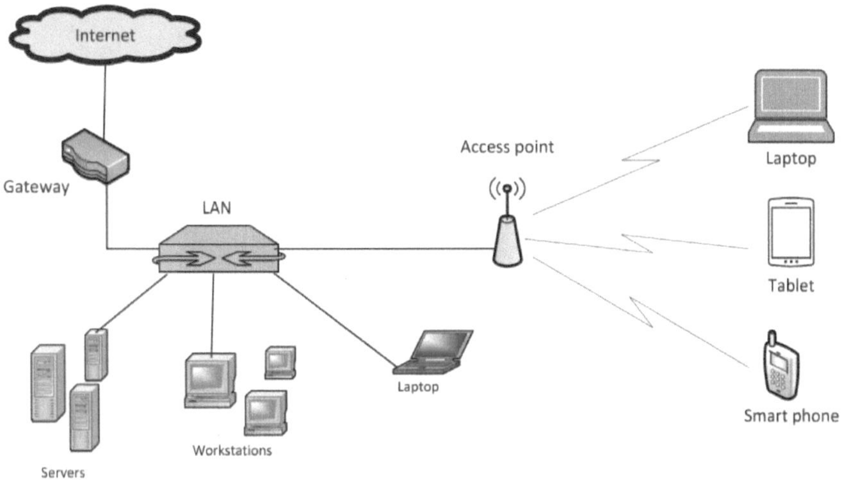

FIGURE 2.3 Wi-Fi network setup.

2.3.3 Wi-Fi

Wi-Fi stands for Wireless Fidelity, and this network is almost present in every educational institute, railway station, airport, organization, shopping mall. The communication of Wi-Fi networks is based on low power, which is practiced by different devices such as laptops, tablets, smartphones, etc. In a Wi-Fi network, a router is the main component, and it operates as a data transmission hub. The users present within a certain range are permitted by the Wi-Fi network to connect with the router. The Wi-Fi network is password protected. Figure 2.3 shows the Wi-Fi network setup.

2.3.4 Infrared Communication

IR radiations are exploited in infrared communication systems for transmitting the information (Kwon et al. 2021). IR communication is used for short-range transmission like in the remote controls of electronic devices such as TV, AC, projector, etc. This communication is also utilized for security control. Photo LED works as a sender, and photodiode works as a receiver in an IR communication system. The LED uses a nonvisible light to send the IR signal that is received by the photodiode. This system operates only when there is a direct line of sight between sender and receiver. Figure 2.4 shows the scenario of infrared communication.

2.3.5 Radio Broadcast

Radio broadcasting is a well-known type of wireless transmission. Radio transmitters use radio waves to transmit information. Radio waves are comparatively narrow, and they can be transmitted over different frequencies. Therefore, a car radio is capable of receiving radio signals from various radio stations. Figure 2.5 shows the scenario of infrared communication.

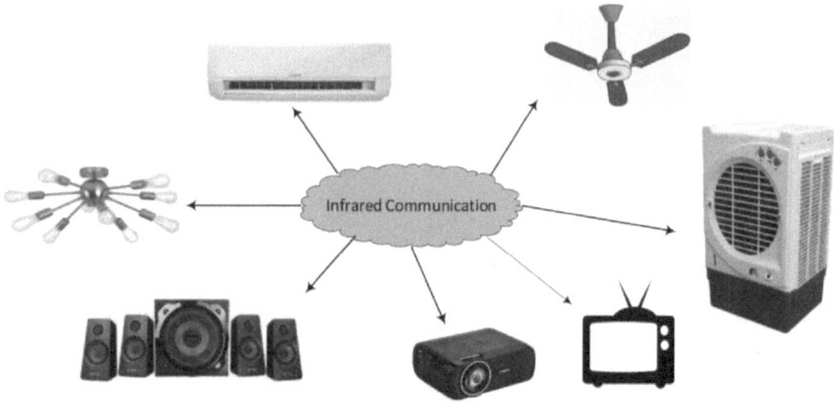

FIGURE 2.4 Infrared communication.

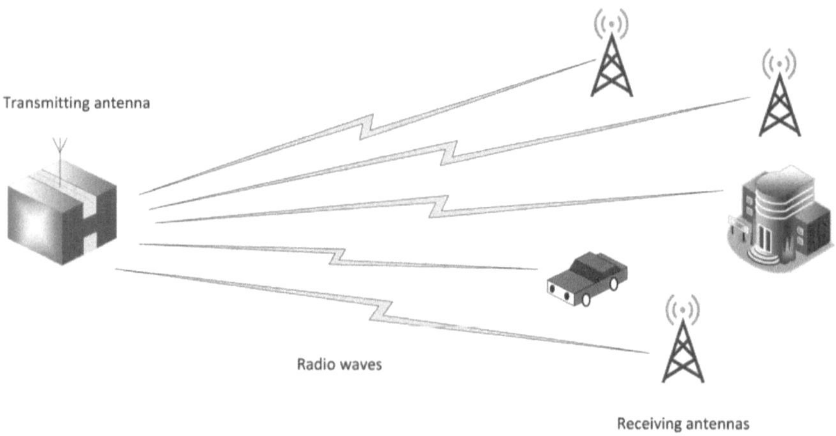

FIGURE 2.5 Radio broadcast communication.

2.4 FEATURES OF THE GREEN WIRELESS COMMUNICATION SYSTEM

Nowadays, improved energy efficiency is considered as the fundamental element of all 5G project solutions and technologies (Elprocus 2015). Energy control aims to minimize the operational expenses of the mobile operator networks, as well as higher customer satisfaction to extended battery life. The base station (BS) is the main component of a wireless system, and it utilizes the highest amount of power for working. For providing a better quality of experience (QoE) to the mobile subscribers the number of BSs is continuously expanding (Osuwa et al. 2014). Therefore, the consumption of energy will also rise proportionally. In the present scenario, it is quite an essential topic for analysis to search for the remarkable techniques to decrease energy consumption and preserve the quality of service (QoS) (Parasher et al. 2019). There are multiple strategies that have been suggested in the literature for converting a base station into a green base station.

2.5 ARCHITECTURE OF THE GREEN WIRELESS COMMUNICATION SYSTEM

In this chapter, a multi-level architecture has been suggested for achieving high EE in the wireless system. In the suggested multi-level architecture, various techniques for gaining high EE have been employed. The primary section of this multi-level architecture is the proper planning and development of the network (Singh et al. 2019). The future green wireless communication system will consist of various devices, which will make an ultra-dense network (Qun et al. 2018). To maintain EE will be the prime concern of future systems, because of the presence of a large number of devices. Therefore, various types of techniques are involved in the multi-level architecture to make the system reliable and environmentally friendly. These techniques are discussed as follows.

2.5.1 Ultra-Dense Network (UDN)

In this technique, a much larger density of radio resources (base stations, access points, etc.) are deployed. In this scheme, the power conservation mode of resources is utilized for decreasing the consumption of energy. In UDN, subscriber density is changeable in the network. But the antennas broadcast the beam radiations in all directions with the same intensity and power. Therefore, the multi-beam carrier aggregation (MBCA) technique is adopted for energy consumption optimization (Xu et al. 2018). In the MBCA technique, the large-power beam is radiated towards the large subscriber density area, and the low power beam is radiated towards the low subscriber density area. Figure 2.6 represents the architecture of the ultra-dense network.

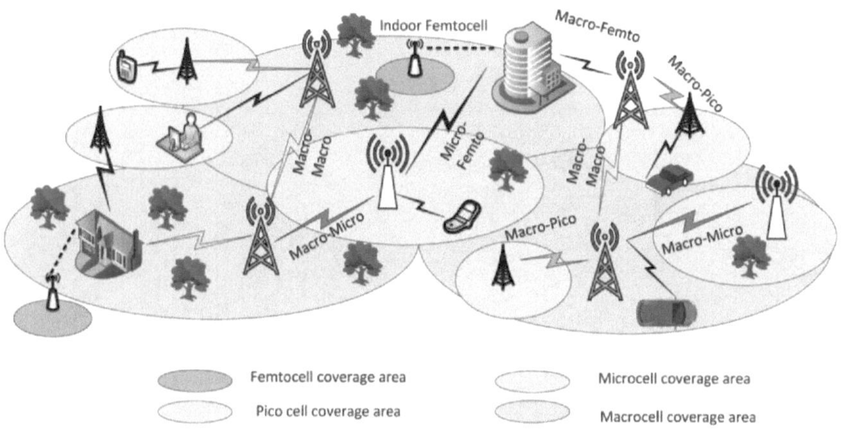

FIGURE 2.6 A scenario of the ultra-dense network.

2.5.2 Device to Device (D2D) Communication

This type of communication technique supports mobile subscribers to communicate with each other with low transmission power (Kaur et al. 2019). Hence, this technique can save more energy in radio resources. In the situation of data congestion and signal unavailability, D2D communication allows alternative data transmission options (Cai et al. 2019). In the upcoming wireless communication techniques, the D2D communications' method is identified as a public security network due to the great data rates and low cost. Figure 2.7 shows the packet transmission through the device-to-device communication method.

The D2D communications' approach is classified into two classes. The first one is inband D2D and the second one is the outband D2D. In inband D2D, the identical licensed band is utilized by both the cellular networks and D2D network (Gupta et al. 2019). The main drawback of this communication is the interference experienced by the D2D link. In the outband D2D communications, D2D links utilize the unlicensed spectrum. In this way D2D communication supports data transmission with low energy usage.

2.5.3 Massive MIMO

At present, the multiple-input multiple-output (MIMO) technique is observed as the most promising solution for green wireless communication systems due to its high energy efficiency and spectrum efficiency (Sachan et al. 2018). Figure 2.8 represents the idea of the massive MIMO technique, which is obtained by employing multiple antennas at the transmitter and receiver end (Yadav et al. 2019). This technique provides high spatial diversity and improves the performance of the system. In the massive MIMO technique, the base station contains a greater number of antennas as compared to MIMO (Gupta et al. 2021). Therefore, this technology is considered, as a more powerful technology for green systems. This technology has numerous benefits like better EE, high throughput, minimum delay, and large capacity. This technique improves the overall experience of the users. Users can send large information files and download high-quality videos, movies, or use real-time services at very high speed.

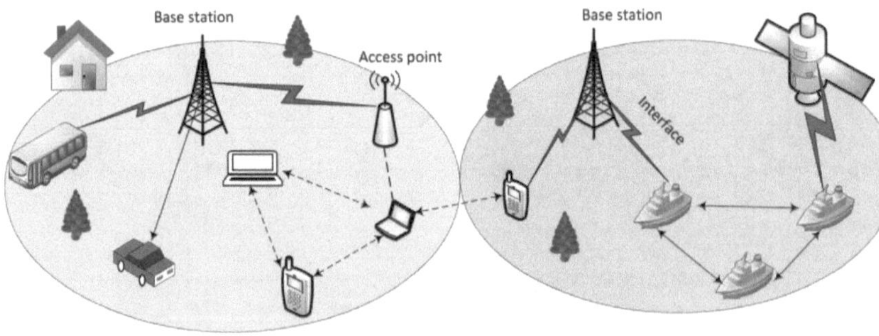

FIGURE 2.7 Scenario of device-to-device communication.

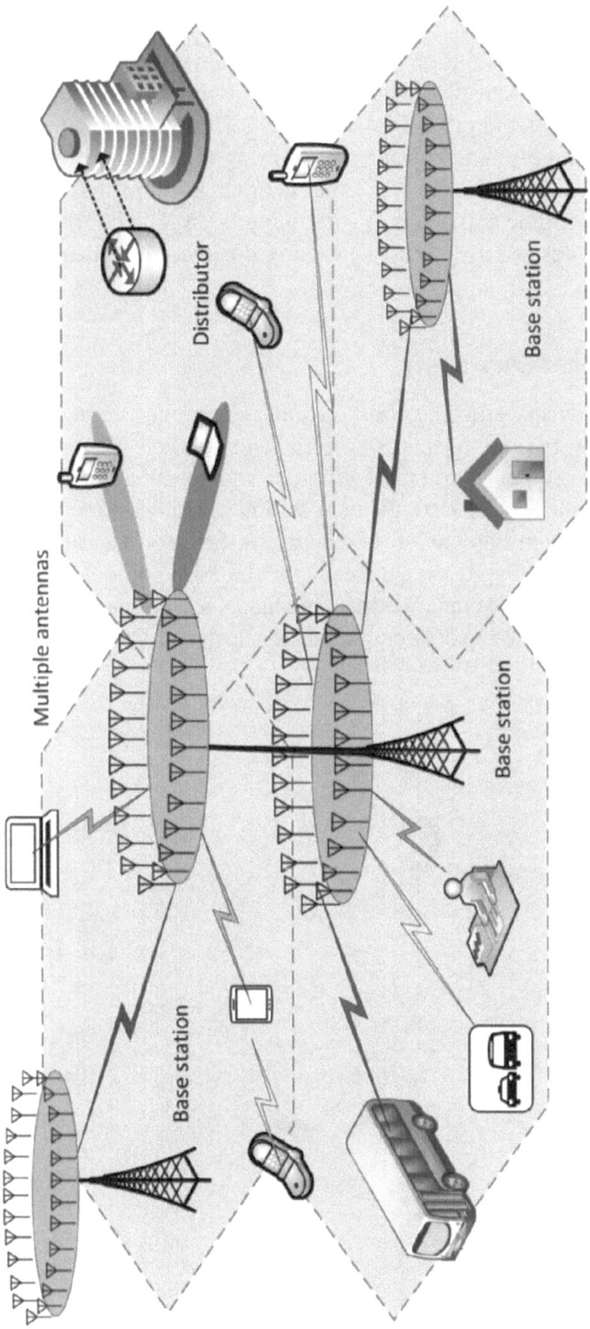

FIGURE 2.8 Massive multiple-input multiple-output technique.

2.5.4 Beamforming

Beamforming is a direct method for minimizing energy expenditure from the communication network. In this technique, an array of antennas focuses the electromagnetic beam radiation towards the desired cell phone user area, as represented in Figure 2.9. The exact position of the desired cell phone user is found by the digital signal processing (DSP) algorithms. In this method, the complex weights are multiplied with the transmitting signals, and these weights are flexible (Srivastava et al. 2021). The other characteristic of these weights is they have signal's magnitude and phase in a particular area. The figure shows the beam generation in the desired user area. Therefore, this technique can preserve a huge amount of energy.

2.5.5 Hardware Improvement

To maximize the energy efficiency of the wireless communication system, improvement in the hardware components is an important solution. Figure 2.10 shows the percentage of energy consumption at different components of the communication system. According to Figure 2.10, the base station consumes almost 58% of energy, which is the maximum amount of the energy. In Srivastava et al. (2020), various techniques for improving EE associated with the base station, power amplifiers, cooling components, cables, and feeder components are discussed. High SE and EE are essential demands for the 5G communication systems. The nonlinearity property

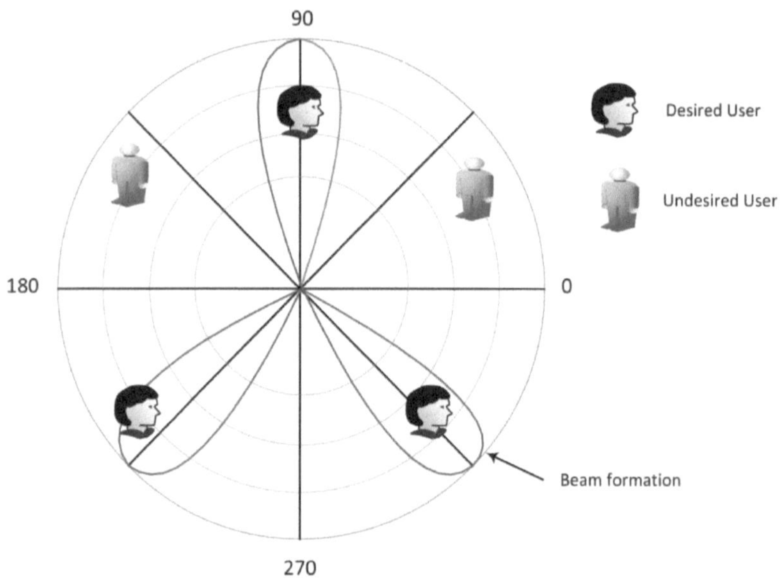

FIGURE 2.9 Beamforming technique.

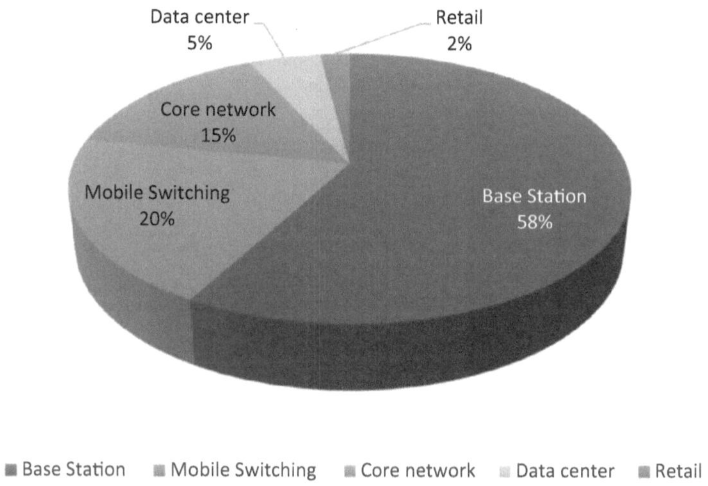

FIGURE 2.10 Percentage of energy consumption at different components of the communication system.

of the power amplifiers (PA) presents vital challenges to achieve these specifications, mainly for multicarrier modulation signals with a large peak to average power ratio. Doherty architecture, switch mode power amplifier (SMPA), and GaN-based amplifier methods are used for improving the quality of PA. Through the hardware improvement, 60% energy can be saved but the drawback is this technique is less flexible and expensive.

2.5.6 Cognitive Radio Network

The cognitive radio (CR) networks' technique is able to make the wireless networks more energy and spectral efficient. The deficiency of the bandwidth is mitigated by utilizing unused extra bandwidth for data communications. Energy-efficient CR networks enhance network performance and thus improve the emerging applications of the network. It is a challenging task to design green CR networks. There are two types of spectrum usage policies. The first one is static spectrum allocation policy, and the second is the dynamic spectrum allocation policy. The static allocation policy is also known as the fixed spectrum allocation policy (Naveen et al. 2020). For designing an energy-efficient CR network, the optimization of the dynamic spectrum allocation policy is required. A four-step process is involved in the dynamic spectrum allocation policy as shown in Figure 2.11.

1. Dynamic spectrum sensing
2. Spectrum analysis
3. Spectrum sharing
4. Spectrum management

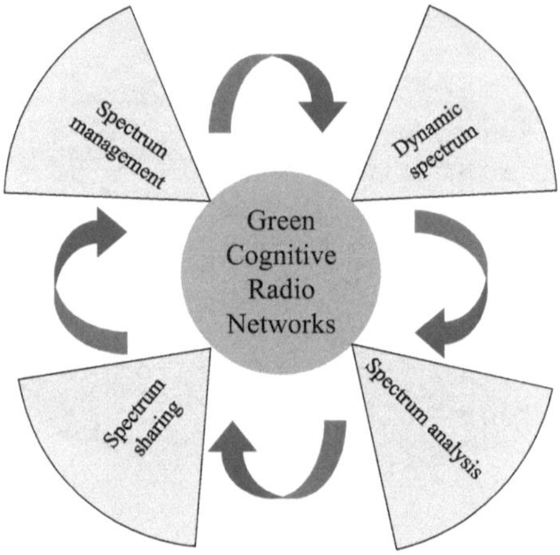

FIGURE 2.11 Key functions of dynamic spectrum allocation.

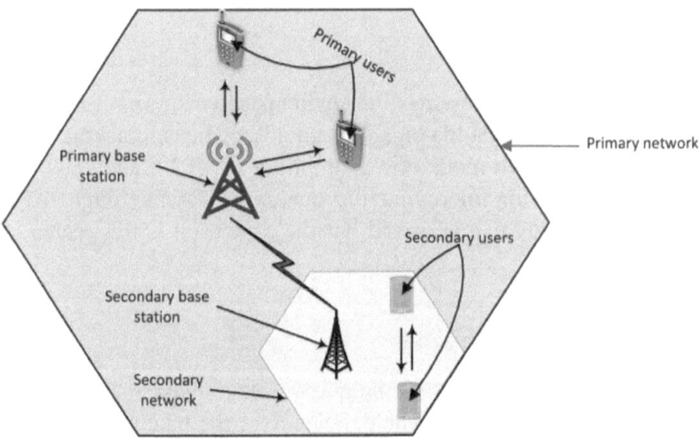

FIGURE 2.12 Scenario of cognitive radio networks.

In the CR network, the primary network (PN) is called a licensed network, and the users associated with this network are called primary users (PUs) or licensed users (Gupta et al. 2020). On the other hand, the secondary network (SN) is called an unlicensed network, and the users associated with this network are called the secondary users (SUs) or cognitive users (Srivastava et al. 2021). Therefore, the SUs can utilize the spectrum holes of PN when the PUs is absent. The coexistence of two different networks is shown in Figure 2.12.

2.6 CHALLENGES OF GREEN WIRELESS COMMUNICATION SYSTEMS

Various possible figures and facts are discussed in this chapter for maximizing the energy efficiency of the wireless communication system. Numerous challenges associated with green communication systems are listed below:

- There must be a proper trade-off between QoS and EE of the system. For example, in the base station ON/OFF scheme, and in the cell zooming technique huge energy can be saved when the base station is off. But this situation influences the execution and QoS of the system.
- Appropriate manufacturing of a communication system is a significant task in measuring and diminishing energy consumption.
- The reduction of energy consumption at one step of the communication system may increase energy consumption at another step of the system, such as in the cognitive radio (CR) technique, appropriate utilization of the bandwidth may decrease the transmission energy of the system. But a large amount of energy is required for the spectrum hole sensing process.

2.7 RESEARCH OPPORTUNITIES OF GREEN WIRELESS COMMUNICATION SYSTEMS

Several energy-efficient strategies explored in prior subsections are important to make a green wireless communication system. Combining these methods and designing a multi-level architecture will start a revolution in the domain of green communication. The future wireless communication system is supposed to recognize the requirements of the users and also address green communication. Figure 2.13 illustrates the summary of future wireless and networking technologies like terahertz communication, software defined networks, supermassive MIMO, NOMA, Mm-wave, and quantum communication. Hence, the research opportunities associated with green wireless communication systems are also discussed in Figure 2.13.

CONCLUSION

This chapter presents the recent trends of the wireless communication system, which envisions a step towards green communication to design next-generation networks. In this chapter an energy-efficient system, types of wireless communication systems, features, merits, demerits, future research direction, and open issue challenges have been reviewed. The necessity to promote green wireless communication systems has become more popular as wireless networks are becoming omnipresent. The energy-efficient green wireless communication systems produce fewer radiative devices and also provide more economic benefits to the users and service providers. The government must also set laws and rules to declare a service supplier as a green service supplier.

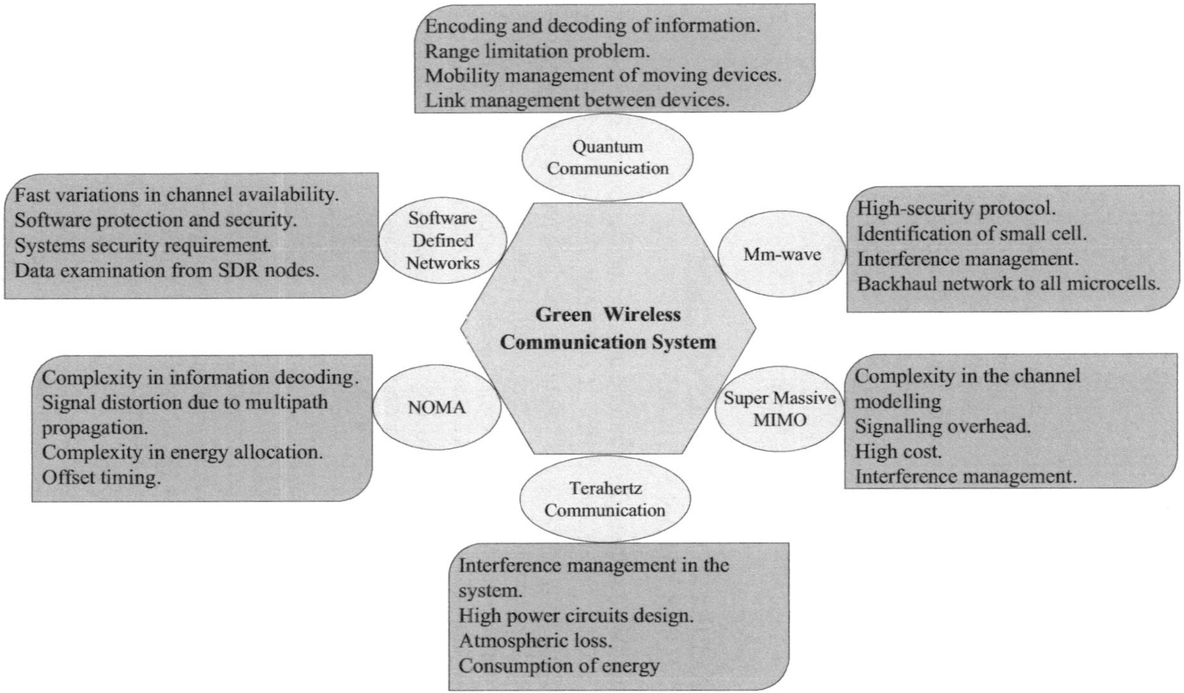

FIGURE 2.13 Challenges of the green wireless communication system.

ACKNOWLEDGEMENT

This work is supported by Women Scientist Scheme-A, Department of Science and Technology, New Delhi, Government of India under Grant File No: SR/WOS-A/ET-154/2017.

REFERENCES

Cai, Yan and Ni, Yiyang and Zhang, Jun and Zhao, Su and Zhu, Hongbo, (2019) "Energy efficiency and spectrum efficiency in underlay device-to-device communications enabled cellular networks" in 2019 China Communications IEEE Transactions on Wireless Communications, vol. 16, no. 4, pp. 16–34,. Publisher (IEEE).

Cayamcela, Manuel Eugenio Morocho and Lim, Wansu, (2018) "Artificial Intelligence in 5G Technology: A Survey", International Conference on Information and Communication Technology Convergence (ICTC), IEEE 2018 pp. 860–865.

Chaturvedi, P.K. (2018) "Microwave, Radar & RF Engineering", Springer.

Homayoun, H., Gang, Y., Kavehrad, M., Behbahani, F., and Galko, P.A. (1994) "Indoor propagation measurements at infrared frequencies for wireless local area networks applications," IEEE Transactions on Vehicular Technology, IEEE, vol. 43, no. 3, pp. 562–576.

G. X. Xu, C. Yuan, W. Chen, X. Tao, and Y. Sun, (2018) "Adaptive cell zooming and sleeping for green heterogeneous ultradense networks," IEEE Transactions on Vehicular Technology, vol. 67, no. 2, pp. 1612–1621.

Ghosh, (2012) www.livemint.com/Politics/zI8QSv9IvLLD01Co3os9PI/Telecom-industry-aims-to-reduce-diesel-consumption.html

Gupta M. S., Kumar K, "AFSOS: An auction framework and Stackelberg game oriented optimal network's resource selection technique in cognitive radio networks," IEEE Transactions on Network and Service Management, 2021.

Gupta M. S., Kumar K., (2019) "Progression on Spectrum Sensing for Cognitive Radio Networks: A Survey, Classification, Challenges and Future Research Issues", Journal of Network and Computer Applications, Elsevier, vol. 143, pp. 47–76.

Gupta M. S., Srivastava A., Kumar K., (2019) "Seamless Vertical Handover for Efficient Mobility Management in Cooperative Heterogeneous Networks," 2nd International Conference on Data Engineering and Communication Technology (ICDECT-2017), Springer, pp.145–153, Pune, India.

Gupta M.S. and Kumar K., (2020) "Application Aware Networks' Resource Selection Decision Making Technique using Group Mobility in Vehicular Cognitive Radio Networks", Vehicular Communications, Elsevier, vol. 26, pp. 100263, 2020.

Kaur, G., Tomar, P. and Tanque, M. (2019) "Integrating Artificial Intelligence to Solve Pervasive IoT Issues, Elsevier, Cambridge, MA, USA.

Kodheli, Oltjon and Lagunas, Eva and Maturo, Nicola and Sharma, Shree Krishna and Shankar, Bhavani and Montoya, Jesus Fabian Mendoza and Duncan, Juan Carlos Merlano and Spano, Danilo and Chatzinotas, Symeon and Kisseleff, Steven and others (2020) "Satellite communications in the new space era: A survey and future challenges", IEEE Communications Surveys \& Tutorials, IEEE.

Kwon, Jung-Hyok and Zhang, Xue and Kim, Eui-Jik, (2021) "Scalable Wi-Fi Backscatter Uplink Multiple Access for Battery-Free Internet of Things" IEEE Access. pp. 30929–30945.

Li, Qun and Wang, Longqi and Xu, Ding, (2018) "Resource Allocation in Cognitive Wireless Powered Communication Networks under Outage Constraint," 2018 IEEE 4th International Conference on Computer and Communications (ICCC). IEEE, 2018, pp. 683–687.

Malmodin Jens and Bergmark Pernilla (2015) "Exploring the effect of ICT solutions on GHG emissions in 2030" in EnviroInfo and ICT for Sustainability, pp. 37–46.

Meraj, M and Kumar, S. (2015) "Evolution of mobile wireless technology from 0 G to 5G", in 2015, International Journal of Computer Science and Information Technologies, vol. 6, no. 3, pp. 2545–2551.

Naveen, R. R., Nayak, A. and Kumar, M. S. (2020). "A survey and performance evaluation of reinforcement learning based spectrum aware routing in cognitive radio ad hoc networks," International Journal of Wireless Information Networks, vol. 27, pp. 144–163.

Osuwa, E. Bala, L. Kazakevich and R. Yang, (2014) "Techniques to improve power amplifier energy efficiency for 5G," 1st International Conference on 5G for Ubiquitous Connectivity, Akaslompolo, pp. 104–109.

Parasher, Y., Singh, P., and Kaur, G. (2019) Green Smart Security System. Green and Smart Technologies for Smart Cities, 165–184.

Sachan A. and Nigam S. and Bajpai A., (2018) "An Energy Efficient Virtual-MIMO Communication for Cluster Based Cooperative Wireless Sensor Network," 9th International Conference on Computing, Communication and Networking Technologies (ICCCNT), pp. 1–6, IEEE.

Singh, P., Dixit, V., and Kaur, J. (2019). Green Healthcare for Smart Cities. Green and Smart Technologies for Smart Cities, 91–130.

Srivastava A, Gupta M. S. and Kaur G., (2021) "Impact of Artificial Intelligence to Solve Pervasive Issues of Sensor Networks of Internet of Things", Artificial Intelligence to Solve Pervasive Internet of Things Issues, Academic Press, ISBN: 978-0-12-818576-6, pp. 367–376.

Srivastava A. and Gupta M. S. and Kaur, G. (2020) "Energy efficient transmission trends towards future green cognitive radio networks (5G): Progress, taxonomy and open challenges," in 2020 Journal of Network and Computer Applications. Elsevier, 2020, pp. 102760.

Srivastava A., and Kaur G, "Resource management for traffic imbalance problem in green cognitive radio networks", Physical Communication. Elsevier, vol. 48, pp. 101437, 2021.

Srivastava A., Gupta, M. S. and Kaur G., "A Game Theory based Approach for Opportunistic Channel Access in Green Cognitive Radio Networks" in 2019 IEEE International Conference on Advanced Networks and Telecommunications Systems (ANTS), pp. 1–6, 2019. Publisher (IEEE).

Statista 2019, Internet of Things (IoT) connected devices installed base worldwide from 2015 to 2025 (in billions): www.statista.com/statistics/471264/iot-number-of-connected-devices-worldwide/US.

Yadav, Vishakha and Kumar, Lalit and Kumar, Pradeep, (2019) "Evolution and Development of Wireless Communication System," International Conference on Computing, Power and Communication Technologies (GUCON) in 2019 IEEE, pp. 53–57.

3 Green Transmission Technologies and Network Protocols

Mani Shekhar Gupta, Akanksha Srivastava, and Ashok Kumar

CONTENTS

3.1 INTRODUCTION OF GREEN TRANSMISSION TECHNOLOGIES

The fast evolution of energy consumption by mobile nodes and wireless networks has posed thoughtful issues; for example, the greenhouse effect due to extreme emission of carbon dioxide and radiation due to extreme emission of high-power

DOI: 10.1201/9781003264477-3

electro-magnetic (EM) waves. According to a report (Hansen et al. 2007, Williams 2008), the emission of carbon dioxide due to human industrial activities is several times larger than the natural absorption capability of nature. The telecommunication industry holds 3% of the global total carbon dioxide to emissions (Kelly 2007, Karl 2003). Particularly, 60% of the energy utilization of the telecommunication industry is attributed to subscribers and communication instruments in mobile node and communication systems (McLaughlin 2008). Therefore, the telecommunication industries and governments have recently shown keen worries on the serious problems correlated to EE in the telecommunication area. This provides a new path in the direction of "Green transmission technologies and networking", which is the topic of interest in this work. The limitations of existing wireless technologies can be illustrated as follows:

- Most of the existing approaches in wireless networking try to maximize measuring parameters, such as reliability, QoS, and throughput, but do not pay even slight attention to the consumption of energy by communication instruments and networking devices. The design of communication instruments is based on performance but not based on EE awareness (Gupta et al. 2019).
- The communication instruments are only fully utilized in peak time hours. In normal hours, their demand of energy is much lower than the given power. Hence, the distribution of power in not dynamic and does not depend upon traffic conditions.
- Most of the existing green approaches achieve EE transmission at the cost of compromising with QoS. Therefore, the trade-off between consumption of energy and performance should be carefully analyzed.

3.1.1 Green Transmission Technology-Based Network Architecture

Figure 3.1 represents the proposed green transmission technology-based network architecture for next generation networks (5G and beyond). The key operational blocks of the proposed green network architecture are as follows.

Macro base station: Macro base stations are installed for minimizing ping-pong effects and enhancing mobility performance for high-speed networking environment.

Micro base station: Micro base stations consist of a remote mobile nodes' controller and baseband signaling unit. The micro base station is linked with mobile nodes and is responsible for measurements of handover parameters related to the target network.

AI enabled data processing unit: This is responsible for signal processing and intelligent resource allocation. It consists of application server and data base used for training and tracking purposes.

Cloud network: The cloud network helps in virtualization and computing techniques for information centers. Cloud-computing-based networking has coordinating and computing abilities to facilitate the exchange of learned data

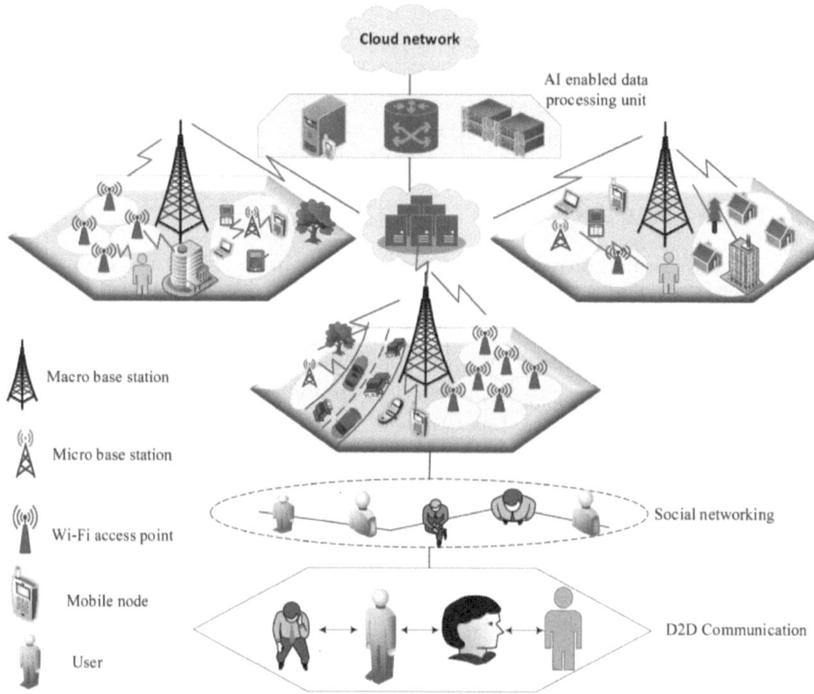

FIGURE 3.1 An artificial intelligence (AI)-based multi-tier green network architecture.

among base stations and mobile nodes, which tiles the path for green transmission networking with intelligent resource optimization.

D2D communication: This helps to establish the direct link between base station and mobile node without consumption of energy and transmission power from the base station.

Social networking: The mobile nodes willingly participate in data delivery and share their information with each other, and follow the protocols of underlying wireless networking.

3.1.2 Green Transmission Technologies

Nearly one-third of the process outflow of a typical wireless communication network is for the consumed energy, so operative green transmission technologies are vital to lessen the consumption of energy associated with complete communication system and networking.

Green transmission technologies are expected to comprehend the requirements of the mobile nodes as well as provide a solution for EE-based green networking. There are various next generation green transmission technologies such as massive MIMO, green IoT, software defined networks, ultra-dense networks, beamforming, D2D communication, and NOMA (non-orthogonal multiple access) (Kumar et al. 2020, Kumar et al. 2020). However, we discuss and summarize the key challenges associated with technologies with future research direction. The consumption of energy in wireless networking devices and circuits is a critical problem in the utilization of renewable energy for designing a green transmission technology based wireless communication system and networks (Okegbile et al. 2020).

3.2 ENERGY-EFFICIENT NETWORK TOPOLOGY

The green transmission approach has emerged as a promising technology with many applications, where energy efficiency is one of the critical necessities (Awasthi et al. 2019). In a mobile ad hoc network (MANET), each mobile node can act as a coordinator so there are several possible network topologies. To minimize the energy consumption, in this work, we try to form network topologies with a smaller number of coordinators while still continuing link connectivity. By minimizing coordinators, the battery life of the mobile node is enhanced and the average duty cycle is minimized.

3.2.1 System Model for Green Network Topology

There are five mobile nodes (MN_1, MN_2, MN_3, MN_4, and MN_5), which participate to form a MANET as represented in Figure 3.2. The MN_1 has a route to sink MN_5. If MN_2 is just a networking device, it is connected with either MN_1 or MN_3. On the other hand, if the coordinator is MN_1, the MN_2 cannot synchronize directly with MN_3. When the data for MN_1 comes from sink MN_5 to MN_3, the MN_3 cannot forward this data to MN_2 since it does not have any information about when the receiver of MN_2 will be ready to receive the data. Symmetrically, if MN_2 is related with MN_3, a data from the MN_1 cannot arrive sink at MN_5. This topology also applies to other intermediate mobile nodes such as MN_3 and MN_4.

3.2.2 Mechanism for Energy-Efficient Green Network Topology

In this work, all the communication channels are assumed to be bidirectional and acknowledgement for each data packet is exchanged after successful reception. Since the non-communicating state consumes nearly the same energy as in the

FIGURE 3.2 System model for green network topology.

communication state (either transmission or reception), we will utilize the duty cycle considered as the fraction of time duration that a mobile node is in communication state or non-communicating state to represent the energy consumption of a mobile node. It is assumed that the data forwarding task is consistently distributed among all mobile nodes. Since each mobile node does not require to forward the data from the other mobile nodes, the duty cycle of each mobile node is denoted as $T_d : (T_d < 1)$. The measurement for the value of T_d associated with duration of wake-up time period, wake-up frequency and the length of the forwarded data. For a green wireless network that consists of i coordinators and j mobile nodes, an average duty cycle is determined as

$$D_j = \frac{i + (j - i)T_d}{j}$$

$$\Rightarrow D_j = T_d + \frac{i}{j}(1 - T_d)$$

In order to minimize the energy consumption of green wireless networks, the number of coordinators should be minimized.

3.3 ENERGY-EFFICIENT ROUTING PROTOCOLS

The energy saving approaches are commonly applied in the hardware design of a wireless communication system. Since the interface among networks is an important part of energy consumption, an extensive literature has been ardent about the low-energy design of a complete network protocol stack of the communication system in order to improve EE (Srivastava et al. 2020). The following subsection discusses the protocol layers with their research directions and transmission control protocol (TCP) for green wireless networks.

3.3.1 Protocol Layers

The programmes used in applications do not interface directly with the hardware part used in the network. Instead, the protocol (set of rules and regulations) software is used to interface among different layers to execute complete communication. The complexity in a set of protocols is associated with the design of real-time hardware implementation and works together with the hardware to achieve green transmission. Figure 3.3 summarizes a conventional OSI protocol stack with layer wise research directions in the field of green networks.

3.3.2 Protocols for Green Networks

Recently, several approaches have been suggested to improve the effects of non-congestion-related losses on TCP performance over communication system. There are mainly two approaches to support a minimum number of retransmissions and hence save energy. These approaches are the divided link protocol and link layer

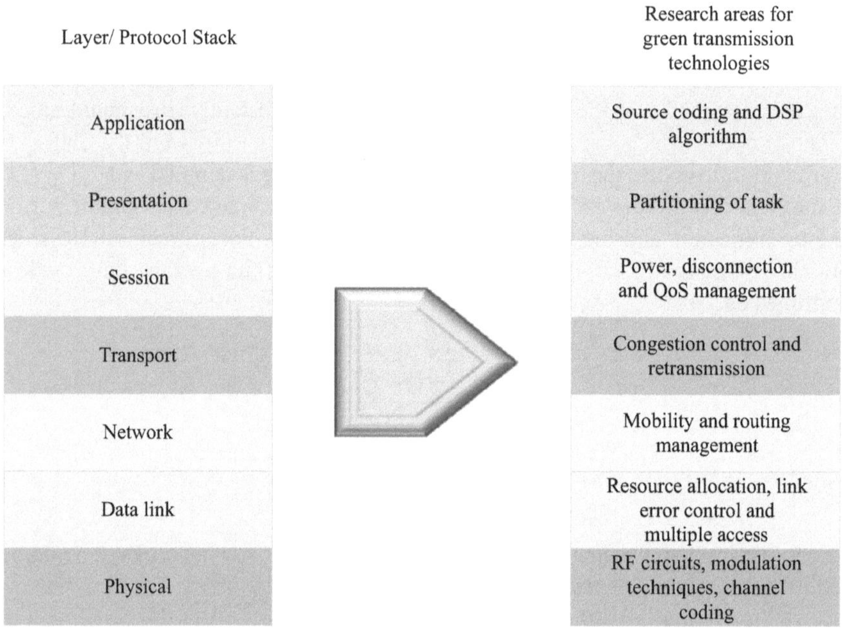

FIGURE 3.3 Protocol stack for green networks and the corresponding research area.

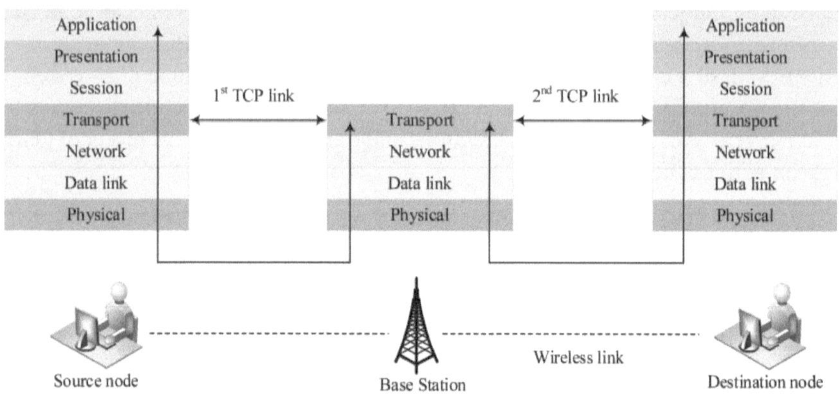

FIGURE 3.4 Divided link protocol approach.

protocol. Figure 3.4 represents the divided link protocol in which TCP links are terminated at the base station. This is performed by dividing each TCP link between the source and destination into two distinct links at the base station.

The link layer protocol approach is represented in Figure 3.5. In this approach, a grouping of forward error correction and local retransmissions is utilized to transmit the data from the TCP source to the destination over the wireless link. The parameters related to wireless link are tuned with the help of local retransmissions.

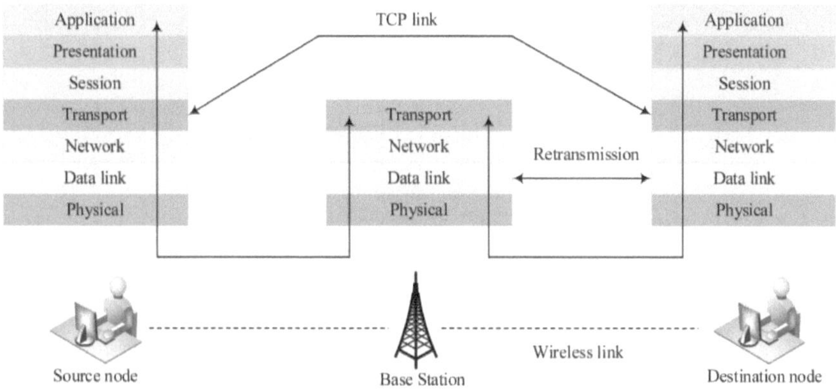

FIGURE 3.5 Link layer protocol approach.

3.3.3 Analysis of Protocol for Energy Consumption

The EE related to a protocol is calculated as the average number of completed transmissions per energy unit. Mathematically,

$$EE = \frac{Average\ number\ of\ completed\ transmissions}{Energy\ (unit)}$$

It is also realized that EE may be quite associated with the kind of TCP applied and the option of tuned parameters related to the protocol. The discussed protocols usually attain higher throughput rates over the communication link than typical TCP. The reason is that the protocols have the capability to acclimatize to a dynamic network scenario. However, the other factors such as frequency, overhead costs, disconnections' handling and mobility also affects the performance of a protocol. Therefore, energy conservation and performance may vary broadly for these protocols reliant upon both external environmental and internal algorithm aspects.

3.4 ENERGY-EFFICIENT DATA TRANSMISSION

Green network-based wireless communication will be a core fragment of the interacted society. It will help to appreciate the idea of unbounded access to networks and partition of data anywhere at any time for anyone and everything. In current wireless networks the performance of energy-based communication plays an important role in next generation networking on the mobile node side. While mobile nodes have empowered elongated battery life in high-energy wireless networks and has been an important portion after the telecommunication revolt. The main objective in green networks is to minimize the total energy consumption of the network. Figure 3.6 summarizes different energy-efficient data transmission approaches. Consequently, in the green network-based wireless communication system, energy performance has an important role to play.

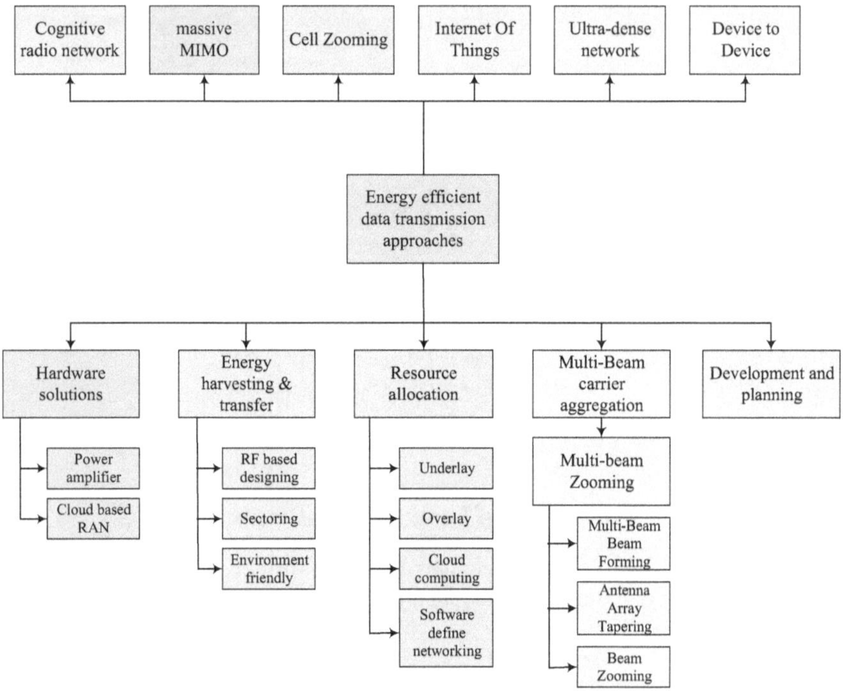

FIGURE 3.6 Different energy-efficient data transmission approaches.

3.4.1 Hardware-Based Solutions

Hardware-based solutions indicate a common class of practices to encompass the green strategy of the RF chain for EE. The intermodulation mutilation is occurred when a high peak to average ratio (PAPR) signal is processed via a nonlinear power amplifier (Elmachkour et al. 2014). To minimize this nonlinear effect from the power amplifier, high linearity is essential to halt contortion. Figure 3.7 represents a hardware-based approach to achieve the objective of green networks. The virtual power is swapped, exploiting the intermingling of measured current or voltage supply at the hardware aspect and controlled voltage or current supply inside the network's base station.

3.4.2 Software-Based Solutions

Recently, the software-based solutions for green networks pulls the attention of the core wireless networking approaches collectively (Rehman et al. 2017). The cloud computing introduces computational mobile nodes designated as edge servers. The edge server (at the base station) will process the data transmission with low intensity. Both the edge and cloud servers have controllers named as edge controller and cloud controller, respectively. Figure 3.8 represents an architecture for cloud computing over a wide area network (WAN), which can offload the transmission of data from

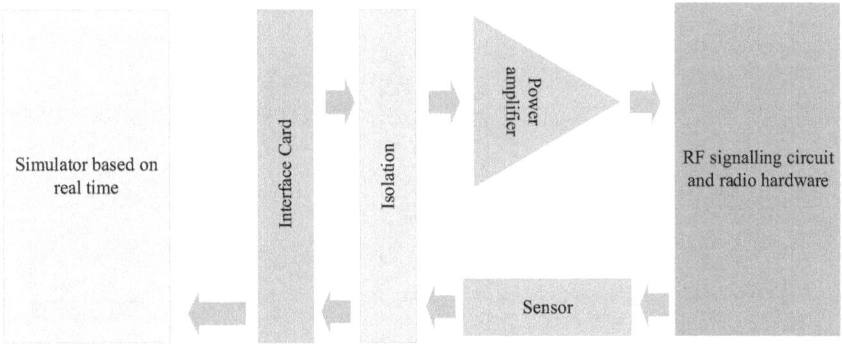

FIGURE 3.7 RF circuit-based hardware approach for low-power transmission.

wireless backhaul to core network. It is ideal design to regulate the cloud data utilization amongst all participating candidate networks in a green transmission-based cloud model.

3.4.3 Energy-Harvesting-Based Approach

The requirements for energy are continuously growing, with unparalleled development in wireless networks' application and services, causing a battery exhaustion issue for mobile nodes and network devices. An energy-harvesting-based approach provides us numerous promising gifts and notable adoptions for next generation wireless networks, i.e., green networks (Srivastava et al. 2021). Thus, the several real-time applications like remote natural monitoring, remote medical solutions, live TV, consumer electronics etc., are considering an energy-harvesting-based approach as a better choice in green communication (Singh et al. 2019). Additionally, for different classes of wireless networks very astonishing measures of energy is essential (Xu et al. 2018). To fulfil energy demands, various types of energy sources are used. The solar system, wind, EM wave, mechanical vibration, wind etc., are available options for energy harvesting in a green wireless network (Parasher et al. 2019). Figure 3.9 represents the communication system for energy-harvesting-based green networks (applied to each of the source and destination ends). For several sources of energy, this approach considers some energy-harvesting circuits such as RF wave and photo-voltaic circuits. An amount of stored energy can be easily estimated due to stochastically harvesting of energy. While to obtain the peak channel coding rate in an energy-harvesting-based approach, the parameters of transmitter antenna are difficult to tune to manage its energy optimally. The energy-harvesting approach is mainly applied to substitute the battery to extend a transmitter's lifetime.

3.5 GREEN ULTRA-DENSE NETWORKS' DESIGN

The exponential evolution in the call for real-time services and higher data rates in communication systems needs a denser installation of base stations within coverage

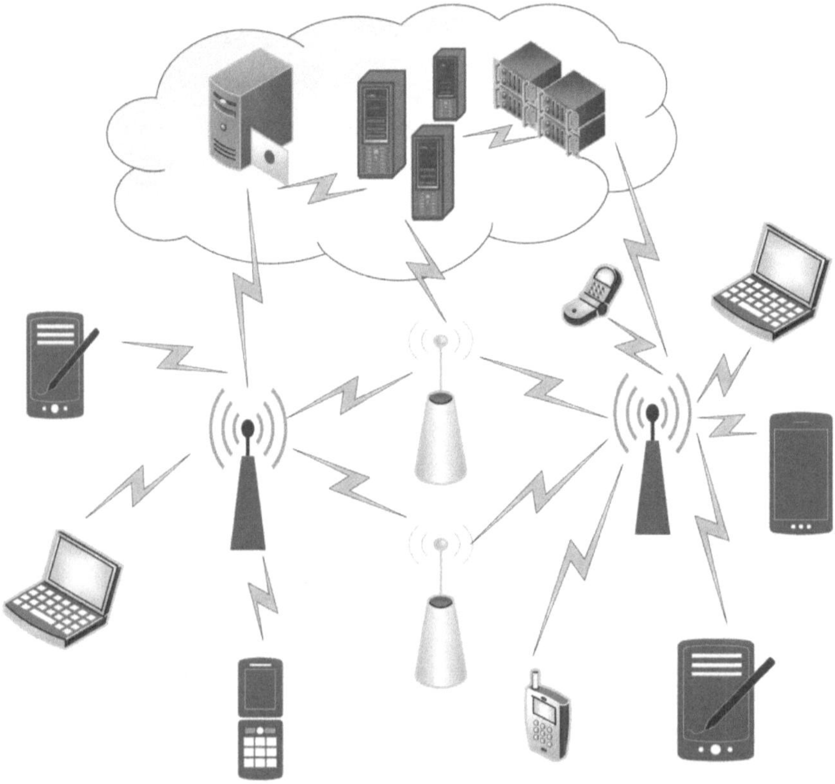

FIGURE 3.8 Cloud-computing (software-based) approach for green networks.

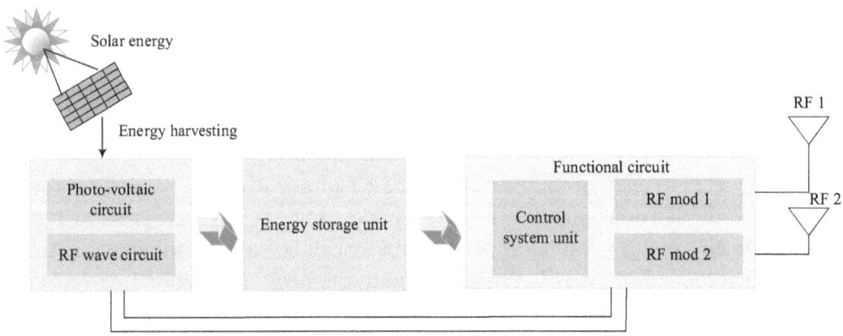

FIGURE 3.9 Energy-harvesting-based approach for green networks.

areas of network. The macro cells are usually installed to cover a large region and are not capable to offer high data rates. This problem can be resolved by minimizing the propagation separation between the mobile nodes, hence minimizing the transmission energy (Hawasli et al. 2017). Therefore, green network integrated solutions based on

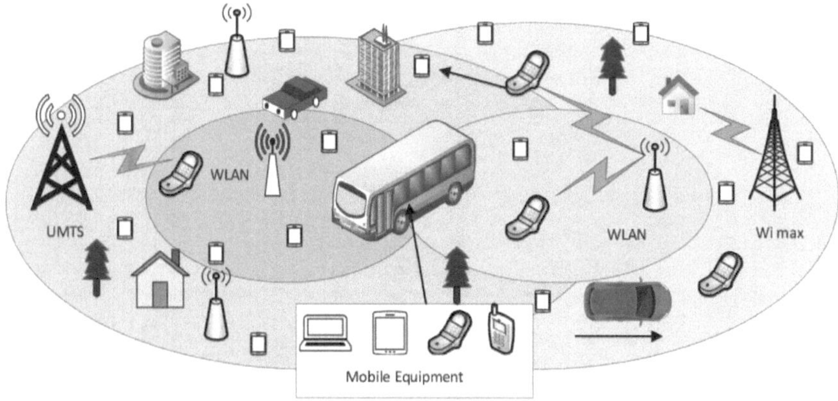

FIGURE 3.10 Green ultra-dense network architecture.

small size cells like femtocell, microcell, and picocell are very hopeful in next generation networking. A typical green communication-based heterogeneous wireless network positioning is represented in Figure 3.10. A picocell/microcell is a cell in the wireless communication system served by a low-energy base station that covers the small region with populated traffic areas like airports, shopping malls, railway platforms, schools, residential areas, bus stations and hotels etc.

The coverage area of the picocell/microcell is in the order of a few hundred meters. These small cells are designed to provide service in much smaller areas. The coverage of femtocells is usually in the order of a few meters. A small cell-based ultra-dense network is more energy efficient in offering a higher data rate to the mobile nodes. As an example, a small cell uses a nearly 150 mW power amplifier and consumes 10 W total energy compared to a macrcell-based deployment (nearly 10 kW) (Kumar et al. 2020). However, to make communication green in an ultra-dense network environment, it is essential to advance energy efficiency affected by inter-cell interference (Gupta et al. 2017). The integration with learning techniques, the ultra-dense network can be used to establish EE green networks and decision-making communication systems.

3.6 GREEN NETWORK PERFORMANCE MEASUREMENTS

Various metrics have been studied so far in order to analyze the energy efficiency related to a wireless communication network and consumption at different stages, i.e., the base station, an internal component, the radio access technology etc. (Piran et al. 2020). These metrics may be classified as energy consumption metrics.

Table 3.1 represents a taxonomy used to analyze the energy metrics. Most current mobile nodes are fortified with multiple radios with dissimilar magnitudes of energy consumptions.

TABLE 3.1
A Taxonomy for Energy Metrics in Green Networks

Metrics	Calculation	Unit	Domain	Creator
Telecommunications' Equipment Energy-Efficiency Ratio	$TEER = \frac{\sum_{i=1}^{n} D_i}{\sum_{j=1}^{n}\left(\frac{P_{0j} + P_{50j} + P_{100j}}{3}\right)}$	log (Watt / Gbps)	Router, Switch, Amplifier	Verizon
Power Usage Effectiveness	$PUE = \frac{\text{Total Facility Power}}{\text{Equipment Power}}$	Ratio	Information Center	Green Grid
Energy Proportionality Index	$EPI = \frac{P_{max\_overload} - P_{idle}}{P_{max\_overload}}$	Percentage	Network Devices	HP Labs
Energy Consumption Rating	$ECR = \frac{\text{Energy Consumption}}{\text{Effective System Capacity}}$	Watt / Gbps	Internet service provider	ECR Initiate
Power Consumption per Line of Broadband	$P_{Bline} = \frac{\text{Power Consumption of broadband Equipment}}{\text{Number of sublines}}$	Watt / subscriber line or Watt / port	Equipment used in broadband networks	ETSI
Data Center Productivity	$DCP = \frac{\text{Useful work}}{\text{Total facility power}}$	Ratio	Information Center	Green Grid
Consumer Consumption Rating	$CCR = \frac{E_{consumer\_device}}{\sum_{j=1}^{n} A_j}$	Rad (dimensionless)	Consumer network devices	Juniper
Normalized Power Consumption	$NPC = \frac{1000 \times P_{Bline}}{\text{Bit rate} \times \text{Line Length}}$	Watt / (Mbps x km)	Equipment used in broadband networks	ETSI
ECR-Weighted	$ECRW = \frac{0.35 \times E_{full\_load} + 0.4 \times E_{half\_load}}{\text{Effective Throughput}}$	Watt / Gbps	Internet service provider	ECR Initiate
Data Center Infrastructure Efficiency	$DCiE = \frac{\text{Equipment Power}}{\text{Total Facility Power}} \times 100\%$	Percentage	Information Center	Green Grid

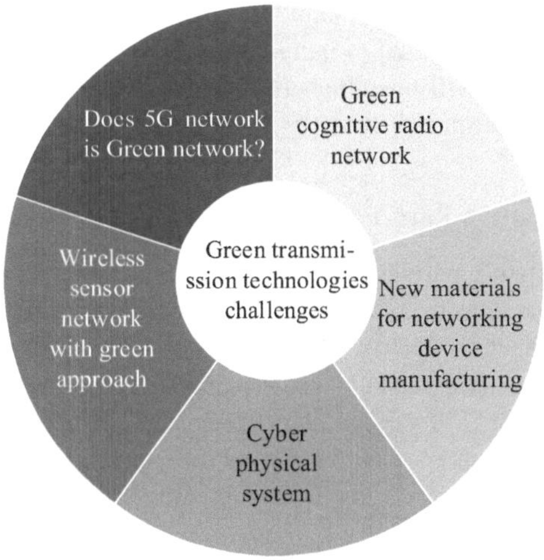

FIGURE 3.11 Green transmission technology challenges.

3.7 CHALLENGES OF GREEN TRANSMISSION TECHNOLOGIES

Green transmission technologies will play a vital role in empowering the novel stage of the energy-efficient ICT industry, but several challenges relating to green networks still continue to be addressed. Figure 3.11 summarizes the implementation issues and is discussed in the following.

3.7.1 5G Network As a Green Network

The proposal of 5G wireless networks is assumed to follow two main standards such as massive MIMO and LTE-Advanced. The greenness of the telecommunication system in the upcoming decade will be extremely subject to the EE-based transmission technologies. Due to use of multiple antenna arrays in m-MIMO technology, a large amount of energy is required. This energy-consuming problem becomes more challenging when multiusers are connected in the network. More attention needs to be paid to the apprehensions of the system complexity and energy cost of link characterization. Future research directions still fall within the sensitivity of network traffic load, interference avoidance, and the adjustment of small cell (pico/femto) power due to dynamic environment (Kaur et al. 2018).

3.7.2 Wireless Sensor Network with a Green Approach

Towards the next generation green wireless networking environment of communication systems, the incorporation of a wireless sensor network and another wireless network, e.g., Wi-Fi, LTE, WiMAX, UMTS, will offer intelligent and ubiquitous services to mobile nodes anywhere and anytime (Chen et al. 2016). For a wireless sensor network, the energy consumption problem becomes more challenging for a sustainable positioning. The sensor visualization, data aggregation diagnosis are some approaches that can be considered to handle this issue. Also, large-scale positioning of green sensor networks will meaningfully aid observation of the greenness of the earth and telecommunication industry.

3.7.3 Green Cognitive Radio Network

One major issue of existing wireless networks is that most mobile nodes are overfilled in a very limited RF spectrum, while the spectrums of other networks are not used opportunistically (Xin et al. 2016). The cognitive radio (CR) technology, permits a mobile node to adopt and change the transmission parameters to communicate through different spectrums in a dynamic manner. The CR networks provide benefits and opportunities to both green network and mobile nodes, but due to the restraint of real-time CR hardware advance, most of the existing work is only based on theory and simulation.

3.7.4 Cyber Physical System

The device-to-device (D2D) communication technology permits intelligent connecting of networking nodes, and mainly the green network-based D2D, will additionally realize energy-efficient communication among end nodes in a wider range (Eappen et al. 2020). This technology supports more dynamic real-time multimedia services and applications such as automation, telematics, smart monitoring etc. The key challenge in EE green D2D communication is the sustainable power supply for the mobile node. The cyber physical system is proposed to coordinate and tightly combine between the physical devices used in networking and communication system's computational control with feedback loops. While carrying out wide research directions on topics regarding the security, reliability, and structural designs of the cyber physical system, the EE challenge will be complex due to the bounded energy distribution of physical actuators and cyber sensors.

3.7.5 New Materials for Networking Device Manufacturing

While various research works are put into energy-saving wireless networking techniques, the new potential to develop a new energy source are very much inspired in recent years. This provides a path to the development of energy-efficient transmission, i.e., green networks. Efficient acceptance of natural energy resources such as solar and wind to assist existing wireless communication systems can be one promising research area. However, scholars are still focusing on improving the conversion efficiency (Srivastava et al. 2019). Also recycling wasted materials and utilization of renewable materials can also pave the way for green networks.

CONCLUSION

This chapter studied the current trends of the various transmission technologies and network protocols, which envision a move towards green communication networks. The green wireless communication system considers a group of recent technologies. For a green communication system, features, advantages, drawbacks, future research direction, and open issue challenges have been discussed. The requirement to develop energy-efficient communication systems has become more popular as wireless networks are becoming omnipresent. The energy-efficient green wireless communication system will produce fewer radiative devices and also provide more economic benefits to the mobile subscriber and service providers. The government must also set laws and rules to declare a service supplier as a green service supplier.

ACKNOWLEDGEMENT

This work is supported by Women Scientist Scheme-A, Department of Science and Technology, New Delhi, Government of India under Grant File No: SR/WOS-A/ET-154/2017.

REFERENCES

A.U. Rehman, L.-L. Yang, L. Hanzo, (2017) Delay and throughput analysis of cognitive go-back-n harq in the face of imperfect sensing, IEEE Access 5, 7454–7473.

Chen, Ruibiao and Shu, Fangxing and Lei, Kai and Wang, Jianping and Zhang, Liangjie, (2021) "User Clustering and Power Allocation for Energy Efficiency Maximization in Downlink Non-Orthogonal Multiple Access Systems" Applied Sciences, Multidisciplinary Digital Publishing Institute.

Eappen, Geoffrey and Shankar, T., (2020) "A Survey on Soft Computing Techniques for Spectrum Sensing in a Cognitive Radio Network", SN Computer Science, no. 1, pp. 1–36.

Elmachkour M., Daha I., Sabir E., Kobbane A., and Ben-Othman J., (2014) "Green opportunistic access for cognitive radio networks: A minority game approach," in IEEE International Conference on Communications (ICC). IEEE, pp. 5372–5377.

Gupta M. S., Srivastava A., Kumar K., (2017) "Seamless Vertical Handover for Efficient Mobility Management in Cooperative Heterogeneous Networks," 2nd International Conference on Data Engineering and Communication Technology (ICDECT-2017), Springer, pp.145–153, Pune, India.

Gupta M.S., Kumar K., (2019) "Progression on Spectrum Sensing for Cognitive Radio Networks: A Survey, Classification, Challenges and Future Research Issues", vol. 143, pp. 47–76, Journal of Network and Computer Applications, Elsevier.

Hansen J., Sato M., Kharecha P., Russell G., Lea DW., Siddal M., (2007) Climate change and trace gases. Philos Trans R Soc 365(1856):1925–1954.

Karl H (2003) An overview of energy efficient techniques for mobile communication systems. Technische University Berlin, Tech. Report.

Kaur, G., Tomar, P., & Singh, P., (2018) Design of cloud-based green IoT architecture for smart cities. In Internet of Things and Big Data Analytics Toward Next-Generation Intelligence (pp. 315–333). Springer, Cham.

Kelly T (2007) ICTs and climate change. ITU-T Technology, Tech. Report.

Kumar A. and Kumar K. (2020) "Relay sharing with DF and AF techniques in NOMA assisted Cognitive Radio Networks,", Physical Communication, Elsevier.

Kumar A. and Kumar K., (2020) "Multiple access schemes for Cognitive Radio networks: A survey,", Physical Communication, Elsevier.
Kumar A., Kumar K., Gupta M. S. and Kumar S., (2020) "A survey on NOMA techniques for 5G scenario," Available at SSRN 3573579.
L. Chen, F. R. Yu, H. Ji, B. Rong, X. Li, and V. C. Leung, (2016) "Green fullduplex self-backhaul and energy harvesting small cell networks with massive MIMO," IEEE Journal on Selected Areas in Communications, vol. 34, no. 12, pp. 3709–3724.
Li, L., Xu, K., Li, T., Zheng, K., Peng, C., Wang, D., Wang, X., Shen, M., and Mijumbi, R., (2018) A measurement study on multi-path TCP with multiple cellular carriers on high speed rails. In Proceedings of the 2018 Conference of the ACM Special Interest Group on Data Communication (pp. 161–175).
M. Awasthi, M. Nigam, V. Kumar, (2019) Optimal sensing and transmission of energy efficient cognitive radio networks, Wirel. Pers. Commun.1–12.
M. Hawasli, S.A. Çolak, (2017) Toward green 5G heterogeneous small-cell networks: power optimization using load balancing technique, AEU-Int. J. Electron. Commun. 82 474–485.
Mani Shekhar Gupta, Krishan Kumar, (2020) "Application Aware Networks' Resource Selection Decision Making Technique using Group Mobility in Vehicular Cognitive Radio Networks", Vehicular Communications, Elsevier, vol. 26, p. 100263.
McLaughlin S (2008) Green radio: the key issues programme objectives and overview. Wireless World Research Forum.
Okegbile, Samuel D and Maharaj, Bodhaswar T and Alfa, Attahiru S., (2020) "Stochastic geometry approach towards interference management and control in cognitive radio network: A survey", "Computer Communications", Elsevier.
Parasher, Y., Singh, P., & Kaur, G., (2019) Green Smart Security System. Green and Smart Technologies for Smart Cities, 165–184.
Piran, Md Jalil and Pham, Quoc-Viet and Islam, SM Riazul and Cho, Sukhee and Bae, Byungjun and Suh, Doug Young and Han, Zhu, (2020) "Multimedia communication over cognitive radio networks from QoS/QoE perspective: A comprehensive survey" "Journal of Network and Computer Applications", Elsevier, vol. 172, p. 102759.
Singh, P., Dixit, V. & Kaur, J., (2019) Green Healthcare for Smart Cities. Green and Smart Technologies for Smart Cities, 91–130.
Srivastava A, Gupta M. S. and Kaur G., (2020) "Energy efficient transmission trends towards future green cognitive radio networks (5G): Progress, taxonomy and open challenges", Journal of Network and Computer Applications, Elsevier, vol. 168, p. 102760.
Srivastava A., and Kaur G., (2021) "Resource management for traffic imbalance problem in green cognitive radio networks", Physical Communication. Elsevier, vol. 48, p. 101437.
Srivastava A., Gupta M. S., Kaur G., (2019) "A Game Theory based Approach for Opportunistic Channel Access in Green Cognitive Radio Networks," IEEE International Conference on Advanced Networks and Telecommunications Systems (ANTS-2019), IEEE, pp.1–6, Goa, India.
Williams F (2008) Green wireless communications. eMobility, Tech. Report.
Y. Xin, D. Wang, J. Li, H. Zhu, J. Wang, and X. You, (2016) "Area spectral efficiency and area energy efficiency of massive mimo cellular systems," IEEE Transactions on Vehicular Technology, vol. 65, no. 5, pp. 3243–3254.

4 Green Ad-Hoc Communication System Based on NOMA Technique

Ashok Kumar, Mani Shekhar Gupta, and Akanksha Srivastava

CONTENTS

4.1 INTRODUCTION OF AD-HOC NETWORKS

In the field of networking, the ad-hoc network defines a network without infrastructure, which is also called a spontaneous network. The comparison of ad-hoc networks with infrastructure-based wireless networks is defined as the other way to understand ad-hoc networks, i.e., cellular networks and wide local area networks (WLANs) (Gupta et al. 2019). The generalized ad-hoc networks consisting of the nodes featured with a transceiver and the created independent network from an infrastructure (Gupta et al. 2020). Thus, the nodes have to design their own networks to communicate with other nodes that lie in their range. Hence, the absence of a centralized structure means the ad-hoc network's nodes act as a router to transmit and receive the information.

DOI: 10.1201/9781003264477-4

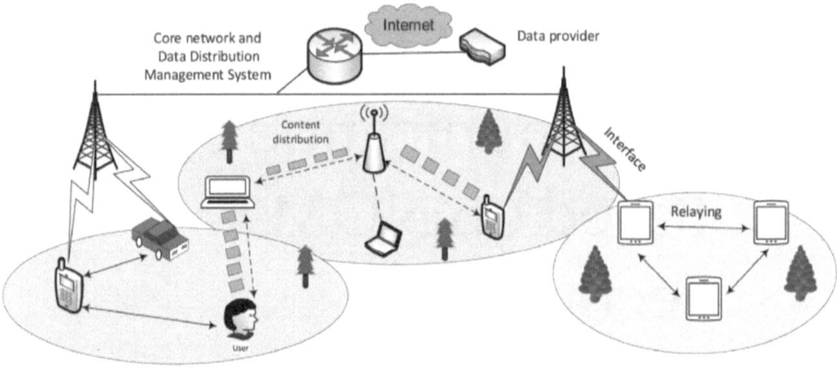

FIGURE 4.1 Ad-hoc networks for the green communication system.

Ad-hoc networks are non-static in nature so that they avoid the single point of failure and make the network more robust (Hoang et al. 2019). Whereas, in an infrastructure-based communication system, the various nodes communicate to the destination node situated in different networks by sending data to an associated access point (AP) (Kumar et al. 2020). It implies that these networks are not powerful enough. In this system, every node has its own networking range, and they have the ability to have a large networking area after combining these small ranges. Additionally, nodes can use one or many hops to transmit data to the desired destination. Figure 4.1 shows the ad-hoc networks for the green communication system.

4.2 INTRODUCTION OF NON-ORTHOGONAL MULTIPLE ACCESS (NOMA) TECHNIQUE

Designing energy-efficient ad-hoc networks is a key requirement of the present scenario. For this, promisingly, the non-orthogonal multiple access (NOMA) (Kumar et al. 2020) technique has become the future for futuristic (5G and beyond) green energy-efficient (Srivastava et al. 2020) communication systems. It also improves the spectral efficiency over existing orthogonal multiple access (OMA) in cognitive radio networks (CRNs) (Kumar et al. 2020).

Nowadays, power consumption of users (mobiles or terminals or moving nodes) has emerged with significant issues faced by researchers, with the need to eliminate or achieve optimal reduction in power consumption to make networks green and suitable for future communication. In other words, it needs to enhance the communicating device lifetime by providing energy from some external factor, i.e., solar panel, energy from received radio frequency (RF) NOMA signal to reenergize the battery. Regrettably, these renewable energy sources are not compatible with small devices like mobiles (Liu et al. 2016). Sometimes, these sources cannot be used for sensor and health care monitoring with real-time needs, but RF signals' energy improves availability and power density and does not depend on climatic conditions. Hence, energy-harvesting (EH) is commonly named simultaneous wireless information and power transfer (SWIPT) (Makki et al. 2020). SWIPT in relaying an ad-hoc network increases the EH and area of coverage for green communication CRNs.

The integration of NOMA and SWIPT prolongs network efficiency and its lifetime. In fairness, EH and NOMA system sum rates are tradeoffs (Moltafet et al. 2018). RF energy is harvested by a strong user and used for further signal forwarding to a weak user. This results in an optimum choice time switching ratio with SWIPT in the NOMA system to avoid impairment of diversity gain (Liu et al. 2016). Nguyen et al. (2019) discuss the impact of the time switching ratio and power allocation over NOMA systems with multiple users. Additionally, for two system users, the authors derived the probability of outage expression with a fixed power in CRNs. This shows if the adopted power is allocated to users that can improve system performance over existing ones.

The literature motivates the above issues to be resolved through the current chapter, which discusses the NOMA system with EH relaying under ad-hoc green CRNs (Kumar et al. 2020). The proposed scenario is a workout in two phases, namely EH and transmission phases, respectively. The key contribution of this chapter is as follows:

- NOMA is being introduced along with EH relay to make the system cooperative and green with a vision of great futuristic ad-hoc CRNs.
- We derived an outage probability expression for both users to determine system performance. Also, we compared it with existing OMA relaying CRNs.
- Next we compare the results as the time switching (between signal transmission and EH times) ratio over the minimum outage probability validated through simulations.
- The advantage, challenges, and applications of ad-hoc CRNs have been comprehensively investigated.

4.3 PROPOSED SYSTEM MODEL AND ASSUMPTIONS FOR DESIGNING OF NOMA-BASED GREEN AD-HOC NETWORK

In this section we have represented an EH relay-based NOMA for green ad-hoc CRNs. Let a NOMA-assisted system model combine a BS as a source, multiple EH relays as a booster, and two destinations as users (U1, U2), as shown in Figure 4.2. These users are treated as primary and secondary users, respectively, in ad-hoc CRN systems. This model operates in two phases: EH by the best relay from R and transformation of information.

4.3.1 Energy-Harvesting Phase

During this EH phase β_T, where β refers ratio of time switching $(0 \le \beta \le 1)$, every relay harvests energy through the received NOMA RF signal. The relay that gives maximum EH can be the best (or provides the best links) and can be used as power for further communication. For choosing the best EH relay, we utilized the partial relay selection (PRS) technique by which BS selects the link that provides maximum

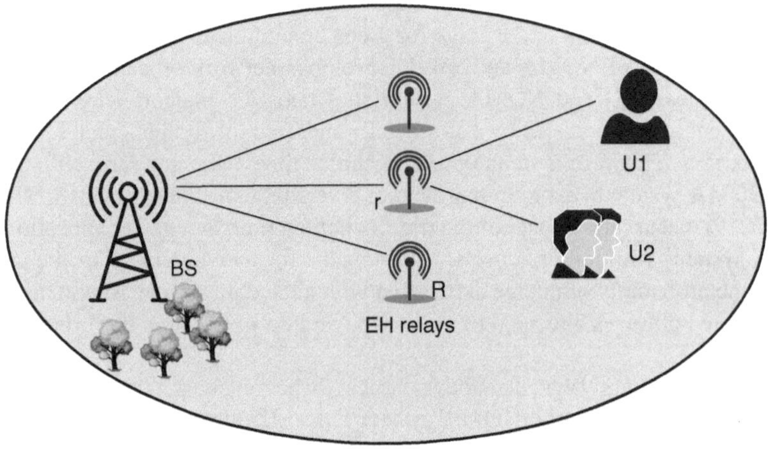

FIGURE 4.2 Proposed EH relays using NOMA scenario.

instantaneous SINR between BS itself and EH relays, i.e., the best relay (r*) can be written by

$$\mathrm{r}^{*} = arg\left[max\left(Y_{b.r}\right)\right] \tag{4.1}$$

Let $\mathbb{E}_h$ be the harvested energy during βT phase, which can be calculated as

$$\mathbb{E}_h = \eta P_b \left|h_{b,r^*}\right|^2 \beta T \tag{4.2}$$

where P_b and η refer to the power transmitted by BS and energy conversion efficiency ($0 \le \eta \le 1$), respectively. $h^*_{b,r}$ shows the fading coefficient between BS to the best EH relay, since the whole harvested energy is utilized for sending information from relay to U_i, $i \in (1, 2)$. Therefore using (4.2) the transmitting the best relay power can be written by

$$P_r = \frac{\mathbb{E}_h}{(1-\beta)T/2} = \frac{2\eta P_b \left|h_{b,r^*}\right|^2 \beta}{1-\beta} \tag{4.3}$$

4.3.2 Transmission Phase

The transmission phase splits into two slots (each having duration (1-β)T/2, one for sending information from BS to all relays and another for further communicating to both users, i.e., U1 and U2 using harvested energy. In first (1-β)T/2, NOMA signal $x_b = \sum_{i=1}^{2} \alpha_i x_i$, aired to $U_i, i \in (1,2)$, respectively. α_i refers the power coefficient with $\alpha_2 < \alpha_1$ and $\sum_{i=1}^{2} \alpha_i = 1$. Hence, the received signal at EH relays can be given by

$$y_r = \sqrt{P_b}\, x_b\, h_{b,r} + w_{b,r} \tag{4.4}$$

where $w_{b,r}$ tends to addition white Gaussian noise (AWGN), i.e., $w_{b,r} \cong CN(0, N_0)$ (Nguyen et al. 2019) between a BS and EH relay, i.e., $r \in (1,2,\ldots,R)$.

Let us assume that the information decoding order moves from U1 to U2 because U1 QoS is not much needed over that in U2. However, it prefers U1 first for decoding and continues. Also SIC is implemented at relay for lower signal detection by rejecting higher value signals. Hence the desired rates for decoding signals x_i at EH relay can be described by

$$R_1^r = \frac{1}{2} log_2 \left(1 + \frac{\rho_b\, \alpha_1^2 \left| h_{b,r} \right|^2}{\rho_b\, \alpha_2^2 \left| h_{b,r} \right|^2 + 1} \right) \tag{4.5}$$

and

$$R_2^r = \frac{1}{2} log_2 \left(1 + \rho_b\, \alpha_2^2 \left| h_{b,r} \right|^2 \right) \tag{4.6}$$

where $\rho_b \cong \frac{P_b}{N_0}$ and $\rho_r \cong \frac{P_r}{N_0}$ are assumed without loss of generality. If $R_1^r \geq R_1$ then SIC is successively workout over U2 for its own signal detection.

On successful operations over signal decoding at EH relay further forwards $x_r = x_b = \sum_{i=1}^{2} \alpha_i x_i$ with the same presumptions towards both users during 2nd half slot $(1-\beta)T/2$. Hence, received signals can be determined through

$$y_i = \sqrt{P_r}\, x_r\, h_{r,i} + w_{r,i} \tag{4.7}$$

where $w_{r,i}$ tends to AWGN, i.e., $w_{r,i} \cong CN(0, N_0)$ between EH relay and ith users.

As per (5), the instantaneous SINR for U1 is given by

$$\gamma_{r,1} = \frac{\rho_r\, \alpha_1^2 \left| h_{r,1} \right|^2}{\rho_r\, \alpha_2^2 \left| h_{r,1} \right|^2 + 1} \tag{4.8}$$

Meanwhile, to decode the U2 signal, SIC needs to be implemented for rejecting the high priority U1 signal, which can be expressed as

$$\gamma_{r,1\to 2} = \frac{\rho_r \alpha_1^2 \left|h_{r,2}\right|^2}{\rho_r \alpha_2^2 \left|h_{r,2}\right|^2 + 1} \tag{4.9}$$

With the satisfaction of $\gamma_{r,1\to 2} = \gamma_{r,1}$, the condition provides the U2 decoding signal as

$$\gamma_{r,2} = \rho_r \alpha_2^2 \left|h_{r,2}\right|^2 \tag{4.10}$$

4.4 PERFORMANCE EVALUATION OF THE PROPOSED SYSTEM MODEL

The system performance examination is done in this section. The performance of the given system can be determined through the outage probability of the users. This subsection examines the behavior of the system.

4.4.1 Outage Probability of U1

This user has the highest priority over the 2nd user. Hence if its data threshold becomes more prominent than its instantaneous SINR, it makes the system in the outage. Mathematically, user U1 probability of outage can be derived by

$$P_{out}^{U_1} = P\left(\gamma_{r,1} < \tau_1\right) \tag{4.11}$$

Substituting values into (11) gets

$$P_{out}^{U_1} = P\left(\frac{\rho_r \alpha_1^2 \left|h_{r,1}\right|^2}{\rho_r \alpha_2^2 \left|h_{r,1}\right|^2 + 1} < \tau_1\right)$$

$$= P\left(\rho_r \alpha_1^2 \left|h_{r,1}\right|^2 < \tau_1\left(\rho_r \alpha_2^2 \left|h_{r,1}\right|^2 + 1\right)\right)$$

$$= P\left(\rho_r \left|h_{r,1}\right|^2 \left(\alpha_1^2 - \alpha_2^2 \tau_1\right) < \tau_1\right)$$

$$= P\left(\left|h_{r,1}\right|^2 < \frac{\tau_1}{\rho_r\left(\alpha_1^2 - \alpha_2^2 \tau_1\right)}\right)$$

$$= P\left(\left|h_{r,1}\right|^2 < \frac{\theta}{\rho_r}\right) \tag{4.12}$$

Therefore, the outage probability of U1 can be taken by

$$P_{out}^{U_1} = \frac{1}{\rho_r} e^{-\frac{\theta}{\rho_r}} \tag{4.13}$$

4.4.2 Outage Probability of U2

As U2 needs to implement the SIC to reject the U1 signal, that first gets its own signal. For this, outage probability can be defined as the probability that a given system lies in an outage or keeps instantaneous SINR down below a predetermined threshold. Hence, the user (U2) outage probability is derived by

$$P_{out}^{U_2} = P\left(\gamma_{r,1\to 2} < \tau_1\right) + P\left(\gamma_{r,1\to 2} \geq \tau_1, \gamma_{r,2} < \tau_2\right) \tag{4.14}$$

where $\tau_1 = 2^{2R_1} - 1$ and $\tau_2 = 2^{2R_2} - 1$ tends to threshold SINR correspondingly associated with 2R1 and R2 minimum data rates of U1 (or PU) and U2 (or SU), respectively.

Nguyen et al. (2019) provides

$$P_{out}^{U_2} = P\left(\frac{\rho_r \alpha_1^2 \left|h_{r,2}\right|^2}{\rho_r \alpha_2^2 \left|h_{r,2}\right|^2 + 1} < \tau_1\right) + P\left(\frac{\rho_r \alpha_1^2 \left|h_{r,2}\right|^2}{\rho_r \alpha_2^2 \left|h_{r,2}\right|^2 + 1} \geq \tau_1, \rho_r \alpha_2^2 \left|h_{r,2}\right|^2 < \tau_2\right) \tag{4.15}$$

Taking the 1st term from (15) as

$$f_1 = P\left(\frac{\rho_r \alpha_1^2 \left|h_{r,2}\right|^2}{\rho_r \alpha_2^2 \left|h_{r,2}\right|^2 + 1} < \tau_1\right)$$

$$= P\left(\rho_r \alpha_1^2 \left|h_{r,2}\right|^2 < \tau_1 \left(\rho_r \alpha_2^2 \left|h_{r,2}\right|^2 + 1\right)\right)$$

$$= P\left(\rho_r \alpha_1^2 \left|h_{r,2}\right|^2 < \tau_1 \left(\rho_r \alpha_2^2 \left|h_{r,2}\right|^2 + 1\right)\right)$$

$$= P\left(\rho_r \left|h_{r,2}\right|^2 \left(\alpha_1^2 - \alpha_2^2 \tau_1\right) < \tau_1\right)$$

$$= P\left(\left|h_{r,2}\right|^2 < \frac{\tau_1}{\rho_r\left(\alpha_1^2 - \alpha_2^2\tau_1\right)}\right) \tag{4.16}$$

Similarly, considering the 2nd term from (15) can be arranged as

$$f_2 = P\left(\frac{\rho_r\,\alpha_1^2\left|h_{r,2}\right|^2}{\rho_r\,\alpha_2^2\left|h_{r,2}\right|^2+1} \geq \tau_1, \rho_r\,\alpha_2^2\left|h_{r,2}\right|^2 < \tau_2\right)$$

$$= P\left(\rho_r\,\alpha_1^2\left|h_{r,2}\right|^2 \geq \tau_1\left(\rho_r\,\alpha_2^2\left|h_{r,2}\right|^2+1\right), \rho_r\,\alpha_2^2\left|h_{r,2}\right|^2 < \tau_2\right)$$

$$= P\left(\left|h_{r,2}\right|^2 \geq \frac{\tau_1}{\rho_r\left(\alpha_1^2 - \alpha_2^2\tau_1\right)}, \left|h_{r,2}\right|^2 < \frac{\tau_2}{\rho_r\,\alpha_2^2}\right) \tag{4.17}$$

Recombining (16) and (17) together gives

$$P_{out}^{U_2} = P\left(\left|h_{r,2}\right|^2 < \frac{\tau_1}{\rho_r\left(\alpha_1^2 - \alpha_2^2\tau_1\right)}\right) + P\left(\left|h_{r,2}\right|^2 \geq \frac{\tau_1}{\rho_r\left(\alpha_1^2 - \alpha_2^2\tau_1\right)}, \left|h_{r,2}\right|^2 < \frac{\tau_2}{\rho_r\alpha_2^2}\right) \tag{4.18}$$

The expression can be rearranged as

$$P_{out}^{U_2} = P\left(\left|h_{r,2}\right|^2 < \frac{\theta}{\rho_r}\right) + P\left(\left|h_{r,2}\right|^2 \geq \frac{\theta}{\rho_r}, \left|h_{r,2}\right|^2 < \frac{\varphi}{\rho_r}\right)$$

Where $\theta = \frac{\tau_1}{\alpha_1^2 - \alpha_2^2\,\tau_1}, \varphi = \frac{\tau_2}{\alpha_2^2}$, and $\eta = \max(\theta, \varphi)$. Resulting

$$P_{out}^{U_2} = P\left(\left|h_{r,2}\right|^2 < \frac{\eta}{\rho_r}\right) \tag{4.19}$$

The outage probability of U2 can be obtained as

$$P_{out}^{U_2} = \frac{\eta}{\rho_r} e^{-\frac{\eta}{\rho_r}} \tag{4.20}$$

4.5 EXTENSION TO MULTIPLE USERS

Extension to multiple users has been given in this section. For multiple users, it needs to calculate the order statistics' function for determining users' outage probability. Therefore, PDF of multiple user extension functions with the help of order statistics, the PDF of $f_{|h_{r,j}|^2}$ can be written as

$$f_{|h_{r,j}|^2}(x)=\frac{U!}{(U-u)!(u-1)!}.f_{|h_{r,j}|^2}(x)\left(F_{|h_{r,j}|^2}(x)\right)^{u-1}\times\left(1-\left(F_{|h_{r,j}|^2}(x)\right)^{U-u}\right) \tag{4.21}$$

where, $|h_{r,j}|$ refers to the channel gain coefficient between the best relay to uth destination. It can be simplified as

$$f_{|h_{r,j}|^2}(x)=\frac{U!}{(U-u)!(u-1)!}\frac{1}{\lambda_{r,j}}\sum_{k=0}^{u-1}\binom{u-1}{k}(-1)^k\times e^{\frac{-x(U-u+k+1)}{\mu_{r,j}}} \tag{4.22}$$

From (22), we have

$$P_{out}^{U_j}=\frac{U!}{(U-u)!(u-1)!}\frac{1}{\lambda_{r,j}}\sum_{k=0}^{u-1}\binom{u-1}{k}(-1)^k\times\int_0^{\frac{\eta}{\rho_r}}e^{\frac{-x(U-u+k+1)}{\lambda_{r,j}}}dx \tag{4.23}$$

After solving, we get multiple user outage probabilities by

$$P_{out}^{U_j}=\frac{U!}{(U-u)!(u-1)!}\sum_{k=0}^{u-1}\binom{u-1}{k}(-1)^k\left(\frac{1-e^{\frac{-x(U-u+k+1)}{\lambda_{r,j}}}}{U-u+k+1}\right) \tag{4.24}$$

where U_j refers to a maximum number of users.

4.6 RESULTS AND DISCUSSION

This section examines the behavior of the proposed system over existing ones based on parameters appended in Table 4.1.

Figure 4.3 compares both users' outage probabilities of both (U1/U2) users with average SINR. The proposed EH relay NOMA-based green ad-hoc CRN system out behaves over existing OMA systems. As we can see, U1 and U2 both improve their outage probabilities in both (NOMA and OMA) cases, but the proposed system shows much more improved performance over OMA with regards to SINR. The outage probability of U1 is sharper (or enhanced) over the outage probability of U2 due to its (U1) higher channel gain than U2. Moreover, the outage probability of U1

TABLE 4.1
Considered Parameters for Analysis

Names	Values	Names	Values
Bandwidth (B)	20.0 MHz	Powers (P1, P2)	0.7 P_b, 0.2 P_b
Data rates	0.3, 0.5 bps	Switching time ratio (β)	0.3
Average gains $\lambda_{b,r}$ ($\lambda_{b,i}$), $i \in (1,2)$	1(2,3)	EH efficiency (η)	0.85

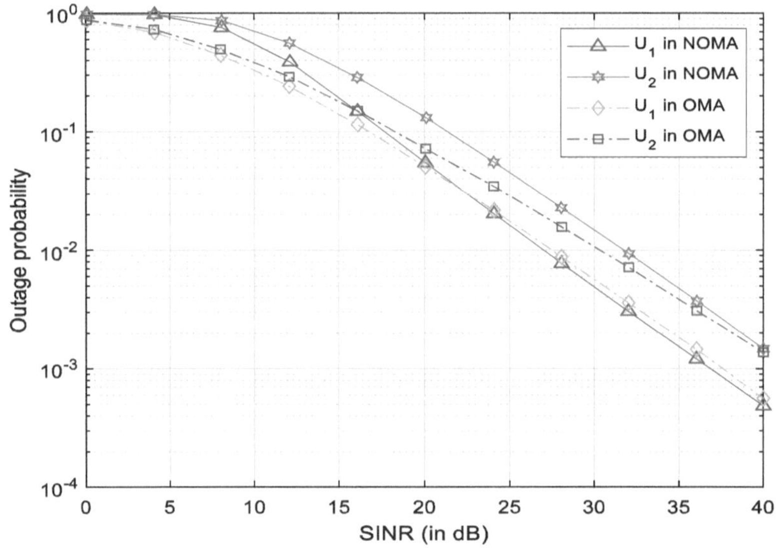

FIGURE 4.3 Outage behavior of proposed and existing systems over SINR at $\rho_r = 0.9$ and $R = 3$.

in the proposed system behaves better than OMA but is the reverse in the case of U2s' outage probability because transmission needs more time slots in OMA over NOMA, respectively, with changing SINR. Hence, the outage is the higher side in OMA than in NOMA systems.

Figure 4.4 describes outage probabilities of U1 and U2 in the proposed NOMA system with regards to SINR at distinct R relays. As Figure 4.4 depicts improved outage behavior (for both users) of a given system with rising relays (in numbers) because more relays give high flexibility for setting links (or connection between BS and relays). This makes signal decoding easier and also increases harvested energy. However, the outage probability of U1 is also significantly improved over the outage probability of U2. Diversity order is not affected anymore by relay numbers. It remains unity.

Figure 4.5 shows the throughput of proposed NOMA with regards to the average SINR with different data rates. When SNR ≤ 10 dB, i.e., low SNR, the system

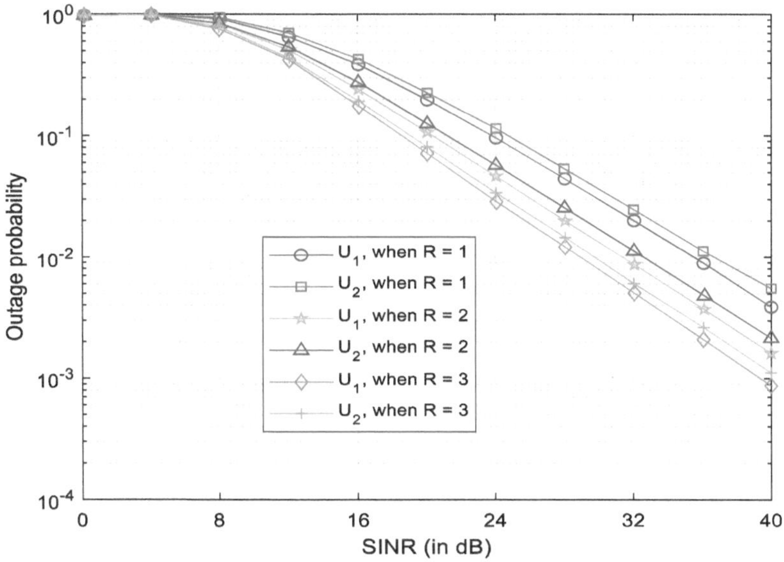

FIGURE 4.4 Impact on users with relays $R \in (1, 2, 3)$ at $\rho_r = 0.9$, $\beta = 0.3$, and $\eta = 0.85$, respectively.

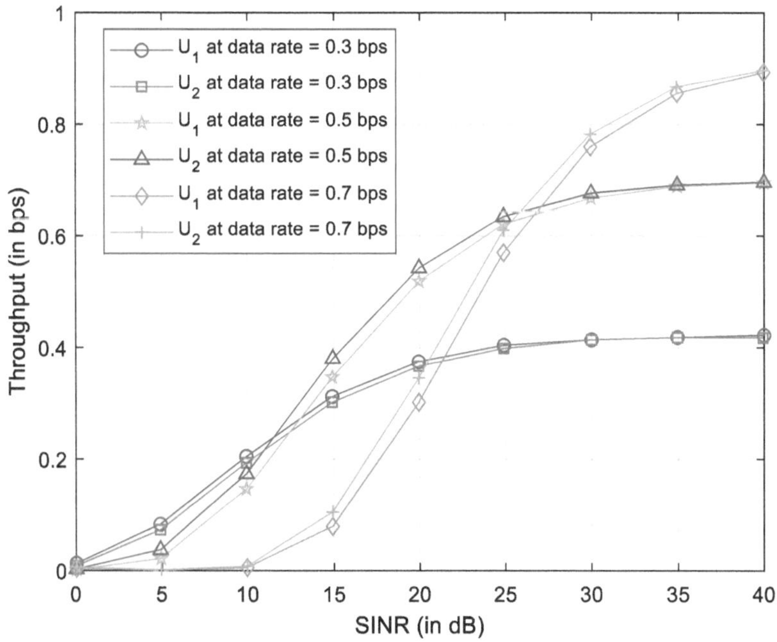

FIGURE 4.5 Comparing throughput of the proposed system at various data rates (i.e., 0.3, 0.5, 0.7).

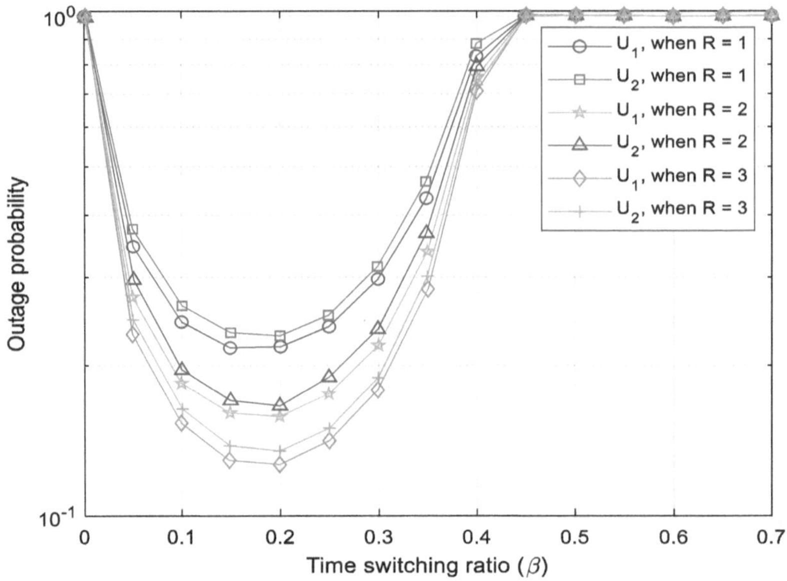

FIGURE 4.6 Outage curves of both users over $R \in (1, 2, 3)$ at $\gamma = 15$ dB and $\eta = 0.85$, respectively.

throughput approached the best-desired value in all cases with low data rate, i.e., 0.3, as given in Figure 4.5. Hence, rising SNR also increases the data rate (i.e., 0.5) to determine the system's optimal throughput (maximum at 0.7). Hence, the optimal throughput of the overall system is enhanced by suitable choosing of the data rates through P_b (based on BS transmitting power).

The impact of the time switching ratio β for EH at different relays on both (U1, U2) users' outage probability is shown in Figure 4.6. In one relay scenario, β approaches an optimal value at which outage probabilities of U1 and U2 are minimized. However, the optimal value of β decreases significantly with increasing relay numbers because good strength of $\gamma_{b,r}, r \in (1,2,\ldots,R)$, as given in Figure 4.6. Moreover, β reached the desired value (i.e., 0.2) irrespective of relay numbers.

4.7 ADVANTAGES OF AD-HOC NETWORK

There are various merits to utilizing ad-hoc CRNs than infrastructure-based networks. Its independence is its strength over the infrastructure (Tran et al. 2018). Thus, the ad-hoc network can possibly be installed in any problematic situation. Hence, it has the following advantages, which are listed below.

- **No infrastructure and lower cost:** In isolated areas where people density is meager, i.e., hilly terrain and deserted areas, it is not feasible or a good deal to establish (or setup) infrastructure. This is because it may be very expensive if it compares how often the residents of that area are utilizing the infrastructure services and what amount per day is consumed/or transmitted at the cost of maintenance, establishment, and repair. Thus, such types of cases requires an infrastructure-free networking system.

- **Mobility:** The 5G or 6G networking systems can support the rapid deployment of independent users. For example, emergency networks are so-called rescue or disaster effort networks and military networks. In such critical situations, no one should be dependent on a centralized infrastructure connection. Here, MANET helps in nodes' mobility.
- **Decentralized and robust:** These networks are inherently very robust, which is its unique feature. This can be understood by taking an example, where one BS (out of many existing) is not operating. Then all associated users with concern BS may lose their connection to different (or other) networks. This may probably happen in a structured-based networking system. But the ad-hoc system has the ability to avoid such problems. Because after the node failing or not operating, you can still connect with the help of some other node through the concept of multi-hop.
- **Easy to build infrastructure:** Impairment of the infrastructure-based networking system sometimes becomes intolerable. It is because it is too hard to replace or repair that faulty network quickly while ensuring the consistent existence of the network. Hence, ad-hoc network deployment is becoming a good choice in such rare cases.

4.8 CHALLENGES OF THE AD-HOC NETWORK

Wireless networking is pretty recognized these days. The ad-hoc networks have low data rates and security (Srivastava et al. 2021). The following are the disadvantages of given networks.

- **Higher error rate:** Unlike wired communication, wireless communication deals with the problem of higher error rate because of the absence of the direct link connection.
- **Low data rate**: It is a central problem of ad-hoc networking systems. The characteristic of the electronic wave prevents communication wirelessly to send more data over the wired system. Thus, a higher frequency, which is more vulnerable towards interference has the ability to send extensive data in the short-range system.
- **Dynamic topology:** Because these networks do not permit the similar aggregation techniques, which are utilized in standard internet protocols, they are vulnerable to scalability problems. If the nodes are movable the routing varies as the nodes move. Current connectivity must be propagated to all network participants. The number of the increased control messages increases traffic load on the bandwidth. This problem is overcome by designing an algorithm for ad-hoc networks based on available information. For this, a suitable algorithm has been required, which is able to examine, compare, and evaluate the relative scalability of the networks in terms of the increased nodes and their movement (Parasher et al. 2019).

- **Energy restriction:** A MANET is a network that permits all movable nodes to send information to others in the absence of a centralized-controlled network. Therefore, they operate with the help of battery power as they do not have other power sources. Because of these restrictions, they must require energy-efficient algorithms, which have the ability to operate with limited power. In such networks, nodes' power is becoming empty while sending information to others. Thus it requires regular charging to the battery source. Hence, MANET cannot be an excellent choice for a centralized network.

4.9 APPLICATIONS OF AD-HOC NETWORKS

Ad-hoc networks are frequently utilized for various scenarios in which an infrastructure-based network is not feasible to construct. This section lists a few significant applications of the above-cited networks.

- **Military use:** Initially, these networks are utilized by defence or military troops/departments at the time of war. For example, many soldiers spread out on an enormous warfield, and they need to convey information. For this, deploying infrastructure and equipping each soldier with a cable in the warfield is not possible. An alternative solution may be in which every soldier is given a transmitter so that they can approach the rest of the soldiers constantly. However, this technique is not feasible here. Because the enemy can intercept constantly established communication and trap them easily. This problem can be overcome if one soldier uses the channel at any instant of time. Thus, ad-hoc networking systems are favored for this situation, where every soldier is facilitating with a small transmitter. It has a smaller coverage area than the above-discussed transmitter. Hence, soldiers can create a network, which is a so-called ad-hoc network. These networks are tough to intercept and more robust. Therefore these are favorites for military troops.
- **Emergency services and rescue operations:** These are the unpredicted and dominant services that most frequently use decentralized ad-hoc networking for communication. For example, any natural calamity, i.e., earthquake, flood, and tsunami, requires an immediate networking system for communication to rescue lives from that affected area and save them. This is because the already established communication structure is entirely damaged or unable to work anymore. That requires deploying a temporary networking infrastructure for as early as possible to influence rescue works. The deployed network is easy to tune and maintain. Additionally, it has the ability to adjust the changes in topology dynamically to support variations in density and node amount. This can be handled by deploying a MANET type of ad-hoc network because its establishment is straightforward. They are temporal and deployed without infrastructure, and they can be removed smoothly as new infrastructure is built (Kaur et al. 2019).
- **Personal area network:** Personal area network (PAN) is a local network, which is created by some other nodes associated with one candidate, i.e., bluetooth.

Bluetooth is nothing but a local wireless network with a minimal range (typically 10 m or 100 m), operating in the 2.4 GHz spectrum, which is an unlicensed, decentralized networking system with no need for a wired link to connect users (Parasher et al. 2019).

- **Wireless sensor networks:** This is the intersection of three different technologies: sensing, communication wirelessly, and networking. It consists of multiple sensor nodes (in number) featured with a transmitter-receiver (transceiver) system. This system has two prominent functions, (i) it is utilized to sense and measure activities of nodes under a cited network. (ii) It is used to relay the collected messages' signal towards the sink nodes, enhancing hop counts. It can help to monitor wild animals or dangerous areas.

4.10 CONCLUSION

The energy-efficient wireless ad-hoc networking for CR has been comprehensively studied. The chapter has also proposed NOMA-based best EH relay selection for futuristic green ad-hoc CRNs, and it performs remarkably superb over existing OMA. It investigated the comprehensive derivation of outage probability for both users to determine system behavior. Also, it calculated throughput and the optimal time switching ratio. Moreover, the outage probabilities of both (U1, U2) proposed system users improved by 9.40% more than OMA systems. Moreover, the advantages and challenges of an ad-hoc network have been comprehensively presented. Significant applications have also been illustrated in the chapter. A great deal is known about their limitations and the trade-offs inherent in EH device selection and implementation. These may include link quality, residual energy, and dynamic energy availability.

REFERENCES

Gupta, M.S., Kumar, K. (2019). Progression on spectrum sensing for cognitive radio networks: A survey, classification, challenges, and future research issues. Journal of Network and Computer Applications 143, 47–76.

Gupta, M.S., Kumar, K. (2020). Application aware networks' resource selection decision making technique using group mobility in vehicular cognitive radio networks. Vehicular Communications 26, 100263.

Hoang, T.M., Van, N.L., Nguyen, B.C., et al. (2019). On the performance of energy harvesting non-orthogonal multiple access relaying system with imperfect channel state information over Rayleigh fading channels. Sensors 19(15), 3327.

Kaur, G., Tomar, P. and Tanque, M. (2019). Integrating Artificial Intelligence to Solve Pervasive IoT Issues, Elsevier, Cambridge, MA, USA.

Kumar, A., Kumar, K. (2020). Multiple access schemes for cognitive radio networks: A survey. Physical Communication 38, 100953.

Kumar, A., Kumar, K. (2020). Relay sharing with DF and AF techniques in NOMA assisted cognitive radio networks. Physical Communication 42, 101143.

Kumar, A., Kumar, K., Gupta, M.S., Kumar, S. (2020). A survey on noma techniques for 5G scenario. Available at SSRN 3573579.

Liu, Y., Ding, Z., Elkashlan, M., Poor, H.V. (2016). Cooperative non-orthogonal multiple access with simultaneous wireless information and power transfer. IEEE Journal on Selected Areas in Communications 34(4), 938–953.

Makki, B., Chitti, K., Behravan, A., Alouini, M.S. (2020). A survey of noma: Current status and open research challenges. IEEE Open Journal of the Communications Society 1, 179,189.

Moltafet, M., Azmi, P., Mokari, N., Javan, M.R., Mokdad, A. (2018). Optimal and fair energy efficient resource allocation for energy harvesting-enabled-PD-NOMA-based HetNets. IEEE Transactions on Wireless Communications 17(3), 2054,2067.

Nguyen, B.C., Hoang, T.M., Tran, P.T. (2019). Performance analysis of full-duplex decode-and-forward relay system with energy harvesting over Nakagami-m fading channels. AEU International Journal of Electronics and Communications 98, 114–122.

Parasher, Y., Singh, P., & Kaur, G. (2019). Green Smart Security System. Green and Smart Technologies for Smart Cities, 165–184.

Parasher, Y., Singh, P., & Kaur, G. (2019). Green Smart Town Planning. Green and Smart Technologies for Smart Cities, 19–41.

Srivastava, A., Gupta, M.S., Kaur, G. (2020). Energy-efficient transmission trends towards future green cognitive radio networks (5G): Progress, taxonomy and open challenges. Journal of Network and Computer Applications p. 102760.

Srivastava A., and Kaur G. (2021). Resource management for traffic imbalance problem in green cognitive radio networks, Physical Communication. Elsevier, vol. 48, pp. 101437.

Tran, H.V., Kaddoum, G., Truong, K.T. (2018). Resource allocation in SWIPT networks under a nonlinear energy harvesting model: Power efficiency, user fairness, and channel non-reciprocity. IEEE Transactions on Vehicular Technology 67(9), 8466–8480.

5 Green Optical Communication Systems

Asbah Masih, Akanksha Srivastava, and Gurjit Kaur

CONTENTS

5.1 INTRODUCTION TO GREEN OPTICAL COMMUNICATION

Over the years, communication technologies have become a necessity for all. With the significant increase in the amount of data transmitted, the growth of optical communication has increased. Optical communication has enabled the transmission of a large amount of data at higher rates. An increase has been recorded in data rate, from a few bits per second to as high as many terabits per second (Tucker et.al. 2011, Prasad et.al. 2005).

With the tremendous increase in optical communication, the concern for high power consumption as well as the environmental impact of these systems has increased. Living in an energy-constrained world the need for efficient energy utilization is necessary. Exploiting the resources will have a large impact on the environment. The carbon footprint of these systems is significant and increasing every day. This increase will not only harm the environment but also have a direct impact

DOI: 10.1201/9781003264477-5

on human health. Thus, to overcome these challenges green optical communication is used.

Green optical communication ensures that needs are met without destroying or depleting our resources. It will reduce the negative effect on the environment by ensuring energy-efficient systems. Green optical communication supports the idea of sustainable development where growth continues with hampering the limited resources we have (Tomar et.al. 2018).

5.2 GOALS OF GREEN OPTICAL COMMUNICATION

The goals of green optical communication are shown below in Fig 5.1. Green optical communication aims to achieve sustainable development by ensuring efficient use of energy along with providing the perks of optical communication. It provides better scalability to improve the quality of services (QoS) of the communication system. Its improved bandwidth, higher flexibility, long-distance transmission, and better security is provided while ensuring less wastage and efficient use of energy. It also focuses on optimization of resources and infrastructure design so it helps in proper load balancing of the data traffic.

5.3 GREEN OPTICAL COMMUNICATION SYSTEMS

The need for improved and energy-efficient systems have led to various green optical communication systems.

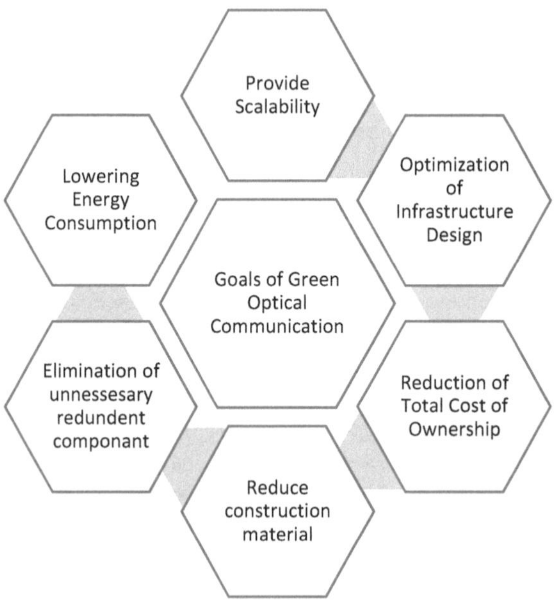

FIGURE 5.1 Goals of green optical communication.

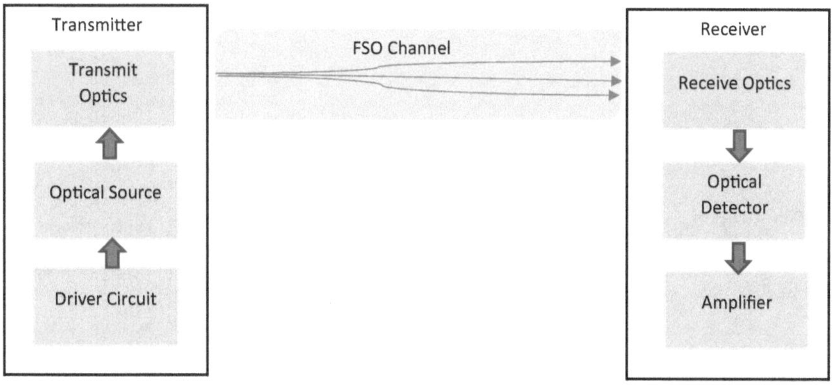

FIGURE 5.2 Block diagram of optical wireless system.

5.3.1 Green Optical Wireless Communication

In optical wireless communication (OWC) wireless communication is carried out using the optical spectrum. Perks of both optical fiber communication and wireless communication, i.e., it has a high capacity of optical fiber communication and the flexibility of wireless communication. Another major advantage of optical wireless systems is that since transmission is using visible light, it is safe for human health, unlike the RF rays whose long-term exposure may cause malignant or benign health issues (Wu et.al. 2017, Cahyadi et.al. 2015). These are also safe for RF link restricted zones such as hospitals, aircraft etc. as they do not interfere with the RF emissions. (Green 2007). The interception disability of signals ensures better security for optical wireless systems. Figure 5.2 shows the block diagram of an optical wireless system.

Since the growth and demand for optical wireless networks have increased, the need for energy-efficient systems is more than before. To achieve this, green optical wireless communication systems are used. Green here stands for lower energy consumption, improving resource utilization, and reducing processes that impact the environment (Jenila et.al. 2021).

The following is a roadmap for green wireless communication systems as shown by Fig 5.3:

In the process of development of optical transmitters and receivers, optical sources like LED with better conversion efficiency should be used. The modulation bandwidth also needs to be increased. Techniques used for transmission should be advanced and there is a need of high responsivity photodetectors. For improvement in performance the transmission data rate needs to be improved and a better spectral efficiency is also needed. Improved energy efficiency and improved reliability also play a significant role.

In uplink, the establishment uplink data rate should be high and the power consumption should be decreased significantly for active and passive uplink. For uplink, reliable techniques should be used. Techniques that would reuse wavelength should be used. For the incorporation of energy harvesting, solar cells should be used to ensure

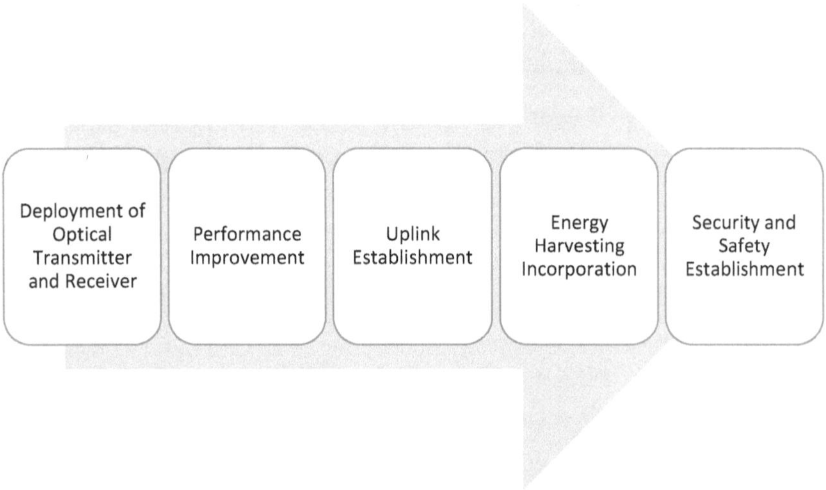

FIGURE 5.3 Roadmap for deployment of green optical wireless communication.

energy harvesting and the harvested energy must be used for the various operations of user devices moreover, battery offload techniques should be included and energy harvesting techniques using incident light should be included.

Security and safety enhancement are possible by inclusion of intentional jamming techniques. In order to ensure secure transmission, position-based transmission should be practiced. Barcodes will be used and skin safety must be included.

Topology control for energy-efficient optical wireless networks (Dong, 2012). The lifetime of a network can be increased by using topology control to assign transmission power per node. This is done to improve the overall energy efficiency of the network. Various algorithms like position-based algorithms, which are used to set up and also maintain the network such that it maintains the minimum energy usage (Rodoplu et.al 1999), algorithms like the max-improvement algorithm (MIA), which works on the principle that each node is selfish and will work to maximize its utility. This algorithm ensured topologies that are globally energy-efficient (Komali et.al. 2008).

5.3.2 Green Optical Wired Communication

Going up on the technology ladder from 2G to 3G to 4G and soon to be 5G, the fiber requirement has been increasing continuously. With 5G in the picture, transmission speeds in excess of 10 Gbps as against 4G technologies with average speeds of 6 Mpbs in India and in order to achieve such speeds fiber connectivity is essential. Figure 5.4 represents the block diagram of optical fiber network.

Various techniques as mentioned below are used to make optical wired communication green and make these systems energy-efficient. These techniques not only make the energy-efficient system but also minimize cost as follows:

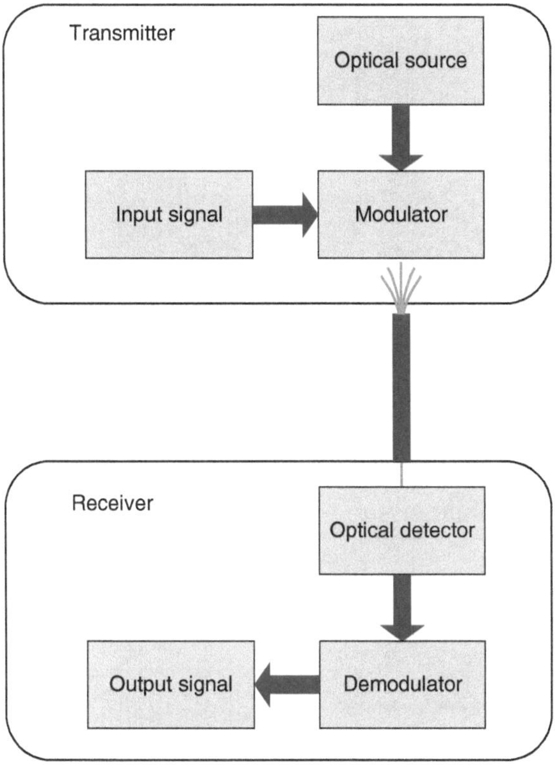

FIGURE 5.4 Block diagram of optical fiber network.

- Dark fibers are often leased from other companies, which reduce the number of lit fibers, and thus it results in minimum network operation cost.
- Cost of hardware and operational cost of amplifiers is reduced when a smaller number of active amplifiers are needed.

Various techniques are practiced to shift optical wired communication to green optical wired communication systems. These techniques have been mentioned below and represented by Fig 5.5.

For maximizing idle network components, we need only those components of the network which are in active stage (using energy) and all the other components are in the resting/idle phase, which helps to reduce wastage of energy as a lower number of fully functional components ensures a lower operational cost. Thus, the hardware cost also goes down. In energy-aware routing and wavelength assignment strategies, the traffic is clustered on a few routes, and thus energy usage is reduced. Renewable energy sources are used to power the sources so the emissions of green-house gases are reduced. In traffic grooming techniques, physical components, which are active, are reused during allocation of requests and total active components are reduced significantly by reusing, thus it limits the expenditure of energy (Masud Hasan et.al.

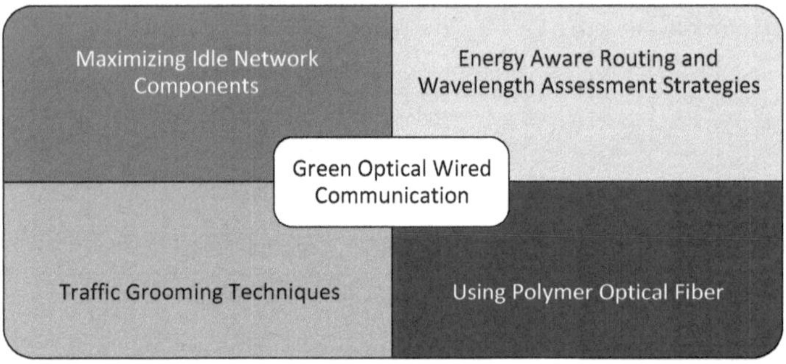

FIGURE 5.5 Techniques for designing the green optical wired communication.

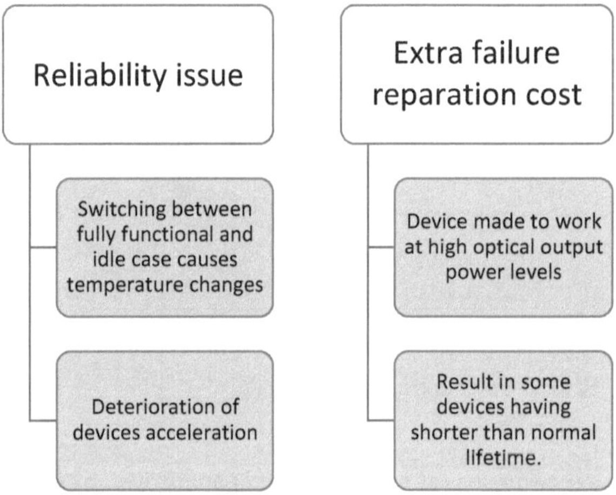

FIGURE 5.6 Negative impact of energy-efficient techniques.

2010). For medium to short haul communication polymer, optical fibers are ideal as they have less loss when light is transmitted and their cost is also lower than glass fibers (Sarah et.al. 2021). It is safe, inexpensive and harmless. Thus it leads towards the green optical communication.

5.3.2.1 Negative Impacts due to Energy-Efficient Techniques

Most of the energy-efficient techniques focus only on reducing energy consumption in the network without considering the overall effect on the network like its reliability etc. Techniques like switching the idle devices in sleep mode are done in order to reduce the consumption of energy may often have a negative impact on various performance metrics of a network (Wiatr et.al. 2015). Various negative impacts on energy-efficient techniques have been mentioned below. Fig 5.6 shows the negative impact of energy-efficient techniques.

5.3.3 Green Deep Space Optical Communication

Free space optical (FSO) communication has emerged as a significant option for communication in space because of its many advantages such as the following

- High data rate
- License free spectrum
- Large bandwidth
- High directivity
- High security
- Fewer power requirements

Communication in FSO is done by establishing links, which can be deep space links or inter-satellite links or links between ground to satellite or vice-versa. Optical subnets are able to provide a full line-of-sight coverage for 24 hours a day of the ecliptic for NASA deep space missions. Figure 5.7 represents the FSO space links.

Despite various advantages, FSO communication in space faces a lot of issues. Since the atmospheric channel is used for propagation, these waves are attenuated by unpredictable weather conditions like rain, fog, snow, haze etc. When FSO is between two satellites then there are various challenges like background noise, pointing and link availability. Space FSO is not influenced by the harsh weather conditions. The following challenges are represented by Fig 5.8.

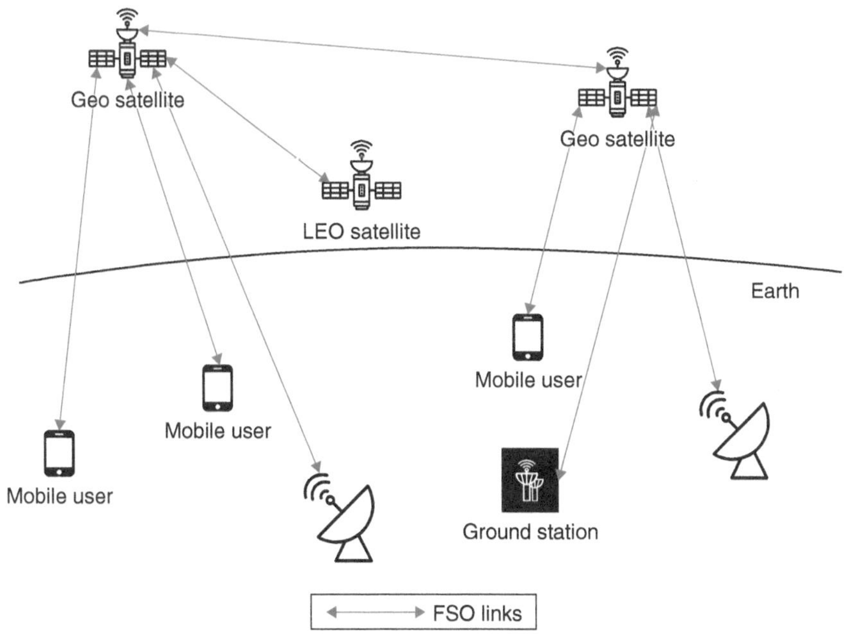

FIGURE 5.7 FSO space links.

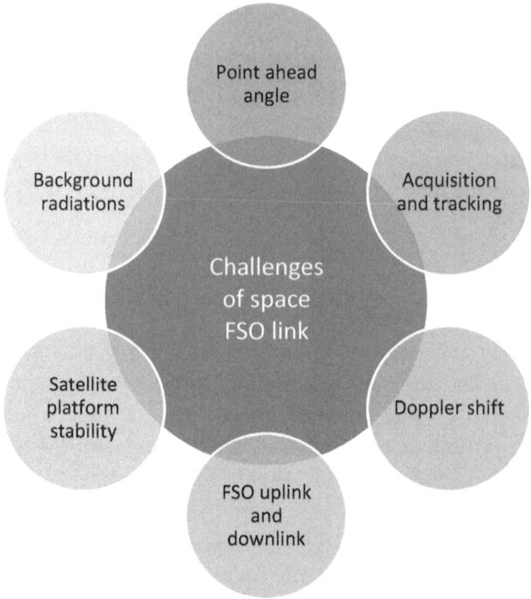

FIGURE 5.8 Challenges of space FSO link.

In order to overcome the challenges and to make the system green, various techniques are used. These techniques ensure efficient utilization of energy, thus making it more useful and safer for the environment.

These turbulence mitigation methods are of two types (Hemani Kaushal et.al. 2016):

- Physical layer methods
- Other TCP layer methods

5.3.3.1 Physical Layer Turbulence Mitigation Methods

Mitigation techniques, which are used at the physical layer, are called physical methods. These are shown in Fig 5.9.

- Aperture averaging is a technique used where channel fading is reduced by averaging fast fluctuations caused by smaller eddies. The size of receiver aperture is increased to do this. (Lyke et.al. 2009, Mendenhall et.al. 2007).
- Effect of turbulence is improved by using diversity technique by operating on frequency, time, and space. In order to make multiple copies of the signal, a small receiver aperture array is used instead of a receiver of one large aperture (Hurd et.al. 2006).
- Relay transmission is beneficial as huge diversity gain is achieved by a single antenna by building a virtual antenna array unlike the multiple aperture arrays used in diversity technique (Wang et.al. 2013, Bhatnagar et.al. 2013).

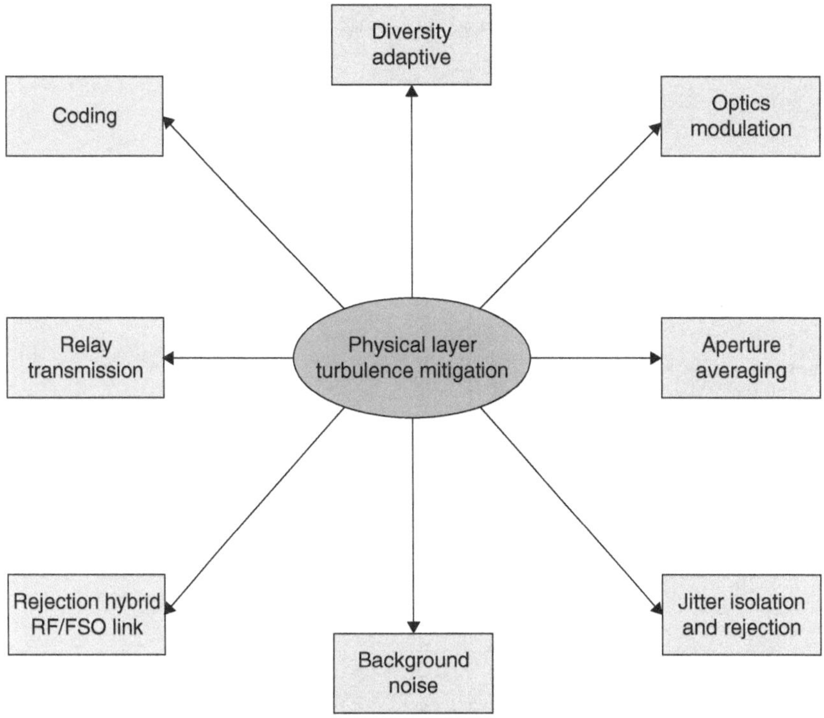

FIGURE 5.9 Physical layer turbulence mitigation techniques.

- Usage of adaptive optics (AO) ensures that the beam is precorrected before it is transmitted. A conjugate of the atmospheric turbulence is added to the beam before its transmission (Zocchi 2005).
- Bandwidth efficiency and optical power efficiency determine the modulation scheme that should be used. For low data rates, power-efficient modulation schemes are easier to implement and are effective (Hemani Kaushal et.al. 2016).
- Error control coding makes use of various FEC schemes like Reed-Solomon (RS) codes, convolutional codes, turbo code etc to ensure improved performance of FSO code (Djordjevic 2011).
- Modulation techniques and spatial filters can be used to mitigate background noise, which is majorly caused by solar radiations during the day (Simon et.al. 2002).
- Hybrid systems ensure that perks of both systems can be used. Since FSO performs poorly in adverse weather conditions it is paired with RF system. This ensures more reliability (Stotts et.al. 2009, Hong-jun 2014).

5.3.3.2 TCP Other Layer Turbulence Mitigation Methods

FSO performance can also be improved by using mitigation techniques at network layer or transport layer. Various techniques are mentioned below in Fig 5.10.

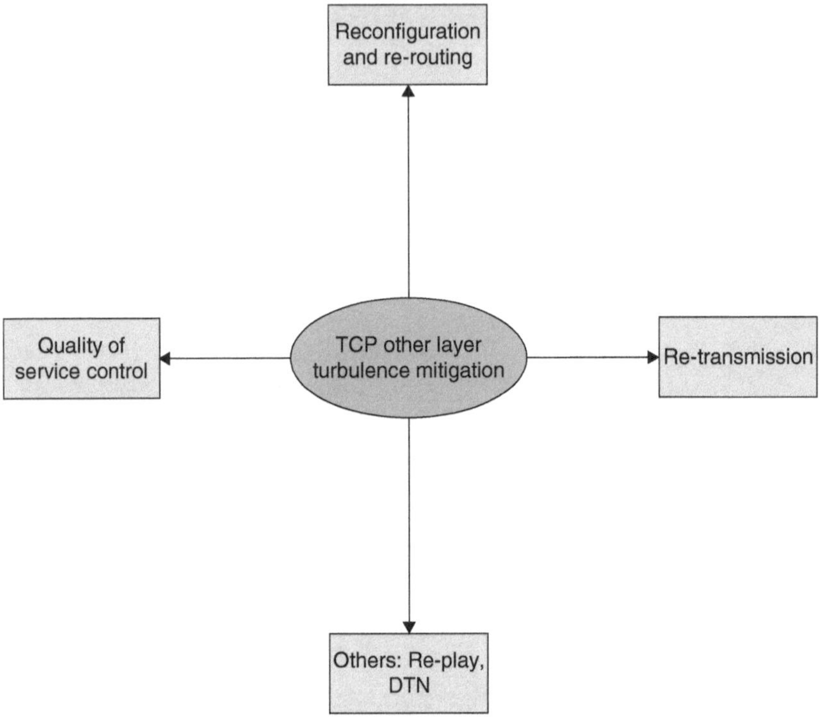

FIGURE 5.10 TCP other layers' turbulence mitigation techniques.

- Retransmission
- This data is transmitted as packets of a particular frame length. Once the receiver gets the packet it sends back an acknowledgement packet. If the transmission was not successful the receiver will not send the acknowledgement packet and thus the transmitter will resend it. To ensure there is no wastage of energy and bandwidth by halting the process to receive the acknowledgement for a data packet, the transmitter continues sending a data packet and if any particular frames' acknowledgement is not received after a certain period, the transmitter assumes it to be lost and retransmits that data packet again. This is a variant of automatic repeat requests (ARQ) (Pathapati et.al. 2010).
- Reconfiguration and rerouting
- If FSO connection is disturbed due to either adverse weather conditions or device failure etc. reconfiguration and rerouting of path is done to increase the reliability as well as availability of FSO link (Son 2010).
- Quality of service control
- For FSO communication, quality of service (QoS) is measured in terms of data loss, energy utilized, data rate, reliability etc. Various routing algorithms can be used, which improves the quality of service of the network (Peach et.al. 2010).
- Others

- In extreme weather conditions, the delay tolerant networking technique (DTN) is applied to FSO. In cases when retransmission or rerouting is not possible then replaying is done. In this case from the edge node up to 5 sec of data will be replayed. This technique is also used for the FSO link to ensure end-to-end connectivity (Burleigh et.al. 2003, Nichols et.al. 2007).

5.4 ENERGY-EFFICIENT ALGORITHMS

- In compliance with the switch off links algorithm (SOLA), according to the threshold value some of the optical network links are disabled in order to save energy when their usage is less (Beletsioti et.al. 2019).
- Reduction in expenditure of energy by reducing the number of components that consume power like amplifiers, regenerators etc. (Manousakis et.al. 2014).
- As an application of the Selene heuristic algorithm, the signal overlap technique is used for power saving in elastic optical networks. Power expenditure is reduced, thus ensuring an energy-efficient (green) system.
- Furthermore, energy expenditure is reduced by saving energy in core networks by selectively powering down cables. A few cables from the bundle are powered down according to the need (Vasic 2008).

5.5 CHALLENGES OF GREEN COMMUNICATION

- Some communication problems are also upgraded and optimized while upgrading communication equipment. To manage the update approach, green communication currently demands for energy consuming equipment. In order to achieve emission reduction and energy saving, communication equipment can be treated in two ways:
    I. The first is the creation of optimization by-products, such as to decrease the density of PCB discs and size of dense PCB boards. It is done by optimizing the circuit design. Energy can be decreased by increasing heat dispersion. Furthermore, by adopting a new design, power can be reduced by employing less equipment.
    II. Another important step is to decrease the energy consumption of the equipment because these systems use the majority of the product's energy.
- In underwater optical communication systems, optimization need is even more because of such a challenging environment. Various modulation techniques in order to make energy-efficient system have been tried and tested (Gabriel et.al. 2012).
- To set a lower bound for energy efficiency of optically amplified systems. This depends on optical transmitters and receivers and also on optical amplifiers (Tucker et.al. 2011). Thus, optimization of all these is required to have green optical communication.
- The problem is figuring out how to create a new structure that balances low energy and high-power processors. In order to expand, communication

systems in the next phase should apply new ways to update the protocols now in use.

- In addition, the quality of skilled facilities and the production of energy consumption services should be increased in order to have a secure and efficient sustainable communication network.
- The indoor OWC systems are evaluated based on these promised properties (high capacity, flexibility, and safety), and the term green is used to describe both safety and security. This concept presents a technology overview and literature study of the OWC system, aids in identifying the obstacles to widespread adoption of green wireless communication systems (Jenila et al. 2021).

5.6 APPLICATIONS OF GREEN OPTICAL COMMUNICATION

Green data centers: The data center is the centralized physical or virtual unit for storage, searching, management, and dissemination of data and information. ITU-T L.1300 defines the best practices for global minimization of the negative consequences of data centers on the environment. The green data center with high-density communication equipment should associate OWC with the communication technology.

Cellular OWC systems: The cellular OWC system works with hexagonal cells that have LED APs as the base station. This is a suitable cellular topology to adopt multiple users in quite large room environments. In these OWC systems, the sources with a wide optical footprint are more appropriate than the narrow transmit beam.

5.7 RESEARCH OPPORTUNITIES ON GREEN OPTICAL COMMUNICATION

An essential trait of the green environment is the generation or conversion of the required energy from the ambient renewable energy sources. As the research on electricity generation from artificial light sources has proceeded, the integration of light-harvesting with the indoor VLC is going to make self-sustainable communication. It will further clear the way to overcome the challenges of wired connection for electricity and battery endurance. Though the researchers have identified this scope, there is still a long way to achieve realistic autonomous communication.

5.8 CONCLUSION

Green optical communications have rightfully gained importance. With an increase in optical communication over the years the only option to continue expanding at such a rate is with green optical communication. It ensures that the needs are met but the environment is not compromised. Green optical communication will not only help in more reliable systems but will also ensure that there is an efficient use of energy thus reducing exploitation of our resources. This will build strong, reliable communication systems along with reducing the carbon footprint.

REFERENCES

C. Gabriel, M. Khalighi, S. Bourennane, P. Léon and V. Rigaud. 2012. Investigation of suitable modulation techniques for underwater wireless optical communication. *International Workshop on Optical Wireless Communications (IWOW)*, 2012, pp. 1–3, doi: 10.1109/IWOW.2012.6349691

C. Jenila, R.K. Jeyachitra. 2021. Green indoor optical wireless communication systems: Pathway towards pervasive deployment. *Digit. Commun. Netw.*, vol. 7, no. 3, pp. 410–444.

F. E. Zocchi. 2005. A simple analytical model of adaptive optics for direct detection free-space optical communication. Opt. Comm., vol. 248, no. 4–6, pp. 359–374.

Georgia A. Beletsioti, Stathis Mavridopoulos, Georgios A. Tziroglou. et.al. 2019. Power-aware Algorithms for Energy-efficient Elastic Optical Backbone and Metro Networks. 10th International Conference on Data Communication Networking. DCNET.

Hemani Kaushal and Georges Kaddoum. 2016. Optical Communication in Space: Challenges and Mitigation Techniques. IEEE Communications Surveys & Tutorials.

I. B. Djordjevic. 2011. Deep-space and near-Earth optical communications by coded orbital angular momentum (OAM) modulation. J. Opt. Soc. Am., vol. 19, no. 15, pp. 14277–14298.

I. K. Son. 2010. Design and Optimization of Free Space Optical Networks. PhD thesis, Graduate Faculty of Auburn University, Auburn, Alabama.

J. A. Mendenhall, L. M. Candell, P. I. Hopman, and G. Zogbi et al. 2007. Design of an optical photon counting array receiver system for deep space communications. Proc. IEEE, Special Issue Tech. Advan. in Deep Space Commun. and Track: Part I, vol. 95, no. 10, pp. 2059–2069.

K. Manousakis, G. Ellinas and E. M. Varvarigos. 2014. Energy-efficient algorithms for translucent WDM optical networks. *16th International Conference on Transparent Optical Networks (ICTON)*, 2014, pp. 1–4, doi: 10.1109/ICTON.2014.6876421

K. S. Pathapati, J. P. Rohrer, and J. P. G. Sterbenz. 2010. End-to-End ARQ: Transport-layer reliability for airborne telemetry networks. Inter. Telemeter. Conf. (ITC).

Komali, R.S., A.B. MacKenzie, and R.P. Gilles. 2008. Effect of Selfish Node Behavior on Efficient Topology Design. Mobile Computing, IEEE Transactions 7(9): p. 1057–1070.

L. B. Stotts, L. C. Andrews, P. C. Cherry, and J. J. Foshee. 2009. Hybrid optical RF airborne communications. Proc. IEEE, vol. 97, no. 6, pp. 1109–1127.

L. Hong-jun. 2014. Overview of U. S. military airborne RF and FSO hybrid communication. Telecom. Eng., vol. 54, no. 2, pp. 1711–1716.

M. K. Simon and V. A. Vilnrotter. 2002. Alamouti-type space time coding for free space optical communication with direct detection. IEEE Trans. Wireless Comm., vol. 50, no. 8, pp. 1293–1300, 2002.

M. Masud Hasan, Farid Farahmand, Jason P. Jue, and Joel J. P. C. Rodrigues. 2010. A Study of Energy-Aware Traffic Grooming in Optical Networks: Static and Dynamic Cases. IEEE

M. R. Bhatnagar and M. K. Arti. 2013. Performance analysis of hybrid satellite-terrestrial FSO cooperative system. IEEE Photon. Tech. Lett., vol. 25, no. 22, pp. 2197–2200.

N. Vasic and D. Kosti´c. 2008. Energy-aware traffic engineering. EPFL Technical Report.

P. Wiatr, J. Chen, P. Monti and L. Wosinska. March 2015. Energy efficiency versus reliability performance in optical backbone networks [invited]. Journal of Optical Communications and Networking, vol. 7, no. 3, pp. A482-A491. doi: 10.1364/JOCN.7.00A482

Prasad, R. S. and Kaur, G. 2005. Optical Communication. Galgotia Publications, ISBN: 9788175154858, 8175154853.

R. A. Nichols, A. R. Hammons, D. J. Tebben, and A. Dwivedi. 2007. Delay tolerant networking for free-space optical communication systems. Proc. IEEE, Sarnoff Symp., Nassau Inn, Princeton, NJ.

R. J. Green. 2007. Secure communications: the infrared alternative. 2007 ICTON Mediterranean Winter Conference, pp. 1–4, https://doi.org/10.1109/ ICTONMW.2007.4446906.

R. Peach, G. Burdge, F. Reitberger, et.al. 2010. Performance of a 10 Gbps QoS-based buffer in a FSO/RF IP network. Proc. SPIE, Free Space Laser Comm. X, vol. 7814.

R.S. Tucker. 2011. Green optical communications – Part I: Energy Limitations in Transport. *IEEE J. Sel. Top. Quantum Electron.*, vol. 17, no. 2. pp 245–260.

Rodoplu, V. and T. H. Meng. 1999. Minimum energy mobile wireless networks. Selected Areas in Communications, IEEE Journal on. 17(8): p. 1333–1344.

S. Burleigh, A. Hooke, L. Torgerson et.al. 2003. Delay-tolerant networking: An approach to interplanetary internet. IEEE Comm. Mag, vol. 41, no. 6, pp. 128–126.

S. D. Lyke and M. C. Voelz, D G.and Roggemann. 2009. Probability density of aperture-averaged irradiance fluctuations for long range free space optical communication links. Appl. Opt., vol. 48, no. 33, pp. 6511–6527.

Sarah, Latifah, Ab-Rahman, Mohd, et.al. 2021. Polymer Optical Fiber Splitter Using Tapered Techniques for Green Technology. 10.5772/intechopen.96645.

T. C. Wu, Y.-C. Chi, H.-Y. Wang, C.-T. Tsai, Y.-F. Huang, G.-R. Lin. 2017. Tricolor r/g/b laser diode based eye-safe white lighting communication beyond 8 gbit/s, Sci. Rep. 7 (1) 11, https://doi.org/10.1038/s41598-017-00052-8.

Tomar, P. and Kaur, G. 2018. Green and Smart Technologies for Smart Cities. CRC Press, Taylor & Francis Group, Dec 2019, Boca Raton: CRC Press, Ist Edition, ISBN 9781138318090 – CAT# K391489, https://doi.org/10.1201/9780429454837.

W. A. Cahyadi, T. Jeong, Y. Kim, Y. Chung, T. Adiono. 2015. Patient monitoring using visible light uplink data transmission, in: 2015 International Symposium on Intelligent Signal Processing and Communication Systems, ISPACS, pp. 431–434, https://doi.org/10.1109/ISPACS.2015.7432810.

W. J. Hurd, B. E. MacNeal, G. G. Ortiz, and R. V. Moe et al. 2006. Exo-atmospheric telescopes for deep space optical communications. Proc. IEEE Aerosp. Conf., (Big Sky, MT).

X. Wang, S. H. Zhao, L. Shi, Y. Li, G. Zheng, G. Zhao, and Z. Zhu. 2013. Performance analysis of high altitude platform multi-hop optical communication with non-regenerative system. Proc. SPIE, Laser Comm. Tech. and Sys., vol. 8906.

Xiaowen Dong, Green Optical Networks, PhD Theseis, The University of Leeds School of Electronic and Electrical Engineering, November 2012.

6 Green Hybrid Wireless-Optical Broadband Networks

Rekha Rani and Gurjit Kaur

CONTENTS

DOI: 10.1201/9781003264477-6

6.1 INTRODUCTION

Information and communication technology have delivered a great chance with environment friendly technologies for the benefit of society by reducing the man power. However, due to customer demands for huge bandwidth multi-media applications like online gaming, video conferencing, HDTV etc. computers, communication devices and networks are increased day by day. Thus, rapidly energy consumption takes place more and more every year. However, traditional access technologies do not support these bandwidth requirements. If energy consumed by these devices continuously increases at an alarming pace, energy deficiency will constitute a hurdle for long term ICT. As a result, developing an energy-efficient network is critical for future ICT advancement. The broadband wireline and wireless access network can meet this expectation. The customer premises are served by a wireline access network, which offers reliable and consistent bandwidth. Coaxial cable, twisted pair, copper wire, and optical fiber can all be used to reach a customer's premises. Optical access network meets the desired requirements such as high bandwidth, reliability, high transmission reach, and wireless access network has less expensive and high flexibility. Many access network technologies for optical fiber access network are cable modem (CM), digital subscriber line (xDSL), fiber to the x (FTTx) and passive optical network (PON). Out of these access technologies, PON is the dominant technology due to being more energy efficient than others (Chowdhury et al. 2010). However, it incurs the significant cost of optical fiber infrastructure.

In contrast to wireline access network, if users are mobile then there are the requirements of untethered access (wireless) instead of fiber. Many different types of wireless access systems exist, each with its own spectrum, bandwidth, range, and applications. WPANs (IEEE 802.15) are wireless personal area networks that connect personal devices over a distance of around 30 feet. Other wireless technologies like Bluetooth, Wireless USB, Infra-red, WiFi, WiMAX etc. are also available. The World-Wide Interoperability for Microwave Access (WiMAX), Wireless Fidelity (WiFi), and cellular networks are the most common wireless access network approaches, however they are all limited in bandwidth.

6.2 OPTICAL ACCESS NETWORK

In this network, the PON technique is most widely used, which offers significant utilization of the bandwidth of optical fiber with high data rate. The main parts of the PON architecture are an optical line terminal (OLT) that resides at the center office (CO) and ONUs. The traffic routes from OLT to ONU in a downstream way, while it is in upstream direction when the traffic routes from ONU to OLT. Based upon the multiplexing techniques, PON is subdivided into three types: time division multiplexing PON (TDM-PON), wavelength division multiplexing PON (WDM-PON), orthogonal frequency division multiplexing PON (OFDM-PON), and hybrid PONs.

6.2.1 TDM-PON

In TDM-PON, the OLT broadcasts the traffic load to the linked ONUs in the downstream direction, and only a single ONU is allowed to provide data in the upstream direction during a preset schedule slot as shown in Figure 6.1. TDM-PON used a frequency having downstream data, which was shared by 'N' number of clients. Upstream data is given by a minimal expense transmitter set in the ONU. The ONUs terminate all data flow and create a help interface for end users using DSL or coaxial cable. OLT, on the other hand, is the gateway interface of the entire PON network to the IP core network, and it is used to transport IP data traffic forward using specific interfaces such as synchronous digital hierarchy (SDH)/synchronous optical networking (SONET) or ethernet. Information is communicated from the OLT to all related ONUs in the downstream fashion, whereas information from ONUs does not cover at a given time allotment by utilizing the arbitration mechanism in the upstream way. Bandwidth allocation criteria is used to schedule the beginning time and length of every time allotment, which is executed at the OLT. The dynamic bandwidth allocation (DBA) approach is utilized to ensure proper bandwidth utilization while also being able to respond to current data dement requirements. But the computation overhead is possessed by the implementation of DBA algorithm in OLT. The architecture of TDM-PON also experiences switchover deficits. To reduce the large propagation delays, interleaved surveying is used. With interleaved surveying, the next ONU to be polled is determined by message-based transmission parameters, whereas the previous ONU is determined by between-ONU transmission guard times. These guard times prevent data transmission from collapsing, just as they allow each switch to turn on and off its laser. Data from the ONUs to OLT reaches at different power stages and OLT should shift its threshold and decision scheme to receive information without any mistakes. So, the OLT would store data of each ONU transmission of all time.

Current deployment of PON systems is generally TDM-PON systems, which incorporate asynchronous transfer mode (ATM) PON (APON), broadband PON (BPON), Ethernet PON (EPON), gigabit PON(GPON), XG-PON, G-EPON, and next generation PON (NG-PON), which allow for a data transfer rate of up to 10 Gbps. A full-service access network (FSAN) launched APON designs to enable multiservice broadband options to business customers. The best hybrid network for reaching that goal was ATM (multiservice) and PON (inexpensive). However, ATM did not satisfy the hope of turning into the universal network protocol. As a result, The ITU-T study group 15 deliberately chose the BPON. But BPON still relies on the ATM networking by nature. BPON defines the line rates as multiple of 8 kHz

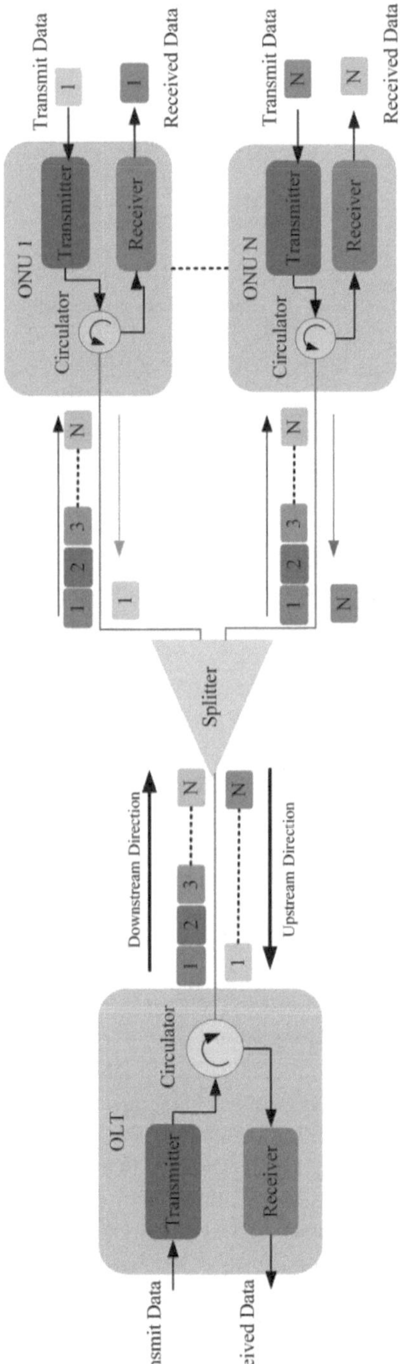

FIGURE 6.1 TDM PON architecture.

for SONET/SDH. An OLT distributes the 8 kHz clock from the OLT to an ONU to support TDM services (voice applications). The ITU-T G.983.3 BPON standard was published originally in 1998 and redefined in 2005 to add high line rate. After BPON, which is developed by ITU-T, the next generation PON technology is GPON. It can cope better with the changes toward Ethernet and IP communication technologies and also meet the quickly developing bandwidth demands. In 2003, the ITU-T G.984 standard corrected the first GPON standard. The Institute of Electrical and Electronics Engineers (IEEE) research committee developed the EPON standard in 2004 as IEEE 802.3ah to extend the core Ethernet LAN MAC to allow new fiber access physical layers. In both directions, EPON uses maximum 1.25 Gbps of data rate and is widely used in Asia and Europe. GPON, which is popular in the U.S, has data rate ranges from 155 Mbps to 2.488 Gbps for the upstream channel and from 1.244 Gbps to 2.488 Gbps for the downstream channel. Downstream wavelength is arranged to 1490 nm with upstream wavelength at 1310 nm for EPON and GPON. Transmission distance for EPON and GPON is up to 20 km but they are having different control message formats, guard times etc. The 10G version of GPON is called XG-PON. It maintains a downstream speed of 10 Gbps and an upstream speed of 2.5 Gbps. Despite the fact that it is identical to GPON, the downstream and upstream wavelengths have been moved to 1577 nm and 1270 nm, respectively. XGS-PON is an enhanced version of XG-PON that employs a wavelength that is quite similar to XG-PON and provides a symmetric data throughput of 10 Gbps in both directions. As, 10G-EPON is standardized by standard IEEE 802.3av in 2009. It supports both symmetric data rate at 10.3125 Gbps for both symmetric channel and asymmetric 10G/1G downstream/upstream operation. 1G-EPON uses 1480 nm–1500 nm band while 10G-EPON uses the 1575 nm–1580 nm band in downstream direction. In the upstream way, 1G-EPON uses 1260 nm–1360 nm band and 10G-EPON uses 1260 nm–1280 nm band that overlap. In 2015, the ITU-T G.989 NG-PON2 standard was published. Previous to XG(S), there is NG-PON2, which uses WDM with multiple 10G frequencies in two ways to transmit 40 Gbps symmetric data rate. TWDM (time and recurrence division multiplexing) is the essential NG-PON2, which is a combination of standard TDM and WDM developments. It likewise upholds highlight point to point WDM overlay, which is appropriate for versatile mobile facility that need low inertness. For upstream transmission, the NG-PON2 wavelength assignment plan uses the 1524 nm–1544 nm band and 1596 nm–1602 nm for downstream transmission. While G.989.2 specifies two different upstream choices with a reduced range band, namely 1528 nm–1540 nm and 1532 nm–1540 nm, the standard G.989.2 specifies two separate upstream alternatives with a reduced range band, namely 1528 nm–1540 nm and 1532 nm–1540 nm. For all PON standards, the idea of allocating wavelengths has been shown in Figure 6.2.

6.2.2 WDM-PON

The OLT and ONUs interact in a point-to-point method in the WDM-PON architecture, which is referred to as point topPoint WDM-PON. It solves the problem of TDM-PON by transmitting many wavelengths on the same optical fiber. In this

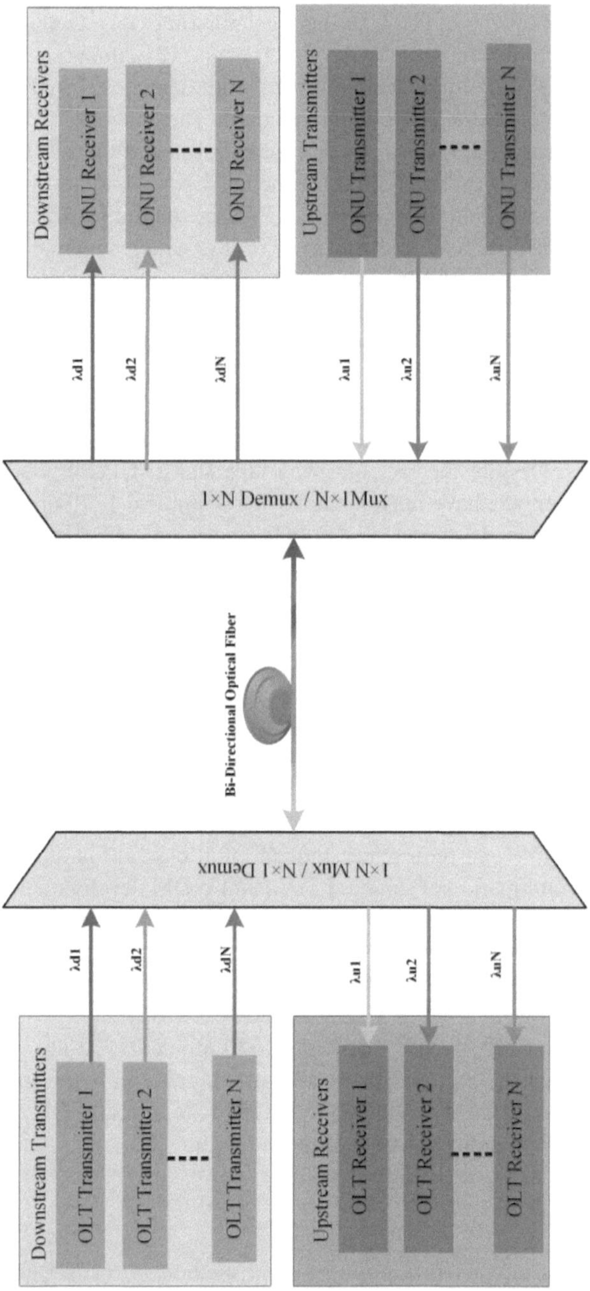

FIGURE 6.2 WDM PON architecture.

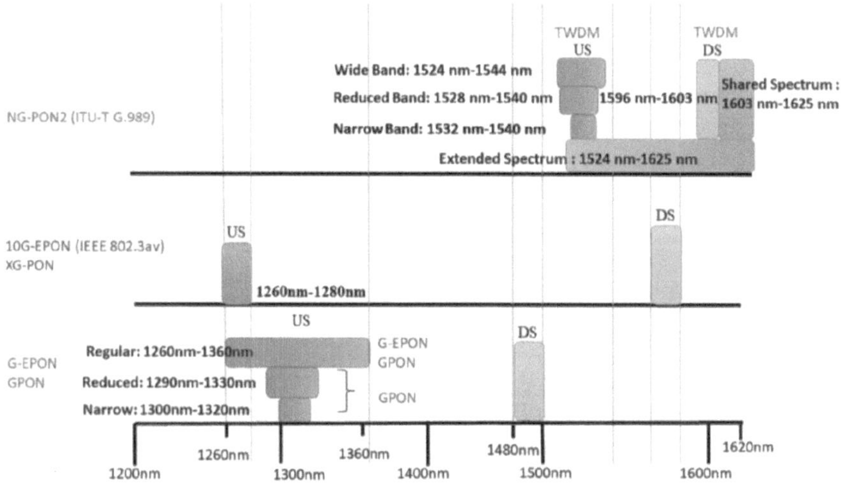

FIGURE 6.3 Wavelength allotted to different PON standard.

framework, ONUs and OLT do not share wavelength capacity, but every ONU and OLT transmitter has a separate wavelength to communicate with each other as shown in Figure 6.3. This multiplexing creates a dedicated communication path between CO and every ONU instead of point to multipoint communication that is in TDM-PON. The WDM-PON network decides adequate wavelengths and spacing between wavelengths. WDM-PONs have the advantage of high speed, as well as scalability in storage and network size, owing to the fact that they combine greater frequency and ONUs. Apart from the increased bandwidth, WDM-PONs provide greater security because information is not shared among ONUs; instead, each ONU accepts its own set of information.

WDM-PON, on the other hand, has the disadvantage of requiring additional WDM active components (array waveguide grating, MUX/DMUX, and tunable laser sources, for example), resulting in higher initial setup costs and higher power consumption. It also increased temperature sensitivity just as raised operational uses in terms of maintenance, production lines, and useless parts. A broad range of solutions have been published in the writing where ONUs is utilized with tunable laser power. One solution is to tune every ONU to the proper wavelength. Another option is to equip each ONU with a wavelength-transparent optical modulator, whereas the OLT has a group of wavelength sources (Kaur et al. 2019).

Two major WDM-PON are used on the basis of wavelength spacing: coarse WDM (CWDM) PON and dense WDM (DWDM) PON. In CWDM PON, wavelengths spacing is more than 20 nm. But in DWDM PON, 3.2 nm wavelength spacing is used, which is much less than that of CWDM. Most recently, wheel-based WDM-PON has been used to support fixed and mobile users with optimal fiber usage and efficient scheduling techniques instead of ring based WDM-PON (Christodoulou et al. 2021).

6.2.3 OFDM PON

OFDM PON utilize numerous orthogonal subcarriers allocated to various ONUs. Its architecture is the same as that of TDM-PON except OFDM modulation and demodulation is used in each time slot and provide a data rate up to 100 Gbps. Thus, OFDM PON has a high spectral efficiency, superior transmission performance, low inter-symbol interference, and high tolerance capability to the fiber chromatic dispersion. Another advantage of OFDM-PON is that bandwidth allocation for each client can be differentiated simply by modifying the subcarrier modulation frequency and scheme. For advanced signal processing, it depends on low-loss ASICs rather than more expensive dynamic elements (tunable laser sources). In Figure 6.4, OFDM-PON uses single wavelength, $\lambda_{downstream}$ and $\lambda_{upstream}$ for downstream channel and upstream channel respectively.

In the downstream direction, OLT generates various radio frequency orthogonal subcarriers by using inverse fast Fourier transform. This function is called radio frequency OFDM modulation. Then, these orthogonal frequencies are up-converted by optical modulator and generate the optical carrier wavelength, $\lambda_{downstream}$ and propagate over the optical fiber. This downstream signal is further divided among the N ONUs by using power splitter. Then, each ONU receiver receives these downstream signals and selects their corresponding assigned subcarriers. After that, fast Fourier transform (FFT) demodulation has been done at the ONU's receiver to get the original transmitted signal. In the upstream direction, each ONU broadcasts a number of orthogonal subcarriers based on available bandwidth, which is then up converted by an optical modulator using same upstream wavelength, $\lambda_{upstream}$. Then, these signals are combined with the help of a power combiner and transmitted along the optical fiber.

6.2.4 Hybrid PON

The hybrid PON network is created by combining any of these approaches, such as TDM-PON, WDM-PON, and OFDM-PON (Shaddad et al. 2014). TDM-WDM and OFDM-WDM are the most common types of hybrid PONs. The hybrid WDM-TDM-PON network makes efficient use of bandwidth while simultaneously improving system performance. In Figure 6.5, each OLT has a different wavelength $\lambda_{d1}, \lambda_{d2} \ldots\ldots \lambda_{dN}$. After multiplexing, it transmits over the bidirectional optical fiber in downstream direction. Then, array waveguide grating (AWG) is demultiplexed according to their wavelengths, which act as a demultiplexer. Further, each wavelength is divided among the number of ONUs by using a splitter. The ratio of splitter is selected according to the number of ONUs. Similarly, this process occurs for each ONU to OLT in upstream direction.

6.3 WIRELESS ACCESS NETWORK

The most famous wireless access technologies are WiFi and WiMAX. WiFi is a wireless local area network (WLAN) based on the IEEE 802.11 standard that was established in the 1990s. It is used extensively in the last 15 years due to cheaper, ease to deployment, high speed data rate. Other standards for WiFi are IEEE 802.11a/b/g/n/y. These standards are different in terms of their data rates and operating frequencies.

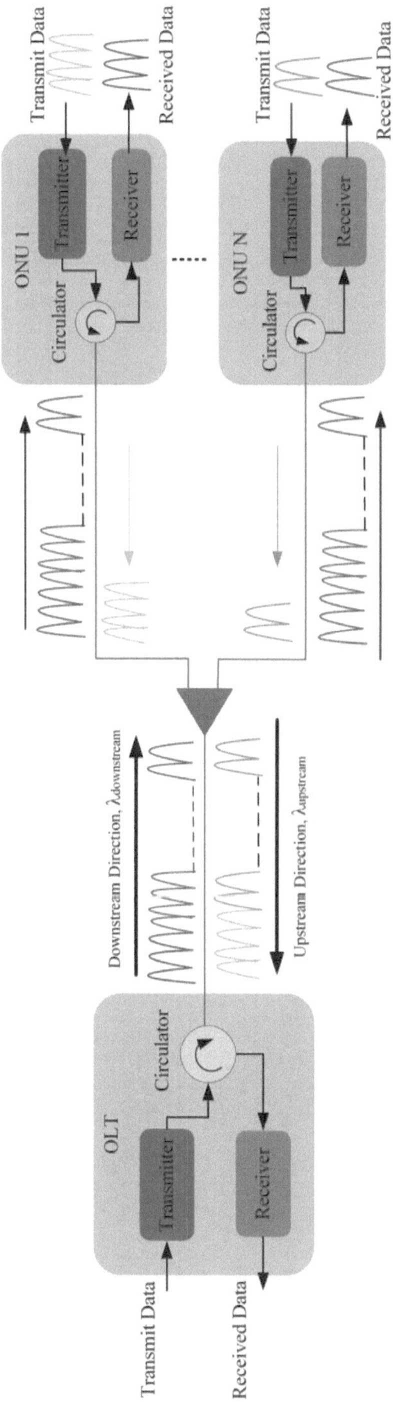

FIGURE 6.4 OFDM-PON architecture.

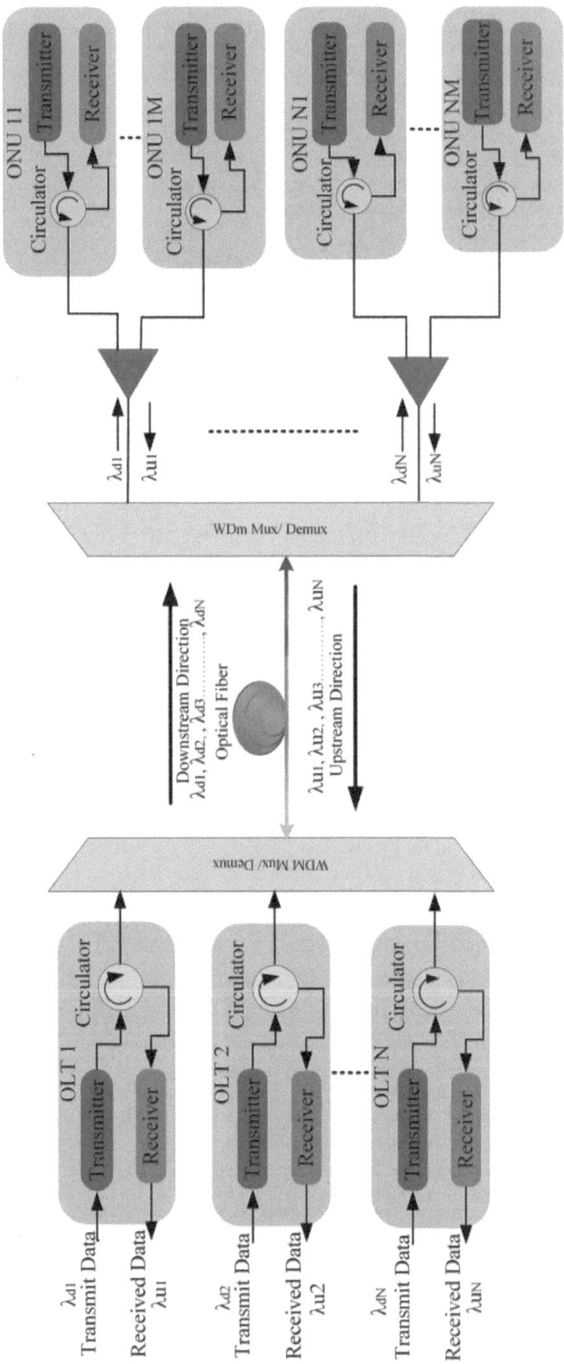

FIGURE 6.5 Hybrid WDM/TDM PON architecture.

TABLE 6.1
IEEE 802.11 Wi-Fi Standards' Comparison

Parameter	IEEE 802.11a	IEEE 802.11b	IEEE 802.11g	IEEE 802.11n	IEEE 802.11y
Operating Frequency (GHz)	5	2.4	2.4	2.4 and 5	3.7
Maximum Data Rate (Mbps)	54	11	54	248	54
Maximum Transmission Indoor Distance (meter)	35	40	40	70	50
Maximum Transmission Outdoor Distance (meter)	**100**	**120**	**120**	**250**	**5000**

TABLE 6.2
Comparison of WiMAX IEEE 802.16 Standards

Parameter	IEEE 802.16	IEEE 802.16a	IEEE 802.16e	IEEE 802.16m
Operating Frequency (GHz)	10–66	2–11	2–6	To be determined
Maximum Data Rate (Mbps)	134	75	15	1000
Cell Size (km)	2–5	7–10	2–5	Microcell (To be determined)

IEEE 802.11a works at an unlicensed band of 5 GHz with utmost 54 Mbps data rate of. IEEE 802.11b/g operate at a 2.5 GHz unlicensed band along with 11 Mbps/54 Mbps data rate (Srivastava et al. 2020). But IEEE 802.11n offers a total 600 Mbps of data rate, which operates at 2.4 GHz or 5 GHz and IEEE 802.11y is currently a work in progress for the 54 Mbps data rate, which operates at 3.7 GHz with maximum transmission reach of 5 km for outdoor environment. The limited transmission range for other standards is from 100 m–200 m.

These standards focused on WiMAX, which is used for a wireless metropolitan area network (WMAN) at a relatively low cost, which can transmit the data over many kilometers across the cities and countries. WiMAX guarantees remote access range up to 50 km when contrasted with the transmission distance of two or three hundred meters for WiFi (802.11 a/ b/ g/ n) as seen in Table 6.1. A solitary base station in WiMAX networks can uphold information rates up to 75 Mbps to the private or business clients. Nonetheless, since numerous clients are served by a single base station, information payload conveyed to end clients is probably going to 1 Mbps for private clients and not many Mbps for business customers. WiMAX works on a 2.5 GHz, 3.5 GHz, and 10.5 GHz licensed band and 2.4 GHz and 5.8 GHz for an unlicensed band. It offers a maximum 75 Mbps of data rate (Rehman et al. 2019). Data rate of up to 1 Gbps is supported in the newest standard IEEE 802.16m but it is used for shorter transmission distance as shown in Table 6.2.

Architecture of mobile WiMAX network as shown in Figure 6.6, in which mobile WiMAX base station covers a cell with the radius of 1 km in city areas, 1 km to 3 km in suburban areas, and more than 10km in a rural area. It can support up to 120 km/h with

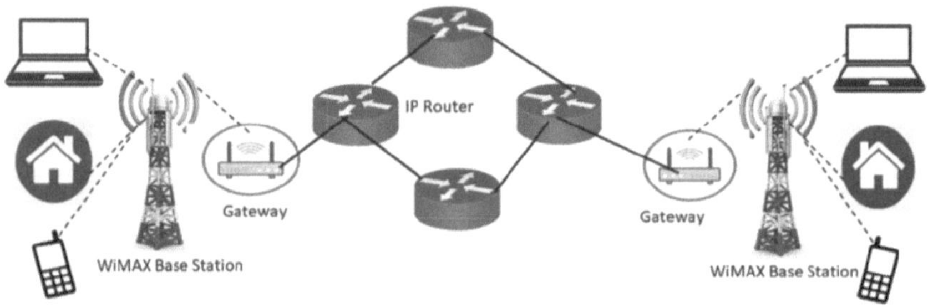

FIGURE 6.6 WiMAX architecture.

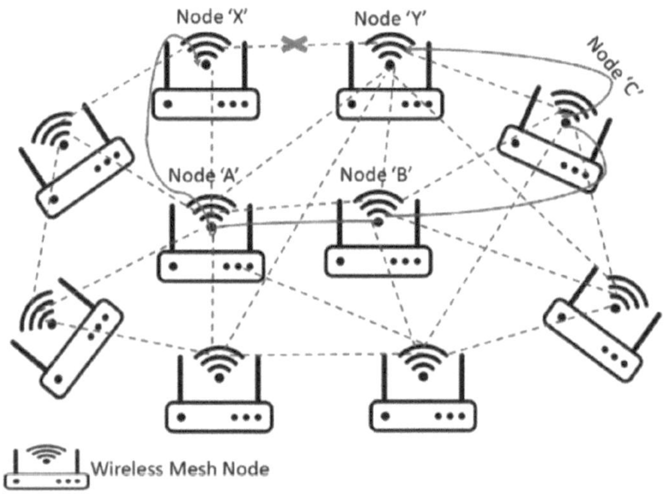

FIGURE 6.7 Wireless mesh network architecture.

handoff for mobile users. The capacity of WiMAX depends on the geographical distribution of users. Suppose if the users are close to the WiMAX base station and with line-of-sight wireless link, then maximum throughput occurs. If users are at the edge of cell, then downlink throughput decreases to 2 Mbps and the uplink to 400 kbps.

The applications of a WiMAX network include voice over internet protocol, instant messaging video conferencing, streaming media, web browsing, interactive, multimedia downloading, and real time gaming.

A wireless mesh network consists of multi-hop wireless communication links to forward the traffic to and from the wireless nodes (Srivastava et al. 2021). In this, each node can make a contact with the other node through a route comprising multiple wireless links. WMNs are dynamically self-configured and automatically maintain the mesh connectivity between nodes in the network.

The main advantage of WMN is having the self-healing capability. Suppose if the link between node 'X' and node 'Y' fails then the traffic between node 'X' and 'Y' can go through the route formed by nodes 'A', 'B', and 'C' as shown in Figure 6.7.

WMNs enhance network coverage cost effectively, improve load balancing across the network, robustness to node failures, fast network deployment, ease of maintenance, and low initial deployment cost.

6.4 HYBRID WIRELESS OPTICAL BROADBAND ACCESS NETWORK (HWOBAN)

In the early 1990s, a hybrid wireless optical broadband access network was developed to provide efficient last-mile connectivity at a low cost (Sarkar et al. 2007). Because optical fiber does not go everywhere, but where it does, it provides a large amount of available bandwidth, the concept of HWOBAN is very appealing because it combines the capacity of optical fiber networks with the ubiquity and mobility of wireless networks. Optical fiber does not go everywhere, but where it does, it provides a highly bandwidth constrained transmission channel, whereas wireless access networks go almost everywhere but provide a highly bandwidth constrained transmission channel. As a result, the combination of these two technologies has the potential to deliver anytime, everywhere broadband access to meet the needs of users in the future. Fiber Wireless (FiWi), grid based reconfigurable optical wireless network (GROWNet), and metro-access are some of the different names for HWOBAN (Bhatt et al. 2016).

6.4.1 Architecture of a Hybrid Wireless Optical Broadband Access Network

Figure 6.8 shows the network of hybrid wireless and optical broadband access (HWOBAN). HWOBAN is the combination of optical and wireless access networks, which serve as the backhaul and front end, respectively. It reduces deployment costs by using a less expensive wireless front-end network and improves broadband access network bandwidth performance. The optical backhaul section in HWOBAN starts at the optical line terminals (OLTs) and ends at a number of optical network units (ONUs). Several wireless routers are embedded in such a way that they can form mesh topology in the wireless access network. Some wireless routers, called gateways are straightforwardly connected to the ONUs. Every gateway is joined with ONU through Ethernet port. Therefore, HWOBAN is used to connect stationary and mobile users with a cloud through the wireless routers, gateways, ONUs, and OLT. In the upstream direction (from end users to OLTs), end users transmit their traffic to one of its vicinity routers of the wireless front end in HWOBAN. Further, this traffic propagates through the mesh topology of wireless network over many hops to one of the gateways. After that, gateways can route the traffic to that ONU, which is connected to it. In the optical fiber backhaul network of HWOBAN, traffic is transferred from each ONU to any OLT through a splitter over shared optical fiber.

But in the downstream direction (from OLTs to end users), traffic load is broadcasted to each ONU and then forward to gateways. Then gateways send this traffic to the specified wireless routers. So, in upstream direction the wireless front-end network is an anycast network and the optical backhaul network used a TDM based medium to

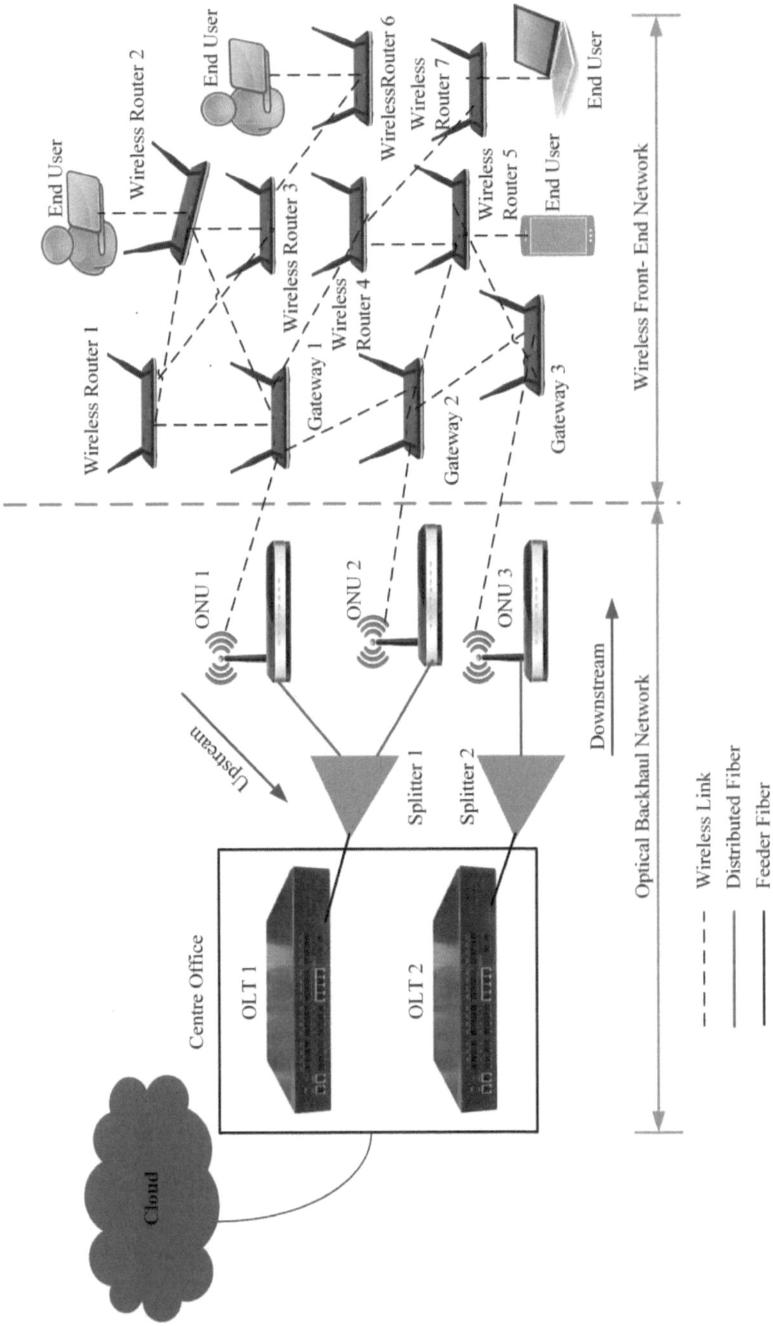

FIGURE 6.8 System Model for HWOBAN.

access the control scheme but in downstream direction, the optical backhaul network is a broadcast network and the wireless front-end network is a unicast network.

6.4.2 Fiber Wireless (FiWi) Access Network

Under the FiWi access network, RoF (radio over fiber) and R&F (radio and fiber) technologies are employed to integrate optical and wireless networks. In the RoF technology-based networks as shown in Figure 6.9, support of a number of wireless applications, radio frequency signals modulated onto an optical carrier in CO are delivered across an optical fiber link to numerous remote antenna units (RAUs). At the RAUs, the received modulated optical signal is detected by photodetector, which converts into electrical signal, further amplified and then directed to the antenna for wireless transmission (Tsagklas et al. 2011). In this a suitable optical element in the charge of managing access to both optical and wireless media is necessary to implement the required medium access control (MAC) protocol in the optical component. R & F based networks bypass the limitations of RoF based networks by using two independent MAC protocols in optical and wireless media to control them separately, and a translation protocol at the optical and wireless network interface, such as the optical network unit base station (ONU-BS) as shown in Figure 6.10.

6.4.3 Grid Based Reconfigurable Optical Wireless Network (GROW-Net)

The GROW-net architecture proposes an innovative technique for providing broadband access in a highly populated and metropolitan area in a flexible manner. To provide actual broadband connectivity to end users, it uses a cost-effective, reconfigurable, and reflecting technology. GROW-net is built on the concept of a hierarchical network as shown in Figure 6.11. As a broadband backhauling network, it consists of a wireless mesh access network segment and a reconfigurable optical grid network (Wong et al. 2009, Shaw et al. 2007). The optical grid network splits the city into grid cells, each of which contains a wireless mesh network. A reconfigurable optical grid is implemented at the fiber intersection in an optical grid network to enable dynamic wavelength routing and add/drop operations. Through optical fiber, wireless mesh gateways are connected directly to reconfigurable grid nodes. The hybrid cell manager (HCM) collects and distributes traffic to and from mesh portals through grid cells to manage the many grid cells in its vicinity.

6.4.4 Metro-Access Ring Integrated Network (MARIN)

The MARIN project, which was introduced in 2005, combines optical access networks into the metropolitan area network (MAN), allowing MAN to share the available bandwidth and resources of the access network. This architecture is built on the integration of various access ring networks to build the MARIN mesh network, which forwards MAN traffic between any two MAN hubs linked to MARIN. DWDM

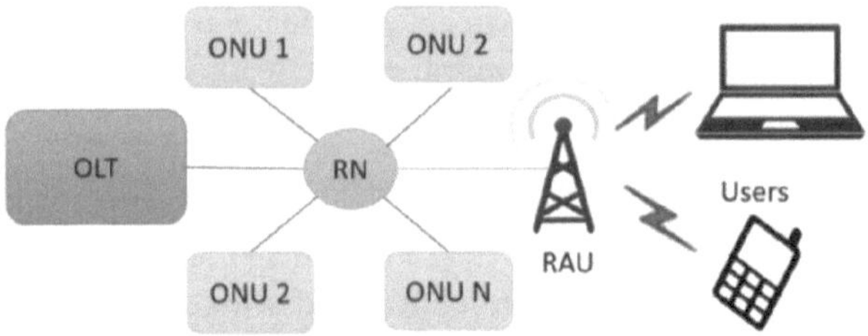

FIGURE 6.9 RoF based FiWi network.

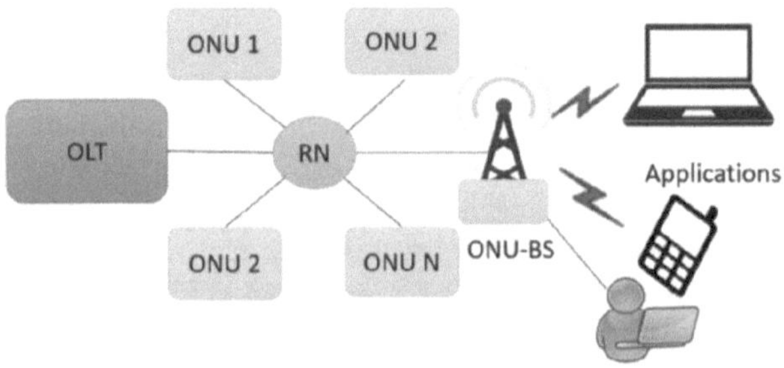

FIGURE 6.10 R & F based FiWi network.

technology is used by the MARIN to improve capacity and traffic routing (Shing Wa Wong et al. 2008).

6.5 FEATURES OF HYBRID WIRELESS OPTICAL BROADBAND NETWORK

1. HWOBAN is the precise cost-controlled network as differentiate to the optical fiber access network. HWOBAN does not require any precious fiber from end-to-end connectivity as well as installing and maintaining the optical fiber is relatively costly. In HWOBAN, end users are connected to their nearest ONUs wirelessly through wireless routers over multiple hops.
2. HWOBAN is much more flexible than the optical fiber access network because end users are connected with each other seamlessly in the wireless front-end network. Therefore, due to their flexibility and multiple hopping capability, WiFi and WiMAX are the most favorable technologies for HWOBAN.
3. A HWOBAN provides self-organizing capability, reliability, and robustness as well as better performance over the traditional optical network due to fault tolerance and better load-balancing capability. In traditional network, if fiber is broken between the splitter to ONUs and the splitter to OLT, congested and wireless routers fail, then the network will be down. But in HWOBAN, all the end users have the capability to form multi-hopping in the wireless front-end

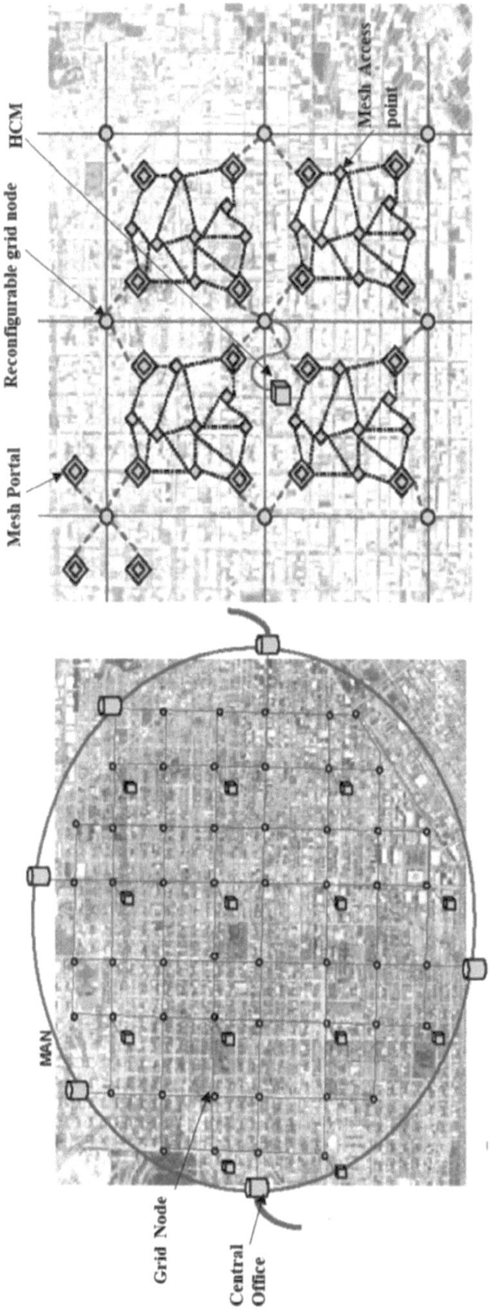

FIGURE 6.11 GROW-Net architecture for HWOBAN("316.Pdf," n.d.).

network and can find the nearest alive ONU. Hence, the users can interact with their ONUs and ONUs will communicate their respective OLTs inside the CO.

4. All the nodes of HWOBAN are having their programmability feature to change its functions. Wireless routers are programmable by open-source firmware, OLTs by using craft port and ONUs by gluing a separate Linux-box.
5. All wireless nodes in the front end of HBOWAN consume very low power. Power consumption of one wireless router is approximately 1 W to 2.5 W.
6. The capacity of HWOBAN is far more than that of the wireless access network. It is due to HWOBAN's high-capacity optical backhaul network (Sarkar et al. 2007).

6.6 PROBLEMS FOR HWOBAN

HWOBANs are thought to be the future of broadband access networks, with a wide range of uses. It becomes critical to focus on concerns such as optimal network device location, survivability, and energy usage, among others. In today's world, the energy efficiency of any network takes precedence over other concerns. In all HWOBANs, there is a large amount of energy wasted owing to idle network parts. We know that the higher the number of awakened devices in a network, the more energy it consumes because the network does not always bear the maximum traffic load. The energy usage of broadband access networks is expected to be roughly 70% of total telecom network energy consumption (Brandão 2010). As a result, the network should use energy-saving management strategies in order to lower its carbon impact.

6.7 SOLUTION FOR ENERGY CONSUMPTION IN HWOBAN

Several factors must be considered in order to reduce network energy consumption and maximize network utilization. HWOBAN architecture, network element deployments, and management methods give flexibility, fault tolerance, reliability, and resilience in the construction of an environmentally friendly green network (Parasher et al. 2019, Singh et al. 2019). As a result, HWOBAN is known as a green hybrid wireless optical broadband access network (GHWOBAN) (Gong et al. 2013). For any network energy management scheme, various aspects need to be considered as explained below.

1. Topologies for HWOBAN should be selected in such a way that it provides a different alternative link to route the traffic from transmitter to receiver.
2. To diminish the power consumption by the network, put some devices of HWOBAN into a sleep mode during low load hours because data varies at different times of the day.
3. Optimal placement of network devices like efficient deployment of ONUs in HWOBAN also play an important role to save energy.

6.7.1 HWOBAN Topologies

In the literature, various important aspects for energy saving have been discussed for HWOBAN. But various topologies also play an important role to enable the energy saving. A capacity imbalance also exists between the optical access network and wireless access network. Thus, redundant capacity for the optical backhaul network supports the extra reliability to the whole network during the failure of any device and traffic can be rerouted through an alternative path. For this, several topologies are used that are able to fulfil the needed capacity and reliability requirements. All these requirements are possible by condense interconnected wireless routers in the wireless access network to divert the traffic through substitute path when the wireless router, gateway, and ONU fail or fiber breaks occur as shown in Figure 6.12. So, this provides extra reliability and flexibility at the time of failure, which helps to enable energy saving in the optical backhaul network.

Moreover, another important aspect of the broadband access network is load unbalancing due to fluctuation of daily traffic that comes directly from the users. So, to deploy the network devices in such a fashion that it can support the maximum traffic load as well as keep the devices in sleep mode during low traffic hours (Mukherjee et al. 2009). As a result, some key considerations must be made when designing the green HWOBN. Network topologies should be chosen in such a way that redundant paths can be provided to ensure that data arrives at its destination on time, as well as traffic profiles that fluctuate throughout the day and the ability to put devices to sleep during low-traffic hours. As a result, lower overall network power consumption is needed to create a green HWOBAN.

6.7.2 Proper Utilization of ONU/OLT in the Optical Ackhaul Network

During the deployment of any network, devices are placed in such a way to hold up peak traffic load, which comes directly from the customer's side. Traffic load changed according to the different times of the day. During low traffic hours, the network's components may be not fully engaged and act as underutilized. Thus, putting those devices into a sleep mode or low power state. Thereby network energy consumption reduces. For example, suppose one OLT fails and put this into sleep mode. Then reroute its traffic through other OLTs, which are connected by different topologies types. Similarly, this can be applied for any ONUs. This method also makes energy-efficient HWOBAN. Power consumed by ONU is approximately 10 W during active hours and less than 1 W during sleep mode (Saeed et al. 2016). In HWOBAN, each OLT tries to manage the power consumption by putting low traffic ONUs in sleep mode and high load ONUs in active mode by using the following Algorithm 6.1.

Algorithm 6.1. Optimal ONU usage

Data: Assign minimum watermark (MinW) and maximum watermark (MaxW), HWOBAN topology
Result: Set of ONUs that can be in sleep mode
Step1: Set the values of MinW and MaxW

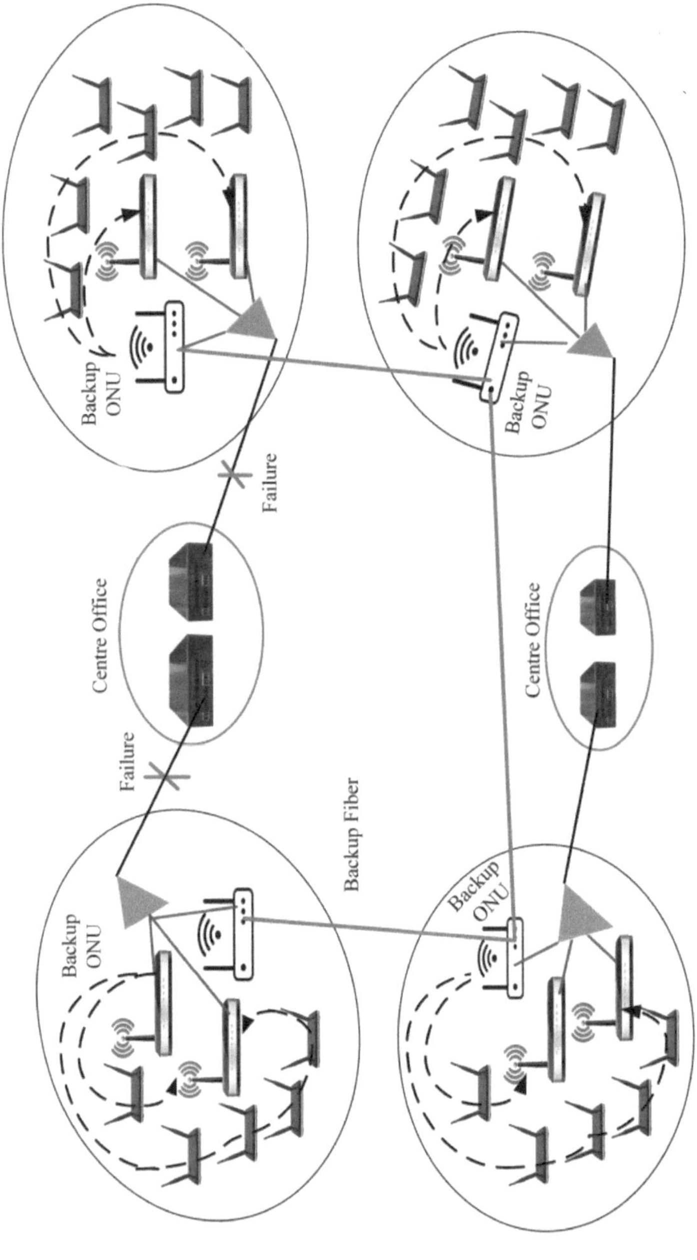

FIGURE 6.12 Segmentation based HWOBAN architecture at the time of fiber failure.

Step2: The OLT assesses traffic load at all ONUs by calculating the length of associated input queues while routing traffic at various times of the day.
Step3: if traffic load < MinW, ONUs to be shut down

else if traffic load > MaxW, ONUs in on mode and switch on other sleeping ONU.
else put ONUs active

6.7.3 ONUs' Placement Algorithms in HWOBAN

Physical resources must be used efficiently in HWOBAN to reduce energy usage. As a result, the placement efficiency of its components is a determining element in regulating energy usage. As ONUs consume the majority of HWOBAN's energy, efficient deployment of the fewest number of ONUs and the best possible site to locate ONUs has an impact on network performance and energy efficiency. As a result, multiple algorithms are used to find the best location for specific ONUs. Ant colony optimization algorithm, random and deterministic method, greedy approach (GA), simulated annealing approach (SAA), and mixed integer programming approach are some of the approaches used (MIP).

6.7.3.1 Ant Colony Optimization (ACO) Algorithm

This method is used to find the optimum placement of ONUs and its minimization. This is a heuristic method that is based on the behavior of a genuine ant colony. To begin, ants look for food in an ad hoc manner around their colony. They leave a pheromone chemical on the ground while searching, which aids other ants in finding the shortest way to the food. They leave a certain amount of pheromone when they return to the nest, which is dependent on the quality and quantity of the food. This entails receiving feedback on pheromone concentration in order to choose the appropriate path. The ACO method is used to locate the ONUs that are the smallest. ONUs is first placed in the center of each grid, each with a different pheromone concentration. After that, ants begin to move across routers in the network. After completion of the full path, ants come back from where they start and provide the feedback about the pheromone concentration of every path. On the basis of pheromone concentration, repeated routers are removed who are having lower pheromone concentration in the ONU's area. This iteration is applied for each router and it is found out that some of the ONUs do not have routers. Those ONUs, which are free from the wireless routers, are removed from the network. By doing this, the number of ONUs are minimized (Saeed et al. 2016).

6.7.3.2 Random and Deterministic Method

It is the simplest method for the best placement of the ONUs. In this method, the whole network is divided into multiple non-overlapping segments where each segment has a separate ONU as shown in Figure 6.13. Each ONU is placed anywhere within its segment. Due to this, the network may not ensure the proper connectivity. If the placements of ONU is predefined or placed in the center of each segment, then this method is called the deterministic method. After that, Euclidean distance

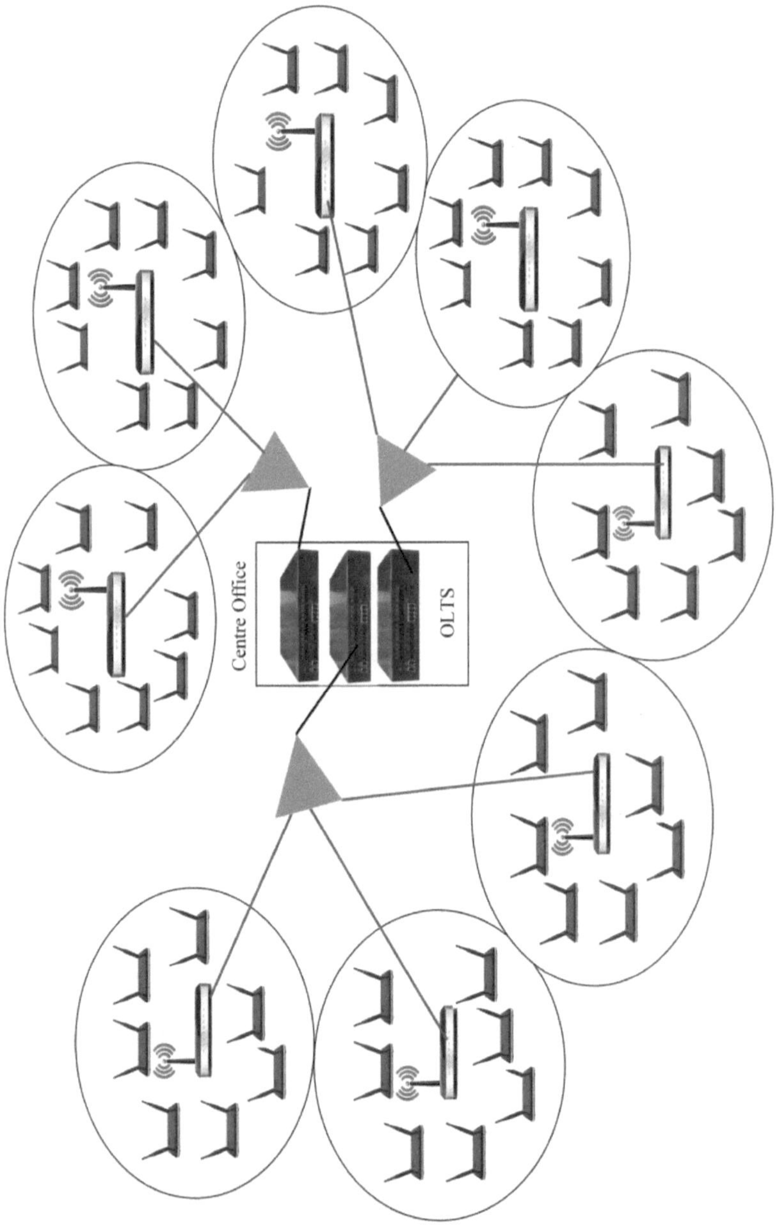

FIGURE 6.13 HWOBAN architecture with optimal ONU placements.

vector is determined from all the ONUs with respect to nodes (wireless routers) and maintains the matrices in an ascending order. Minimum Euclidean distance between ONUs and node, identifies the primary ONUs. Several backup ONUs are also found out according to the 2nd minimum Euclidean distance, 3rd minimum Euclidean distance, and so on. Assume that a total of 'N' ONUs are distributed in such a way that the position of wireless routers is known. Consider (X_i, Y_i) as the coordinates (location) of the *i*th ONU, which has *k* wireless routers and coordinates of the wireless router (x_j, y_j) where $j \in (1, 2, 3...., k)$. So, the cost matrices are used to deploy the *i*th ONU from the wireless routers as the expected Euclidean distance between the ONU and its wireless routers, and the related cost matrices are given as

$$C_{ONUi} = \left(\sum_{j=1}^{k} \sqrt{(x_{j-} X_i)^2 + (y_{j-} Y_i)^2} \right) / k$$

$$C_{overall} = \sum_{i=1}^{N} C_{ONUi}$$

Both approaches are unsuitable for ONU installation since they have lower processing requirements, are not well suited for non-uniform wireless router distribution, and do not provide sufficient network connectivity.

6.7.3.3 Greedy Approach (GA)

This algorithm is based on the divide and conquer method. The entire wireless front-end network is separated into many non-overlapping areas using this strategy. Primary ONUs are also discovered in this method, which is similar to the random and deterministic methods. However, in the Greedy technique, the first ONU is placed in the center region, and each primary ONU's position (coordinates) corresponding to their premium nodes is optimized. So, using the random and deterministic method to identify primary and backup ONUs for each node, obtain a set of premium nodes for primary ONUs, and optimize the placement of each primary ONU with respect to its premium nodes. Place their primary ONU as shown below for a set of premium nodes, (x_j, y_j), $j \in (1, 2, 3...., k)$, and assume that all wireless routers are dispersed independently and identically.

$$(X_i, Y_i) = \left(\frac{\sum_{j=1}^{k} x_j}{k}, \frac{\sum_{j=1}^{k} y_j}{k} \right)$$

The Greedy algorithm-based network performs better as compared to the random and deterministic algorithm-based network. But it finds the local optimal placement of ONUs after detection of premium node rather than global optimization (Sarkar, Mukherjee, and Dixit 2006a).

6.7.3.4 Simulated Annealing Approach (SAA)

This approach is based on the Monte-Carlo method. Instead of premium nodes, the globally optimal placement of ONU will be determined by calculating the minimal expectation of the cost function for all wireless routers with regard to other ONUs. Initialization, perturbation, cost calculation, acceptance, and updated are the five phases of the process (Sarkar et al., 2006). The first deployment of ONUs is done with GA. Furthermore, the perturbation phase moves the ONUs by a small random distance in either an upward, downward, left, or right direction from their current position, and then calculates the new cost of the ONUs' position and determines the correlation between the new and old cost placement, which is C_{ONUi}^{new} and $C_{overall}^{new}$ in the cost calculation phase.

$$C_{ONUi}^{Initial} = C_{ONUi}^{Primary} + \sum_{j=1, j \neq i}^{L-1} \alpha_j C_{ONUi}^{Backup}$$

$$C_{overall}^{Initial} = \sum_{i=1}^{N} C_{ONUi}^{Initial}$$

where $C_{ONUi}^{Primary}$ is the cost of premium nodes for primary ONU, $\sum_{j=1, j \neq i}^{L-1} \alpha_j C_{ONUi}^{Backup}$ is the cost of other (L-1) nodes for backup ONU, and $C_{overall}^{Initial}$ = overall cost of all 'N' ONUs.

Then according to it, SA accepts or rejects the relocation of ONUs. If $C_{ONUi}^{new} \leq C_{ONUi}^{Initial}$ then SAA allows the new placement of ONUs, otherwise it takes the new position with a unique probability. The same process is repeated until there is no additional cost reduction in the updated phase. Thus, this state is called equilibrium state (Sarkar et al. 2006b).

SAA's network deployment cost was lower than GA's. Individual ONUs can save anywhere from 10% to as little as 0.05 percent by switching from SAA to GA. The overall cost savings from SAA versus GA is often less than 3%. As a result, SAA provides a better solution than GA, yet GA provides accurate results (Sarkar et al. 2006b).

6.7.3.5 Moth-Flame Optimization Algorithm (MFO)

As the complexity of the HWOBAN increases, then optimization is required to find a better way to resolve this complexity problem. The moth-flame optimization method (MFO) is used to find the best location for various ONUs. It is based on the movement of moths in a straight line that keep a fixed angle with the moon while they fly at night (Singh and Prakash 2017).

Algorithm 6.2

1. Initialized the conditions for optimization process. Conditions consists of the number of candidate solutions, number of iterations, lower bound, and upper bound of the variables. Moth represents the numerous candidate solutions.

Variables is the location of moths in the space. Here moths represent ONUs and the location of moths denotes the location of ONUs. Then corresponding cost function values are calculated and stored in a matrix.

2. Then another matrix is formed, which consist of the flame's values. While moving around the space, moths dropped the flames by looking at the search space. Each moth comes across the flame and upgrades its position if a better solution is acquired. Both the positions are the solution but flames are the optimization solution and moths are the actual answer. With this process, the moth does not misfortune the best positions with respect to the moon. The transverse orientation of moths is the focus of this algorithm, $M_i = S(M_i, F_j)$ where M_i is ith moth and F_j denotes the jth flame and S is spiral function.
3. Then the logarithmic spiral path is determined based on the distance between moth and flames.

   Logarithmic spiral for the MFO algorithm is given as

   $$S(M_i, F_j) = D_i \exp(bt) \cos(2\pi t) + F_j$$

   It depicts the moth's next position in relation to the flame.
   where 'D_i' denotes the ith moth's distance from the jth flame

   $$D_i = | F_j - M_i |$$

   'b' determines the logarithmic spiral's shape
   't' can be any number between [-1 and 1].
   t = -1 signifies the flame's closest value, whereas t = 1 denotes the flame's farthest value.

6.7.3.6 MIP (Mixed Integer Programming) Approach

A hybrid wireless-optical broadband access network appeals to many people. This algorithm aids in determining the best number of devices for a given traffic load. According to the literature, the cost of laying fiber in WOBAN is more than the cost of devices (Sarkar et al. 2007a). As we know that an OLT can use only a limited or constant number of ONUs. If there are more ONUs in the network then it can lead to the additional cost for the installation of OLTs. The cost of HWOBAN rises as the cost of OLTs rises, and this problem is modeled as a "mixed integer programming (MIP)" problem. MIP focuses on optimizing the pre-deployment of ONUs and wireless routers in the HWOBAN wireless network, as well as the fiber infrastructure between ONU and OLT in the HWOBAN optical network (Sarkar et al., 2007). It reduces the cost of the network by using an optimal number of wireless routers, gateways, ONUs and efficiently usages of fiber. Thus, efficient operation of HWOBAN, MIP approach uses some constraints: traffic flow constraints, wireless capacity constraint, wired capacity constraint, wired directionality constraint, path length constraint, and ONU installation constraint (Chowdhury et al. 2010).

6.7.4 Energy Aware Routing Protocols

The goal of energy conscious routing strategies is to reduce overall network energy consumption. The HWOBAN design employs a variety of routing protocols. Energy aware protocols are classified using the least path routing and link state routing algorithms. The minimum hop routing algorithm (MHRA) and the shortest path routing algorithm (SPRA) are two types of routing algorithms (Ali 2013) based on the smallest path routing schemes whereas link state is used in the predictive-throughput routing algorithm (PTRA), delay aware routing algorithm (DARA), and capacity and delay aware routing (CaDAR) (Ghazisaidi and Maier 2011). Each node uses link state advertisement to routinely advertise link-state (LS) related data to the other nodes (LSA). Following the collection of LSAs from all nodes, each node creates a network map and updates the routing table using Dijkstra's algorithm to transport traffic throughout the network. Dijkstra's algorithm is one of the most widely used routing techniques for determining the shortest path tree between two nodes.

6.7.4.1 Minimum Hop Routing Algorithm (MHRA) and Shortest Path Routing Algorithm (SPRA)

MHRA and SPRA algorithms are easy to implement due to its simplicity. These are mostly used in wireless front-end network of HWOBAN. They are based on the smallest path principle neglecting other traffic demands. In the case of SPRA, the cost metric is inversely proportional to path capacity but it is equal to one in case of MHRA. Both techniques are static routing algorithms, which means they do not need to adapt in response to changing network conditions. As a result, when load balance has variations, the connection conflicts, or network congestion arises, and they cannot be corrected in a timely manner (Ali 2013).

6.7.4.2 Predictive-Throughput Routing Algorithm (PTRA)

It is not the shortest path-based algorithm like MHRA and SPRA. According to the LSA, it selects the route among the set of possible alternative routes between wireless routers and gateways in such a way that it fulfils the throughput needs. Basically, PTRA gets the previous samples of link states continuously across wireless links. Based on them, it gathers the link-state information dynamically and estimates the throughput of each path. Then according to highest throughput, select that path. The major drawback in PTRA is packet delay because the maximum throughput selected path might be the larger path and packets travel the longer path. So, it is not suitable for delay-sensitivity services. So the shortest path should be chosen, which is having a maximum throughput and less delay (Ali 2013).

6.7.4.3 Delay Aware Routing Algorithm (DARA)

The DARA method is based on minimizing the time delay between the wireless router and the gateway in the HWOBAN wireless network. It also tries to avoid congestion and provides the load balancing. This is a proactive routing algorithm where time delay from wireless router to gateway and from gateway to non-wire router are considered. Each wireless router on the way up to the gateway is considered as a systematic order. Then LSAs transmit the advertisement in a timely manner, which confirms the status of the wireless link states. Further, optimal weight is assigned to the links according to the LSAs. The delay of packet in the wireless network of HWOBAN has four different categories: slot queuing delay, synchronization delay, transmission delay, and propagation delay.

Propagation delay is neglected due to all the routers being densely packed altogether. Transmission delay is confined on the successful path capacity. Transmission delay will be lower when link capacity is high. Slot synchronization delay exists with TDM that transmit the packet from one wireless router to its nearby routers within predefined schedule slots. Delay in queuing dependent on router service speed and packet arrival speed. If the packet arrival rate is higher and the service rate is lower, it will be significant.

The sum of three delay slots' synchronization delay, transmission delay, and queuing delay is the average packet transfer delay, Q (system time) of the path. Packet arrived at wireless routers (from wireless router to gateway) and gateways (directed to wireless routers) are in Poisson distribution λ_n and λ_m respectively with exponential interarrival time. The ratio of system time and interarrival time is called traffic intensity of the network, λ_i, where λ_n, λ_m, and λ_i are the packet receiving rate at router $n \in [1, |R|]$, packet arrival rate at gateway $m \in [1, |G|]$ and packet arrival rate at link i, $i \in [1, |W]$, respectively. R, G, and W are the sets of wireless routers, gateways, and wireless paths, respectively (Sarkar et al. 2007b). With mean length, $1/\mu$ and effective path capacity, C_i, data lengths are autonomous and dispersed exponentially. The delay in packet transfer for every path 'i' is Q_i

$$Q_i = \left(\frac{1}{\mu C_i} + \frac{1}{2\mu C_i} + \frac{\rho_i}{\mu C_i - \lambda_i} \right)$$

where $\frac{1}{\mu C_i}$ = transmission delay

$\frac{1}{2\mu C_i}$ = slot synchronization delay

$\frac{\rho_i}{\mu C_i - \lambda_i}$ = queuing delay

$\rho_i = \frac{\lambda_i}{\mu C_i}$ is called link utilization.

So, DARA gets the predicted link delay matrics to provide the link weight.

This algorithm mainly relies on the LSAs but LSAs change dramatically from one LSA to another LSA. Thus, the predicted link weight may go wrong and take too much processing time and decision making. Suppose, if the packet wants the link with least delay (weight), then other links in the mesh get overloaded and due to this, congestion takes place throughout the network. So, load balancing is required for proper utilization of all the components of the network. But during least traffic time, some components of the network lead to underutilization. Thus, it is important to decrease the energy consumption throughout the complete network by just putting underutilization devices or very low load devices to off mode. Then, reroute the data through other 'on mode' devices. When load increases then 'off mode' devices can be turned on to fetch the increased traffic. Thus, some modification is required in the DARA algorithm to balance the load. When data is transmitted through the path, its successor path weight is the remaining capacity on that path.

Remaining Capacity = Original Capacity – Traffic Flow on the path

Here residual capacity is considered as link delay matrics (Chowdhury et al. 2010). To discover the advanced way from the source to objective, we must choose the most reduced lingering capacity link. Suppose there is a small network in which traffic is flowing from the source to destination as in Figure 6.14. The traffic is routed through the link Source-A-B-C-Destination, which has the lowest link weight (2+2+2+1=7). Although this path is not the shortest route because it chooses the path with four hops in the wireless front-end network. It may increase the average path length as well as time delay in the network.

Further, the solution of this situation, hop offset is introduced. The objective of the hop offset is to lower down the average path length. So, add the hop offset to the link weight for each hop. Consider hop offset is 1, then residual capacities are 7+ 4*1=11, 9+3*1=12, and 13+2*1=15 for link Source-A-B-C-Destination, Source-E-F-Destination and Source-G-Destination respectively. Residual capacity is minimum for link Source-A-B-C-Destination and it is selected path. Now suppose hop offset is equal to 3, Source-E-F-Destination will be the selected path. If hop offset is 5 then Source-G-Destination will be the chosen path. Hence, choose the ideal bounce offset so that expected path length ought to not raise disproportionally.

If average path length grows very large, then there are more hops in the path. That means more wireless transmitters and receivers are there. Each transmitter and

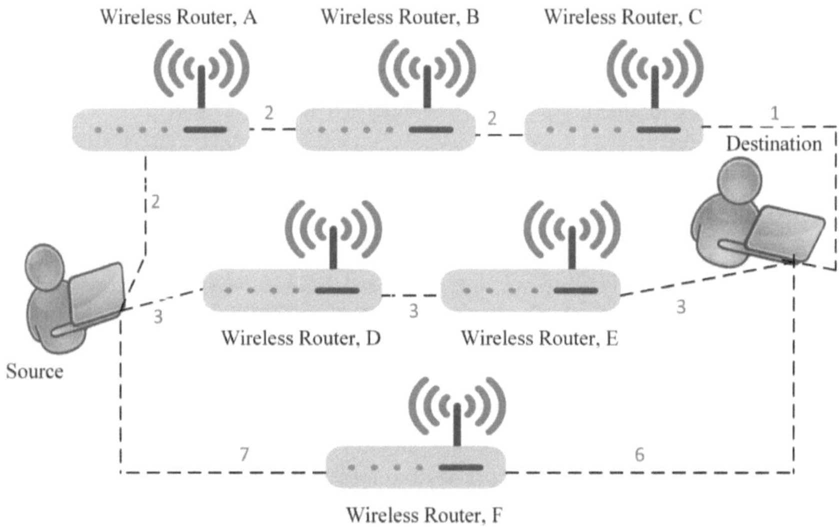

FIGURE 6.14 Wireless mesh network with residual capacity.

receiver need power to work. So, improper selection of average path length may lead to power consumption. Thus, for power saving, appropriate selection of average path length is required in the network.

6.7.4.4 Risk and Delay Aware Routing Algorithm (RADAR)

This algorithm is a DARA expansion. Similar to DARA, all the wireless routers advertise link states periodically by LSA and assign weight to the wireless link. In RADAR, every wireless router updates their risk list to get the records of various faulty devices (gateway or OLT failure). They automatically update the risk list and rerouted the packets. The risk list contains the information about link number, primary gateways, secondary gateways, tertiary gateways, path status, and link delay. It protects multiple wireless access networks and optical access networks from faults and decreases projected packet time delay and packet loss (Reaz et al. 2009).

6.7.4.5 Capacity and Delay Aware Routing Algorithm (CaDAR)

This algorithm not only determines the shortest possible link latency, but it also distributes the link weight to all devices in a very efficient manner. In the wireless front-end network of HWOBAN, due to inefficient routing, packets may require the multiple hops from the source to destination, which results that some links are overutilization and others may be underutilization. But, if the outgoing capacity of nodes is not properly allocated among their paths, then packet delay becomes larger. Hence, capacity assignment along with packet delay are the two most important parameters for any reliable network. To solve this problem, the CaDAR routing algorithm is used with an aim to optimally assign the radio capacity of a node across its outgoing path of wireless front-end network to get minimum average delay (Reaz et al. 2009).

The capacity of a wireless front-end network is limited, as each node has just one radio capacity that is shared across all of the node's outbound links via time division multiplexing (TDM). Predetermined time slots are assigned to each node based on the number of links existing at that node. Each node has a different number of links. So many time slots are allotted to the nodes for various induced paths. Specifically, in CaDAR, every node notifies their nearby nodes so that they know the current traffic load of all their outgoing links by advertising the LSA messages. Furthermore, to enhance the capacity of the wireless front-end of HWOBAN, multiple radios at nodes can be deployed.

Suppose there are four nodes x, y, z and u. Let node x be connected with the other three nodes y, z and u, which operates on three time slots as shown in Figure 6.15. Similarly, z operates on two different time slots. From LSA, optimal outgoing capacity is assigned based on traffic flow on each wireless link and average delay should be as minimum as possible. If any link does not carry any traffic, then treat that path as inactive and assign zero duration to that path. In Figure 6.15, different outgoing traffic from node x and z are λ_{xy}, λ_{xz}, λ_{xu}, and λ_{zx}, λ_{zu} respectively that is $\lambda_{xy} \neq \lambda_{xz} \neq \lambda_{xu} \neq \lambda_{zx} \neq \lambda_{zu}$. Based on their traffic, assign different time slots for different links for the same node. Then capacities need to be assigned according to

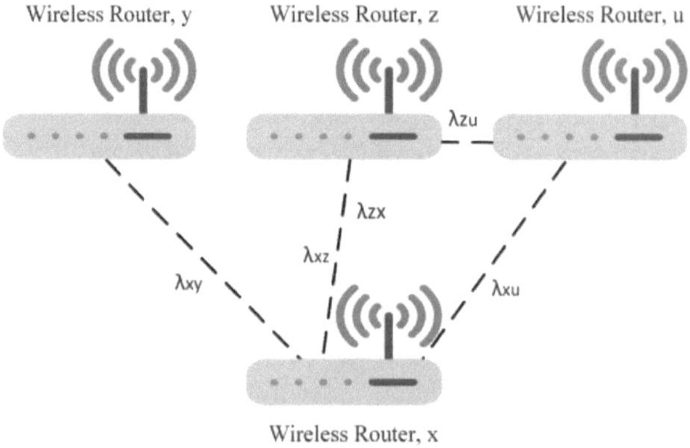

FIGURE 6.15 Wireless mesh network with traffic flow.

λ_{xy}, λ_{xz}, λ_{xu}, λ_{zx}, and λ_{zu}. Capacities for node x are denoted as C_{xy}, C_{xz}, and C_{xu} and for node z as C_{zx} and C_{zu}.

6.8 CHALLENGES OF HYBRID WIRELESS-OPTICAL BROADBAND NETWORK

Here, some research challenges will be discussed from the information gathered from HWOBAN literatures.

1. **MPCP (multi-point control protocol) Based L2 Routing Protocol:** Deployment of current HWOBAN uses different data transfer techniques for the wireless front-end network and optical backhaul network. Optical backhaul network uses Ethernet PON (EPON) technology and function of Ethernet is emulated by Layer-2 signaling mechanism called MPCP-based dynamic bandwidth allocation (DBA) or interleaved polling with adaptive cycle time (IPCT) that permits the OLT to distribute the bandwidth to all the ONUs periodically. If any failure occurs, then different dynamic protection schemes can be used for network survivability.

   But in wireless front-end network, dynamic link-state routing protocol (layer 3), OLSR (optimal link-state routing) is used as proactive to create a self-organizing capability. But, OLSR uses an excess amount of bandwidth for update of the link state dynamically (Chowdhury et al. 2009a). So, static routing can be used to get the better performance as compared to dynamic allocation routing in HWOBAN. So, to provide seamless connectivity through all the segments of HWOBAN, an alternative routing protocol called integrated Layer-2 (L2) is requires to reduce the layer 3 generated overheads. Hence, anyone can research on the MPCP based L2 routing protocol for HWOBAN. End to end compatibility takes place between optical segment and wireless segment with the help of this protocols and OLT can drive multiple ONUs and one ONU can drive multiple gateways for wireless segment.

2. **TDM-OFDM MAC for wireless segment:** As the optical backhaul network of HWOBAN uses a TDM-based MAC protocol, but conventional wireless front-end network uses collision-based MAC protocols. With IEEE 802.11 g standard wireless routers, carrier sense multiple access with collision avoidance (CSMA/CA) MAC protocols are now employed. However, this protocol limits the capacity of the wireless section. As a result, the TDM-based MAC protocol increases the capacity of the wireless front-end network (Chowdhury et al. 2009a). TDM paired with orthogonal frequency division multiplexing (OFDM) MAC protocols can be used to increase the performance of wireless networks.
3. Memory and processing power for wireless routers near gateways should be more as compared to far away routers from gateways. As nearer wireless routers to the gateways have more traffic load than the distant ones. That means a wireless router in the vicinity of the gateways should be optimally placed with a large number of wireless routers. Various technologies, like numerous radios, directional antennas, and multiple input multiple output (MIMO), can be used to increase the wireless router's capacity.
4. Existing HWOBAN do not exhibit adequate self-organizing capability and faulty tolerance. To ensure the flexibility, load shifting/balancing is required between the devices.

6.9 APPLICATIONS FOR HYBRID WIRELESS-OPTICAL BROADBAND NETWORK

1. HWOBAN has attracted significant growth in its applications with emerging technologies. It enables machine to machine (M2M) communications, focusing on how HWOBAN can be used in emerging smart grid applications.
2. HWOBAN can be used from multimedia communication to telepresence to disaster relief including IoT, software defined network, NG-PONs etc.
3. It can be used for inter building communication, smart homes, smart agriculture, smart transportation, water monitoring, waste management, education (Liu et al. 2016).
4. This network is secured, used for distances from nm to more than thousand km, has high capacity and enormous data transmission rate to build a smart city. Smart cities are fast emerging as potential solutions for making the systems sharper and environmentally friendly.

6.10 RESEARCH OPPORTUNITIES FOR THE HYBRID WIRELESS-OPTICAL BROADBAND NETWORK

1. Different technologies, such as multiple input multiple output (MIMO), multiple radios, and directional antenna, can be used to expand the capacity of a wireless router. Theoretical maximum capacity of wireless routers is of order 54 Mbps (Chowdhury et al. 2009b). Capacity achieved from wireless routers is low due to the interference and other surrounding interferences.

2. Performance of HWOBAN degrade due to too many wireless hops especially for video data. Placement of intelligent and more gateways in the wireless front-end network may help to improve the HWOBAN performance (Chowdhury et al. 2009b).
3. HWOBAN should be programmable and configurable to facilitate future networks.
4. Further, calculate the parameters that is packet loss rate, jitter, hop length versus energy cost for HWOBAN.
5. Channel condition, modulation methods can be considered in the energy aware routing algorithm for HWOBAN.

6.11 CONCLUSION

This chapter gives an overview of optical access technologies, wireless broadband, and the hybrid of these two technologies. This combination optimizes the best of twos, which provides a high bandwidth, reliability, high transmission reach, less expensive, and high flexibility broadband access network. This chapter emphasizes the recent energy-efficient schemes for HWOBAN to reduce the power consumption by using topologies, load balancing, optimum placements of ONUs in a HWOBAN, and energy aware routing algorithms. These algorithms are designed to ensure network survivability when wireless router, gateway, and ONU are failed/inactive or fiber breaks. So, this provides extra reliability and flexibility at the time of failure, which helps to save a considerable amount of energy. Review of various energy-efficient mechanisms concludes that energy efficiency in HWOBAN is the foremost problem.

REFERENCES

Ali, A. (2013). Routing in WOBAN: A Review. *International Journal of Computer and Communication Engineering*, *2*(2), 110–114. https://doi.org/10.7763/ijcce.2013.v2.150

Bhatt, U. R., Chouhan, N., & Upadhyay, R. (2016). Energy Efficient Segmentation for Green FiWi network. *Physics Procedia*, *78*, 75–81. https://doi.org/10.1016/j.procs.2016.02.013

Brandão, A. L. (2010). On the energy consumption of relay networks. *IEEE Vehicular Technology Conference*, *1*, 9–11. https://doi.org/10.1109/VETECF.2010.5594366

Chowdhury, P., Mukherjee, B., Sarkar, S., Kramer, G., & Dixit, S. (2009a). crit. In *IEEE Network* (Vol. 23, Issue 3). https://doi.org/10.1109/MNET.2009.4939262

Chowdhury, P., Mukherjee, B., Sarkar, S., Kramer, G., & Dixit, S. (2009b). Hybrid wireless-optical broadband access network (WOBAN): Prototype development and research challenges. *IEEE Network*, *23*(3), 41–48. https://doi.org/10.1109/MNET.2009.4939262

Chowdhury, P., Tornatore, M., Sarkar, S., & Mukherjee, B. (2010). Building a green wireless-optical broadband access network (WOBAN). *Journal of Lightwave Technology*, *28*(16), 2219–2229. https://doi.org/10.1109/JLT.2010.2044369

Christodoulou, C., & Ellinas, G. (2021). Resilient architecture for optical access networks. *Photonic Network Communications*, *41*(1), 1–16. https://doi.org/10.1007/s11107-020-00910-y

Ghazisaidi, N., & Maier, M. (2011). Fiber-wireless (FiWi) access networks: Challenges and opportunities. *IEEE Network*, *25*(1), 36–42. https://doi.org/10.1109/MNET.2011.5687951

Gong, X., Hou, W., Guo, L., & Zhang, L. (2013). Optik Dynamic energy-saving algorithm in green hybrid wireless-optical broadband access network. *Optik – International Journal for Light and Electron Optics, 124*(14), 1874–1881. https://doi.org/10.1016/j.ijleo.2012.05.030

Kaur, G., Tomar, P. and Tanque, M. (2019). "Integrating Artificial Intelligence to Solve Pervasive IoT Issues, Elsevier, Cambridge, MA, USA.

Liu, J., Guo, H., Nishiyama, H., Ujikawa, H., Suzuki, K., & Kato, N. (2016). New Perspectives on Future Smart FiWi Networks: Scalability, Reliability, and Energy Efficiency. *IEEE Communications Surveys and Tutorials, 18*(2), 1045–1072. https://doi.org/10.1109/COMST.2015.2500960

Mukherjee, B., & Chowdhury, P. (2009). Green wireless-optical broadband access network (WOBAN). *Optics InfoBase Conference Papers*, 2–3. https://doi.org/10.1364/acp.2009.thdd2

Parasher, Y., Singh, P., & Kaur, G. (2019). Green Smart Security System. Green and Smart Technologies for Smart Cities, 165–184.

Reaz, A., Ramamurthi, V., Sarkar, S., Ghosal, D., Dixit, S., & Mukherjee, B. (2009). CaDAR: An efficient routing algorithm for a wireless optical broadband access network (WOBAN). *Journal of Optical Communications and Networking, 1*(5), 392–403. https://doi.org/10.1364/JOCN.1.000392

Rehman, S. U., Ullah, S., Chong, P. H. J., Yongchareon, S., & Komosny, D. (2019). Visible light communication: A system perspective—Overview and challenges. *Sensors (Switzerland), 19*(5), 1–22. https://doi.org/10.3390/s19051153

Saeed, K., Chaki, N., Pati, B., Bakshi, S., & Prasad, D. (2016). *Progress in Advanced Computing and Intelligent* (Vol. 2, Issue July 2018). http://dx.doi.org/10.1007/978-981-13-0224-4_19

Sarkar, S., Dixit, S., & Mukherjee, B. (2007). Hybrid wireless-optical broadband-access network (WOBAN): A review of relevant challenges. *Journal of Lightwave Technology, 25*(11), 3329–3340. https://doi.org/10.1109/JLT.2007.906804

Sarkar, S., Mukherjee, B., & Dixit, S. (2006a). Optimum placement of multiple Optical Network Units (ONUs) in optical-wireless hybrid access networks. *2006 Optical Fiber Communication Conference, and the 2006 National Fiber Optic Engineers Conference, 2006*, 2–4. https://doi.org/10.1109/ofc.2006.216015

Sarkar, S., Mukherjee, B., & Dixit, S. (2006b). Towards global optimization of multiple ONU placement in hybrid optical-wireless broadband access networks. *COIN-NGNCON 2006 – The Joint International Conference on Optical Internet and Next Generation Network, Coin*, 65–67. https://doi.org/10.1109/COINNGNCON.2006.4454489

Sarkar, S., Yen, H. H., Dixit, S., & Mukherjee, B. (2007a). A mixed integer programming model for optimum placement of base stations and optical network units in a hybrid wireless-optical broadband access network (WOBAN). *IEEE Wireless Communications and Networking Conference, WCNC*, 3910–3914. https://doi.org/10.1109/WCNC.2007.714

Sarkar, S., Yen, H. H., Dixit, S., & Mukherjee, B. (2007b). DARA: Delay-aware routing algorithm in a hybrid Wireless-Optical Broadband Access Network (WOBAN). *IEEE International Conference on Communications*, 2480–2484. https://doi.org/10.1109/ICC.2007.410

Shaddad, R. Q., Mohammad, A. B., Al-Gailani, S. A., Al-Hetar, A. M., & Elmagzoub, M. A. (2014). A survey on access technologies for broadband optical and wireless networks. *Journal of Network and Computer Applications, 41*(1), 459–472. https://doi.org/10.1016/j.jnca.2014.01.004

Shaw, W. T., Wong, S. W., Cheng, N., Balasubramanian, K., Zhu, X., Maier, M., & Kazovsky, L. G. (2007). Hybrid architecture and integrated routing in a scalable optical-wireless access network. *Journal of Lightwave Technology*, *25*(11), 3443–3451. https://doi.org/10.1109/JLT.2007.909202

Singh, P., & Prakash, S. (2017). Optical network unit placement in Fiber-Wireless (FiWi) access network by Moth-Flame optimization algorithm. *Optical Fiber Technology*, *36*, 403–411. https://doi.org/10.1016/j.yofte.2017.05.018

Singh, P., Dixit, V., & Kaur, J. (2019). Green Healthcare for Smart Cities. Green and Smart Technologies for Smart Cities, 91–130.

Srivastava A., Gupta M. S., Kaur G., (2020). "Energy efficient transmission trends towards future green cognitive radio networks (5G): Progress, taxonomy and open challenges", Journal of Network and Computer Applications, Elsevier, vol. 168, pp. 102760.

Srivastava A., and Kaur G, (2021) "Resource management for traffic imbalance problem in green cognitive radio networks", Physical Communication. Elsevier, vol. 48, pp. 101437.

Tsagklas, T., & Pavlidou, F. N. (2011). A Survey on Radio-and-Fiber FiWi Network Architectures. *Journal of Selected Areas in Telecommunication*, 18–24.

Wong, S. W., Campelo, D. R., Cheng, N., Yen, S. H., Kazovsky, L., Lee, H., & Cox, D. C. (2009). Grid reconfigurable optical-wireless architecture for large scale municipal mesh access network. *GLOBECOM – IEEE Global Telecommunications Conference*. https://doi.org/10.1109/GLOCOM.2009.5426037

Wong, Shing Wa, Shaw, W. T., Cheng, N., Qiao, C., & Kazovsky, L. (2008). Dynamic wavelength allocation in a converged and scalable interface for metro-access ring integrated networks. *OFC/NFOEC 2008–2008 Conference on Optical Fiber Communication/National Fiber Optic Engineers Conference*, 25–27. https://doi.org/10.1109/OFC.2008.4528618

7 Green Underwater Communication Systems

Dushyant Singh Chauhan, Gurjit Kaur, and Dinesh Kumar

CONTENTS

7.1 INTRODUCTION OF UNDERWATER COMMUNICATION

Today's era is of wireless communication and various internet applications rely on Wi-Fi technology. Smart devices, such as laptops, cellphones, and sensor-enabled watches, offer new capabilities that aim to improve users' wellbeing. According to CISCO, nearly 67% of the world's population will have access to the internet by 2023. Today, more than 80% of internet data is consumed in indoor applications, resulting in a significant increase in energy consumption. In the coming decade, demand for

DOI: 10.1201/9781003264477-7

internet connectivity is expected to rise even more; however, this pattern is concerning since the environmental costs of this expansion will be adverse (Cisco 2020). As a result, alternative energy-saving technologies are required to off-load wireless traffic, and visible light communication (VLC) is one of the leading technologies.

As per the survey of US NOAA, oceans cover around 97% surface of the Earth's water (Pearlstine et al. 2020). Nowadays, after exploring the atmosphere, there is a need to discover the oceans, i.e., the oceans' environment for several other applications like, to check the change in climate, surveillance, a checkup of oil rigs, the study of aquatic life, and unmanned operations. Acoustic and radiofrequency are the key underwater communication technologies (UWC). Acoustic communication has already been used in many underwater applications, but due to some drawbacks, i.e., difficulty to achieve high data rate, low information capacity due to multipath and absorption effect. As far as radio waves are concerned, they are widely used for broadband wireless applications, but they too have a disadvantage in that they cannot propagate through seawater. When sound waves propagate a large time-delay appears due to the slow propagation. So, this gives rise to optical communication in the underwater medium.

Since VLC is a stream of optical communication, which is distinctive because the sources which are used for illumination purposes can also be used for underwater communication purposes. In VLC, the LED light intensity is modulated so fast, so that light flickering is imperceptible to the eyes. These LEDs flash too fast so that data can be transferred over a distance. The LED emits lights, which correspond to ones, and zeros in binary. As the blinking rate of the LED increases the data transmission rate increases. This allows the transmission of various types of information like data, images, and videos between the nodes. The light signals are collected and recovered at the receiver side with the help of a photosensitive detector, which is called a VLC receiver (Sticklus et al. 2018). Light signals propagate at a faster rate in water (3×10^8 m/s) as compared to acoustic signals. As a result, for short-range and high-throughput, underwater visible light communication (UVLC) is considered a leading technique in designing a wireless communications system. UVLC has many advantages as compared to acoustic communication, some of them include low latency, better security, and larger bandwidth. Moreover, designing these systems is relatively easier, cost-effective, and installation is easy. Due to these numerous advantages, UVLC can be considered as a substitute to fulfill the necessities of huge data and high data transmission speed in underwater communications. Figure 7.1 shows the arrangement of an underwater wireless sensor network.

LEDs have become the most widely used and highly efficient lighting alternative due to advancements in LED technology. LEDs do not only use less energy and emit fewer greenhouse gases, but they also have a life expectancy of more than ten years or 50,000 hours. The ability to save almost 75% on energy has encouraged the extensive usage of LEDs for both outdoor and indoor applications. LEDs, for example, can switch to different light intensities at a very rapid rate, which is imperceptible to the human eye, allowing information to be transmitted through the light, which enables wireless transmission. LEDs have already been used to achieve rates

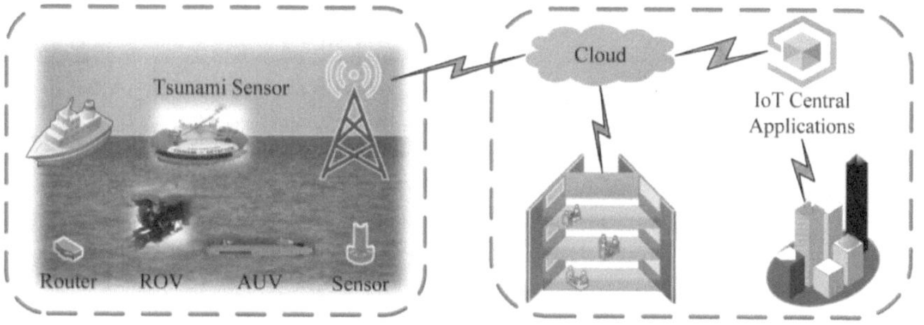

FIGURE 7.1 Arrangement of underwater wireless sensor network.

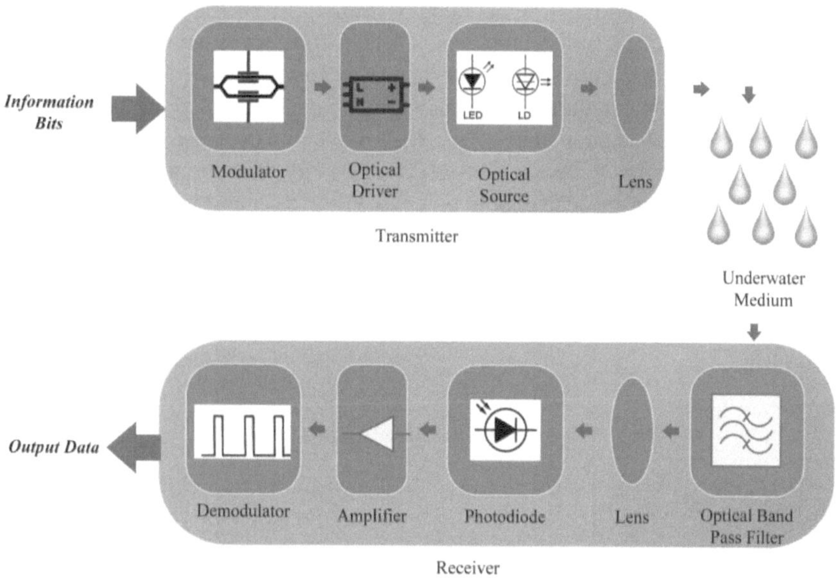

FIGURE 7.2 General block diagram of the UVLC system.

exceeding 10 Gbps, while laser diodes have been used to achieve a tantalizing rate of 100 Gbps (Zeng et al. 2016). LEDs can be employed for communication, navigation, and sensing, according to new research (Kaushal et al. 2016).

Figure 7.2 shows the general block diagram of the UVLC system. First, the data from the source is modulated and passed to the optical transmitter, which can be a laser or LED. It is cost and power-efficient, it is easily available. So, it is commonly used as an optical source. But its modulation bandwidth is low and as a result the data rate will be less. If we want a high data rate then we should use a laser. Therefore, a source that produces a very high coherence like laser diode is used for providing small beam divergence, and large modulation bandwidth was suggested to build high-speed UVLC networks.

A green laser has been employed, which operates at a wavelength of 532 nm to implement an optical wireless communication (OWC) known as an underwater optical wireless communication (UOWC) link at a data rate of 1 Gbps over a length of 2 m long pipe (Hanson et al. 2008). But the background light has a bad effect on BER performance. The unfavorable effect of water turbulence due to relatively small aperture size is another problem. Then optics is used at the transmitter side to focus the beam towards the target. The beam passes through the channel. Then lenses are used at the receiver side to focus it towards the photodetector. The photodetector finally performs the optoelectrical conversion.

Despite various advantages of VLC, several other phenomena, such as turbulence, scattering, and absorption undesirably affect the photons' propagation in UOWC. Optical waves show distinctive characteristics at different wavelengths in underwater propagation as shown in Figure 7.3. In Yi et al. (2015) it has been found that attenuation within the blue and green lights ranging from 450–550 nm wavelengths is very small in comparison with other wavelengths. Due to this fact, interest has been arising in the development of blue-green sources and detectors in UVLC.

When photons come in the influence of water molecules, absorption and scattering attenuate the signal strength in the 532 nm spectrum based on the salinity of the water. Once a photon absorbs it converts into thermal energy, this phenomenon is known as absorption, and it is an irreversible process. Whereas scattering is the process of spreading the photons in random directions in which due to finite aperture size some part of photons is not properly received at the receiver, and some other portions of photons are not received due to the delay produced by various propagation paths. The

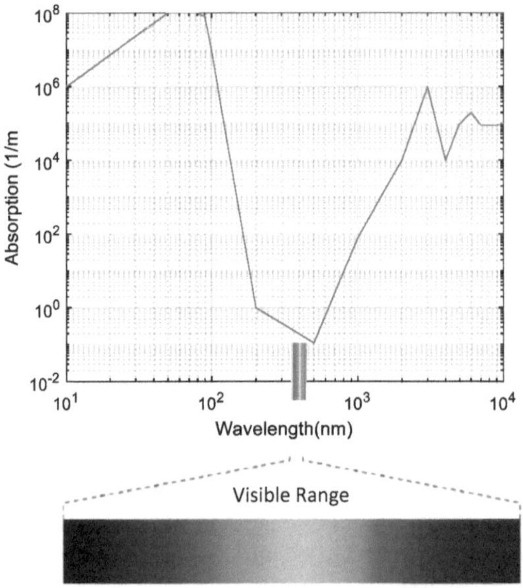

FIGURE 7.3 Absorption of the electromagnetic spectrum in water.

TABLE 7.1
Comparison of UWC Technologies

UWC technologies	Advantages	Limitations
Acoustic	• The most widely used UWC technology • Long communication range	• Lowest data transmission rate (in kbps) • Large communication latency (in a sec) • Costly, bulky, and power-consuming transmitter and receiver
RF	• Easily transmission from water and air boundaries • More tolerant to water salinity and turbulence • Alignment of transmitter/receiver not necessary • Moderate data transmission rate	• Short link range • Costly, bulky, and power-consuming transmitter and receiver
Optical	• The data transmission rate is too high (Gbps) • Immune to transmission latency • Cheaper and small size transceivers	• Difficulty in crossing air/water interface • Severe absorption and scattering • Moderate link range (up to 10 meters)

attenuation of light by water droplets is due to two phenomena, i.e., absorption and scattering, which directly affect the arrival angle, intensity, and phase of the incident light, as a result, time-jitter, multi-path fading, and inter-symbol interference (ISI) arise. The general block diagram consists of one source LED and one APD at the receiver side. Since LEDs in underwater communication are used due to their fast switching, excellent brightness, low power consumption, high tolerance to humidity, low cost, and high data rate (Kaushal et al. 2016). Table 7.1 shows various underwater technologies with advantages and limitations.

7.2 FEATURES OF A GREEN UNDERWATER COMMUNICATION SYSTEM

Recent advancements in underwater communication technologies have driven our attention to various underwater applications. Furthermore, the rapid development of wireless communication enables AOUVs to talk with one another and exchange information with ROUVs. However, a lack of collaboration among these vehicles leads to not only poor communication quality but also imbalances in the resources used in underwater wireless networks, resulting in excessive energy consumption. Figure 7.4 shows various techniques to balance resource consumption and save energy (Obaidat et al. 2012).

By utilizing a photon-counting receiver approach the occupancy of the channel by the UOWC system can be detected. This technique is known as spectrum sensing and can be used to save the energy requirements in the underwater system (Wu 2020). Another technique is known as routing adaptation, which helps the UOWC to select

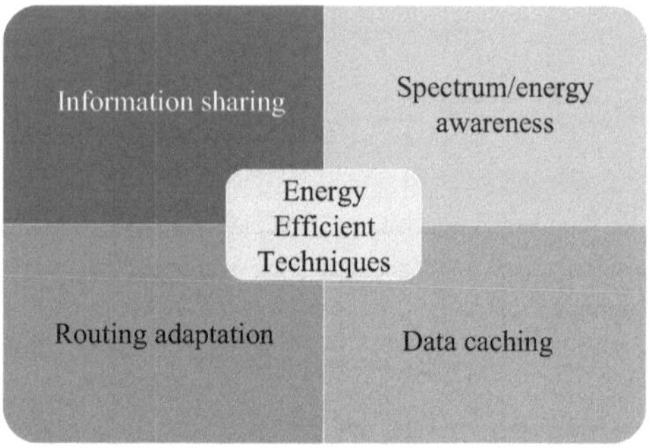

FIGURE 7.4 Techniques to balance resource consumption and save energy.

the shortest path between the underwater sensors within the underwater network. A routing protocol based on the geographic location in which the energy awareness metrics are used to detect the neighbor sensor node and the energy requirements in the network (Almesaeed et al. 2021).

7.3 WAVE TRANSMISSION IN THE AQUATIC MEDIUM

7.3.1 Transmission Waves for Underwater Communication

7.3.1.1 Acoustic Waves

For a long time, underwater transmission has been performed by acoustic waves. For medium distances, this technology enables relatively modest data rates and also does not provide guaranteed link security. In addition, the delay in signal transmission has been significantly enhanced. It is an old technology, and while it can transmit data across long distances, it can only be performed at a moderate data rate. As a result, it is unsuitable for broadband transmission due to the huge demand for data rates in underwater applications. Even though sound travels reasonably well in the air, it travels far more effectively in the water. Sound travels at a speed of 340 m/s in clean air, which is around four times slower than sound travels underwater. The main limitations of acoustic waves regarding underwater communication have been shown in Figure 7.5.

7.3.1.2 Electromagnetic Waves

To achieve faster wireless transmission of data, Electromagnetic (EM) waves in the radio frequency (RF) spectrum can be used. Temperature and depth do not affect them, because of both absorption and scattering, the attenuation of EM waves is considered as the most important effect when propagating through the water. Because seawater is a highly lossy medium, the behavior of EM waves in seawater and freshwater is considerably different. The absorption loss and propagation velocity of EM waves may both be characterized as functions of frequency (Abdou et al. 2011).

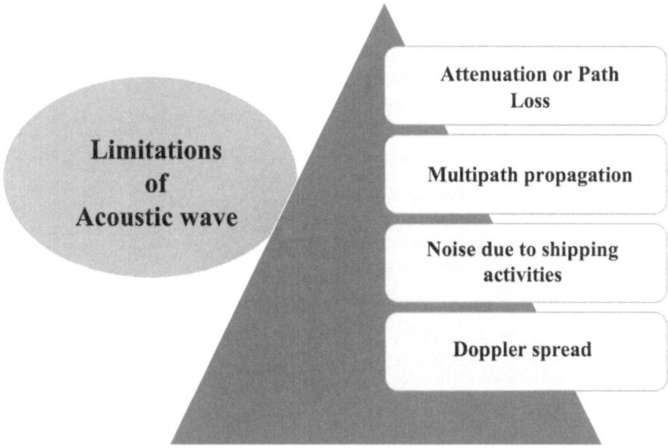

FIGURE 7.5 Limitations of acoustic underwater communication.

Inside water, EM waves are dependent on frequency and it is proportional to the square of the frequency. The use of low frequencies such as very low frequency (VLF) and extremely low frequency (ELF) is primarily for this purpose. VLF radio waves have a frequency range of 3 to 30 kHz, can transmit data at 300 bits per second, and can penetrate salt water to a depth of 18 m. To summarize, EM waves' main drawbacks are the severe attenuation in ocean water and large size antenna requirement in freshwater.

7.3.1.3 Optical Waves

Optical wireless communication offers several significant benefits, including extremely high data transmission rates, small and light transceiver components, secure communication links, and low operational costs (Prasad et al. 2005). To perform communication for short, medium, or long-range, optical wireless technology utilized modulated optical beams. In the visible spectrum (i.e., 400–700nm), optical waves give an alternate method of providing high data rates in the water. That is why optical underwater wireless communication has attracted a lot of attention in recent years as a high-speed (10–100 Mbps), secure, and dependable alternative to conventional acoustic underwater communications systems (Darwiesh et al. 2018).

Due to the water, some optical signals are absorbed in water, which is one of the major drawbacks, the other being scattering of these signals through particles and aquatic bodies present inside the water. Seawater, on the other hand, has lower absorption in the visible spectrums, i.e., blue/green area. Optical links with high data rates can be achieved by utilizing appropriate wavelengths, depending on the type of water. In the clear ocean, the minimum attenuation is around 0.460 m, but in dirty waters, it rises to over 0.540 m for coastal waters (Ata et al. 2014).

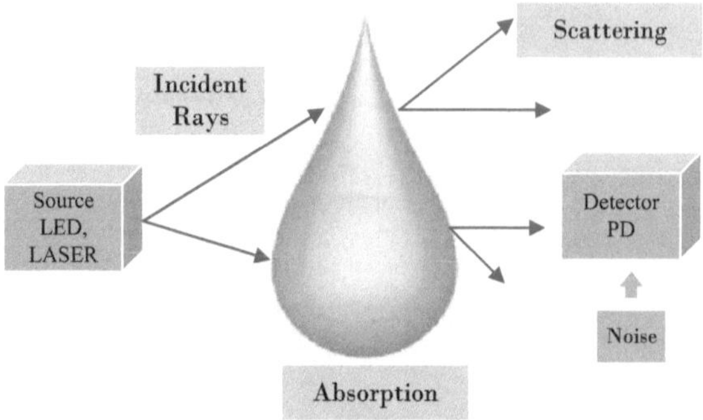

FIGURE 7.6 Absorption and scattering of optical waves due to water.

Beer's law estimates the received power $P(z)$ propagating through a medium, given an initial power P_o. The received power can be expressed as

$$P(z) = P_o e^{-c(\lambda)z} \tag{7.1}$$

where Z is the distance in the medium, extinction coefficient is $c(\lambda)$ in m^{-1}, which represents the overall attenuation caused by water propagation. The Beer-Lambert law is restricted only to mitigation due to absorption. In a practical scenario, there may be many occurrences of multiple dispersions. It also assumes that the source and recipient are in precise alignment and can be employed in the communication contexts of line-of-sight (LOS). In addition, the law of Beer overlooks temporary dispersion (Ata et al. 2014). Figure 7.6 shows the effect of absorption and scattering of an optical wave due to the particle present in the water.

Apart from absorption and scattering the other predominant effect in underwater communication is fading of the signals due to the refractive index variations inside seawater. This effect is also known as optical turbulence, which arises due to temperature and salinity variations. A parameter that defined the degree of fading is known as the scintillation index (SI).

When a Gaussian beam propagates through weak turbulence, the SI is expressed as the sum of two components, without taking scattering into account (Yi et al. 2015).

$$\sigma_R^2(z,\lambda) = 8\pi^2 K^2 z \int_0^1 \int_0^\infty k\phi_n^{ocean}(k)\left[1-\cos\left(\frac{zk^2}{K}\xi\right)\right]dk d\xi \tag{7.2}$$

7.4 UNDERWATER GREEN CHANNEL CHARACTERIZATION

Absorption and scattering, both wavelength-dependent processes, are the two primary mechanisms impacting light propagation in water (Mullen et al. 2009). Absorption is the process in which there is a permanent loss of intensity and it can be determined by the RI of water. The basic intrinsic optical property (IOP) used to represent water absorption is the spectral absorption coefficient $\alpha(\lambda)$, where λ is the light wavelength. The diversion of light from its original path is referred to as scattering.

7.4.1 Effect of Particles on Optical Propagation

When an incident light strikes a particle present in water then some incident power will be absorbed and some incident power scattered out making a solid angle $\Delta\Omega$. Figure 7.7 shows how the remaining flux makes an angle $\bar{\psi}$ with the propagating direction. $\beta(\psi, \lambda)$ is known as the spectral volume scattering function (SVSF).

To model scattering, the SVSF is employed as the primary IOP. The spectral scattering coefficient $b(\lambda)$ can be obtained as (Gabriel et al. 2013)

$$b(\lambda) = 2\pi \int_{\pi/2}^{\pi} \beta(\psi, \lambda) \sin \psi d\psi \tag{7.3}$$

Some part of the incident power is reflected due to reflection, so the parameter associated with this type of reflection is known as the back-scattering coefficient and is represented as $b_b(\lambda)$, which is obtained as

$$b_b(\lambda) = 2\pi \int_{\pi/2}^{\pi} \beta(\psi, \lambda) \sin \psi d\psi \tag{7.4}$$

Channel fading caused by oceanic turbulence can potentially influence the performance of a UOWC system.

7.4.2 Effect of Water Types on Optical Propagation

Because underwater materials and water quality differ from one place to the other, usually four major water categories have been considered (Kandouci 2022).

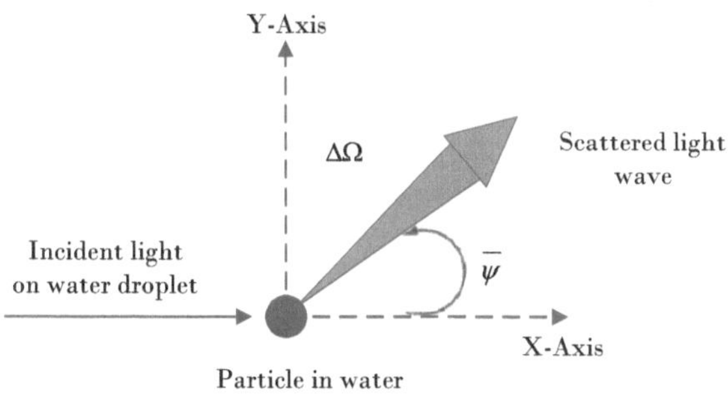

FIGURE 7.7 Incident light strikes the particle present in the underwater medium.

TABLE 7.2
Various Attenuation Constants Associated with Different Types of Water

Water types	C (mg/m³)	a (m⁻¹)	b (m⁻¹)	c (m⁻¹)	b_b (m⁻¹)
Turbid harbor water	5.9	0.295	1.875	2.170	0.0076
Costal ocean water	0.83	0.088	0.216	0.305	0.0014
Clear ocean water	0.31	0.069	0.08	0.151	0.0010
Pure sea water	0.005	0.053	0.003	0.056	0.0006

- Unpolluted seawater: The key limiting factor is absorption. The low spectral scattering coefficient values cause the beam to travel nearly in a straight line.
- Clear ocean waters: This type of water has a larger number of particles present in the ocean, which causes scattering of optical signals during propagation.
- The abundance of planktonic matter, debris, and mineral components in coastal ocean waters gives rise to more scattering.
- Turbid harbor: They include a lot of dissolved and suspended particles.

Table 7.2 shows typical values for the various attenuation constant associated with the different water types. By setting the chlorophyll concentration ***C*** to levels that were near to the attenuation coefficient ***c(m⁻¹)*** given in Kandouci (2022) who used the bio-optical model to determine the parameters.

7.4.3 Light Propagation in Water

When the light beam parameters along with the IOP of the medium were given, the behavior of light can be described by the RTE in the water medium. Let us call $L(z,\theta,\phi,\lambda)$ the light radiance in $Wm^{-3}sr^{-1}nm^{-1}$, where z is the distance from the transmitter θ and ϕ are the polar and azimuthal angles, respectively, $r = z/\cos\theta$ (Gabriel et al. 2013).

$$\frac{dL}{dr} = -cL + L^E + L^I\,(Wm^{-3}sr^{-1}nm^{-1}), \tag{7.5}$$

where L^E and L^I are elastic and inelastic scattering path functions, respectively. The inelastic scattering is referred to as the loss of photons owing to a wavelength change. But this inelastic scattering can be negligibly small due to its small contribution to RTE solution and can be ignored. Most earlier studies have ignored the elastic scattering by assuming that propagation is in a straight line, as specified by Beer-Lambert's law.

7.5 UNDERWATER GREEN TRANSCEIVER DESIGN

Recent advances in blue-green sources and detectors have increased interest in underwater optical communication (UOC). This transceiver utilizes the low attenuation optical window for optical transmission. Extended ranges, large data rates, and low latencies have all been shown using laser-based systems. Low-power, low-cost, and small LED-based systems have been successfully demonstrated in seawater (Ahmad et al. 2017).

Most studies have approached UOC as if it were a point-to-point communication system. Point-to-point linkages, especially on mobile platforms, necessitate precise pointing and tracking. Some systems additionally employ big aperture photomultiplier tubes (PMTs) with a 20-inch aperture to expand the receiver field of view (FOV). Compact systems, on the other hand, are preferred in particular instances. Smaller platforms with small systems lack the volume and large energy is required for advanced tracking and pointing systems. Whereas PMTs, which occupy a large-area, become costly and inconvenient to use. Smart antennas play an important role in making a communication system green. Signal processing capabilities enable smart antennas to give information of transmitting beam and angle of arrival. It is only logical to explore the advantages of applying similar strategies to the underwater environment to make the system green.

7.5.1 Green Transmitters

Transmitters play an important role in any communication system. In UOWC most researchers have demonstrated their systems when a transmitter and receiver are in LOS communication. So, when the receiver deviates from its actual position due to water waves and aquatic bodies present in the ocean, there is a loss of optical propagation, and energy is wasted during this type of communication. So, the transmitter has to be smart enough for no loss in optical signals while transmission (Simpson et al. 2012). The following are the key features of the smart transmitter:

- Increased directionality
- Electronic switching beam steering

One can increase the directionality of the transmitter by increasing the field of view (FOV) of the LEDs attached to it. The smart transmitter consists of seven LEDs arranged in a hexagon form, similar to the smart receiver (Parasher et al. 2019). Every LED has its wide FOV focuses and creates a narrow beam in the direction of the receiver. Every LED is controlled by the controller attached to it using the modulator, which decides the direction of illumination. This is the basis for an electronic switching beam steering system

7.5.2 Green Receivers

The optical transmission in UOWC receivers plays a crucial role in detecting the optical signals transmitted by the LEDs. Receivers have to be strictly aligned with the transmitter FOV to capture the optical signals properly. So, pointing errors have to be minimum to improve the quality of reception. This optical technology could also improve the performance of point-to-point communications by providing angle and spatial diversity. Previous research in this area has demonstrated optimum methods for combining the optical signals using a photodiode array to increase the signal and decrease noise while minimizing loss of bandwidth (Tucker 2010). As a result, the smart receiver possesses the following features:

- Wider FOV
- Estimation of the angle of arrival

7.5.3 Benefits of the Green Optical System

It is important to consider a smart transceiver for UOWC that can estimate water's IOP. The optical signal is being transmitted in the desired direction and it can detect the incident light beam at the receiver's side as the IOP is known. Already known IOP helps in reducing the transmitted power requirements and reduced the wastage of power during detection. Figure 7.8 shows various benefits of employing green optical systems in UOWC.

7.6 ENERGY HARVESTING FOR UNDERWATER SENSOR NETWORKS

While using UOWC a huge amount of data can be collected from various underwater sensors and AUVs, these sensor nodes require some power to operate and these have

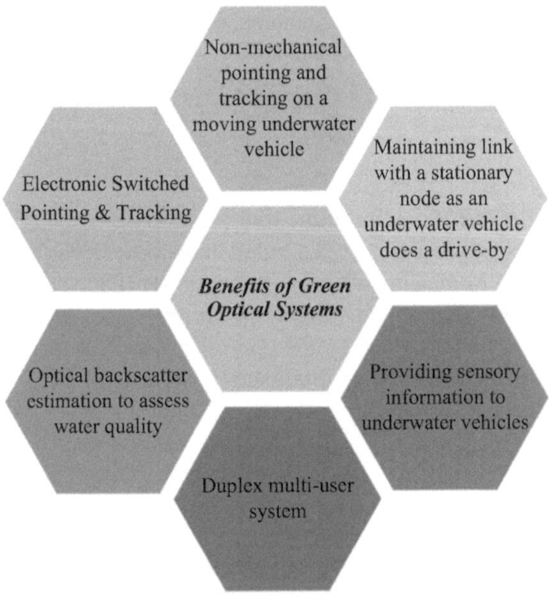

FIGURE 7.8 Benefits of green optical communication systems.

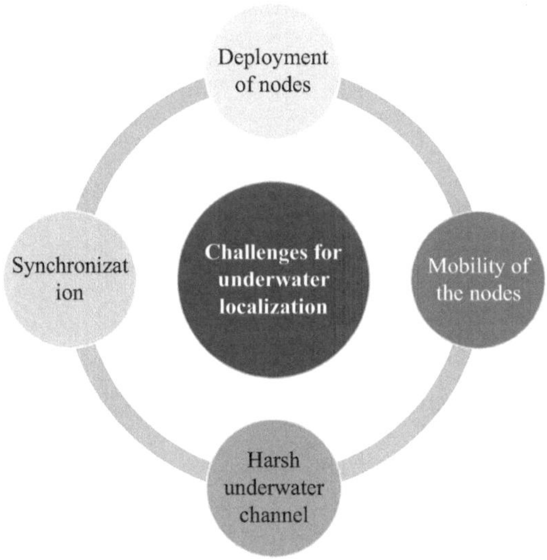

FIGURE 7.9 Challenges associated with underwater localization.

limited charging capabilities. During data transmission from these nodes, there is a need to charge the batteries or there is a need to save the consumption of energy. The saving of energy in any communication system is referred to as energy harvesting. The following are the techniques to harvest energy for UOWC.

- Localization
- Routing protocols
- Cooperative routing

An efficient routing technique based on mobile sinks and cooperative routing is demonstrated (Khan et al. 2019). A cooperative communication-based approach is utilizing broadcast techniques to make an energy-efficient communication system. Based on their depth of knowledge, nodes choose their relays and destination nodes. The message exchange overhead, on the other hand, is substantial, resulting in high network energy usage. It uses less energy by minimizing redundant packet transmission. Figure 7.9 shows various challenges associated with the underwater localization.

7.7 APPLICATIONS OF UNDERWATER WIRELESS COMMUNICATIONS

The underwater environment has been utilized for several applications like, to check the change in climate, surveillance, a checkup of oil rigs, the study of aquatic life, IoT applications and unmanned operations due to its high data rate, low latency, and low power consumption as compared to its counterpart (Kaur et al. 2018). So, this section describes the applications and various challenges that occur in designing an energy green optical communication link.

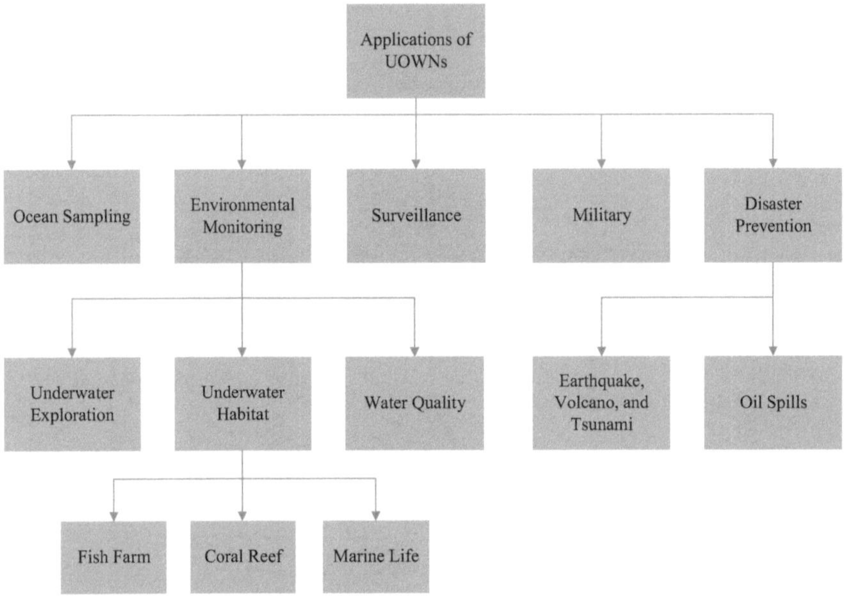

FIGURE 7.10 Applications of UOWNs.

7.7.1 Application Deployment

The degree of portability and thickness varies depending on the organization of submerged sensor systems. Single hubs attached to docks, AUVs, ROVs, and submerged sensors are typical examples of underwater devices. Semi-portable submerged systems, on the other hand, can be suspended from floats left to set up for several hours or days. These arrangements of devices are static over extended ranges, allowing for the construction of a system topology to progress the network (Zhang et al. 2015). Figure 7.10 shows various underwater networks' application.

7.7.2 Challenges of Underwater Wireless Communication

Even though there has been a lot of research on UOWC in the last few years, there is still a lot of work to be done to overcome some of the problems. We offer some suggestions for upcoming research on these systems below.

7.7.2.1 Channel Modeling

Due to the development in various simulation tools and realistic models have been created so far. However, channel modeling for horizontal links has been resolved. But modeling vertical linkages, on the other hand, has received little attention due to challenges that occur during the vertical communication in UOWC. Another factor to consider is the effect of turbulence. In most circumstances, turbulence may be overlooked in deep underwater oceans, but its effects cannot be neglected in shallow waters. Most of the researchers have included the free-space optical wireless model to model underwater channels, but these models are unable to accurately account for turbulence in UOWC. There have been challenges also for temperature and salinity variations in building such an underwater model.

7.7.2.2 Energy-efficient Transmission Techniques

Another point to investigate is the creation of appropriate signaling methods that are developed for the aquatic channel. Up till now, UOWC systems have used generally traditional modulation and coding algorithms. Considering the unique characteristics of the underwater channel while designing transmission systems can improve connection performance and reliability. At the data link and physical layers, optical data transmission must employ powerful channel coding and energy-efficient modulation techniques for relatively different types of water conditions.

7.7.2.3 Reduces Link Misalignment

Since power consumption by the underwater application is a major concern in UOWC, minimizing optical intensity losses by lowering beam divergence is critical. The use of smart transceivers must be capable of self-adapting in operating conditions. The transmitter utilizes increased directionality and electronic switching beam-steering techniques to save energy, make UOWC smart. The receiver must adaptively take care of the pointing and alignment issues by the knowledge of IOP of the various water conditions.

7.8 CONCLUSION

This chapter focuses on the significant role of green technologies in underwater wireless communication systems. The chapter illustrates the important features of a green underwater communication system utilizing various techniques to balance resource consumption and save energy. Various challenges have also been discussed while designing a green optical communication link. By utilizing a green transceiver and low attenuation optical window for optical transmission, extended ranges, large data rates, and low latencies can be achieved. Low-power, low-cost, and small LED-based systems have been demonstrated for underwater communication, which makes the system green. The increasing demand for underwater devices and energy-efficiency concerns are further increasing the demand for a green communication system.

REFERENCES

Abdou, A.A., Shaw, A., Mason, A., Al-Shamma'a, A., Cullen, J., and Wylie, S., 2011, October. Electromagnetic (EM) wave propagation for the development of an underwater Wireless Sensor Network (WSN). In *SENSORS, 2011 IEEE*: 1571–1574.

Ahmad, Z., Rajbhandari, S., Salih, O., and Green, R., 2017, July. Demonstration of a multi-hop underwater visible light communication system. In *2017 19th International Conference on Transparent Optical Networks (ICTON)*: 1–4.

Almesaeed, R. and Jedidi, A., 2021. Dynamic directional routing for mobile wireless sensor networks. *Ad Hoc Networks, 110*: 102301.

Ata, Y. and Baykal, Y., 2014. Structure functions for optical wave propagation in underwater medium. *Waves in Random and Complex Media*, *24*(2): 164–173.

Cisco, U., 2020. Cisco annual internet report (2018–2023) white paper. Online](accessed March 26, 2021) www.cisco.com/c/en/us/solutions/collateral/executive-perspectives/annual-internet-report/whitepaper-c11-741490.html.

Darwiesh, M., El-Sherif, A.F., Ayoub, H.S., El-sharkawy, Y.H., and Hassan, M.F., 2018, April. Hyper-Spectral Laser Imaging of Under-Water Targets. In *The International Conference on Mathematics and Engineering Physics* (Vol. 9, No. International Conference on Mathematics and Engineering Physics (ICMEP-9): 1–10.

Gabriel, C., Khalighi, M.A., Bourennane, S., Léon, P., and Rigaud, V., 2013. Monte-Carlo-based channel characterization for underwater optical communication systems. *Journal of Optical Communications and Networking*, *5*(1): 1–12

Hanson, F., and Radic, S., 2008. High bandwidth underwater optical communication. *Applied Optics*, *47*(2): 277–283.

Kandouci, C., 2022. Investigation of Jervol water types properties effects on underwater optical wireless OCDMA system performances for different modulation techniques. *Optical and Quantum Electronics*, *54*(1): 1–13.

Kaur, G., Tomar, P., and Singh, P., 2018. Design of cloud-based green IoT architecture for smart cities. *Internet of Things and Big Data Analytics Toward Next-Generation Intelligence*: 315–333. Springer, Cham.

Kaushal, H., and Kaddoum, G., 2016. Underwater optical wireless communication. *IEEE Access*, *4*: 1518–1547.

Khan, A., Khan, M., Ahmed, S., Abd Rahman, M.A., and Khan, M., 2019. Energy harvesting based routing protocol for underwater sensor networks. *PloS one*, *14*(7): 0219459.

Mullen, L., Laux, A., and Cochenour, B., 2009. Propagation of modulated light in water: implications for imaging and communications systems. *Applied Optics*, *48*(14): 2607–2612.

Obaidat, M.S., Anpalagan, A., and Woungang, I. eds., 2012. *Handbook of green information and communication systems*. Academic Press Waltham, Massachusetts.

Parasher, Y., Singh, P., and Kaur, G., 2019. Green Smart Security System. *Green and Smart Technologies for Smart Cities*. CRC Press: Boca Raton, London, New York, 165–184.

Pearlstine, L.G., Beerens, J.M., Reynolds, G., Haider, S.M., McKelvy, M., Suir, K., Romañach, S.S., and Nestler, J.H., 2020. Near-term spatial hydrologic forecasting in Everglades, USA for landscape planning and ecological forecasting. *Environmental Modelling & Software*, *132*: 104783.

Prasad, R. S. and Kaur, G., 2005. *Optical Communication*. Galgotia Publications, ISBN: 9788175154858, 8175154853

Simpson, A., B., Hughes, L., and Muth, J. F., 2012. Smart Transmitters and Receivers for Underwater Free-Space Optical Communication. *IEEE Journal on Selected Areas in Communications*, *30*(5): 964–974.

Sticklus, J., Hoeher, P.A., and Röttgers, R., 2018. Optical underwater communication: The potential of using converted green LEDs in coastal waters. *IEEE Journal of Oceanic Engineering*, *44*(2): 535–547.

Tucker, R.S., 2010. Green optical communications – Part I: Energy limitations in transport. *IEEE Journal of selected topics in quantum electronics*, *17*(2): 245–260.

Wu, W. C., 2020. Spectrum sensing in optical wireless communication network. *Transactions on Emerging Telecommunications Technologies*, *31*(4): 3716.

Yi, X., Li, Z., and Liu, Z., 2015. Underwater optical communication performance for laser beam propagation through weak oceanic turbulence. *Applied Optics*, *54*(6): 1273–1278.

Zeng, Z., Fu, S., Zhang, H., Dong, Y., and Cheng, J., 2016. A survey of underwater optical wireless communications. *IEEE Communications Surveys & Tutorials*, *19*(1): 204–238.

Zhang, X., Cui, J.H., Das, S., Gerla, M., and Chitre, M., 2015. Underwater wireless communications and networks: theory and application: Part 1 [Guest Editorial]. *IEEE Communications Magazine*, *53*(11): 40–41.

8 Green Optical Networks Based on Machine Learning Techniques

Shilpa Jindal

CONTENTS

8.1 INTRODUCTION OF MACHINE LEARNING FOR OPTICAL NETWORKS

Since the inception of the internet, users worldwide are utilizing its services in all domains of life. To name a few technologies like virtual reality, smart machines, remote access, drones, artificial intelligence, cloud computing, internet of things, energy efficient, quick mobile charging, advent of fast microprocessors that could process data in a fraction of seconds, optical fiber cable's tremendous capacity of data transmission and distribution in Mbps and Tbps, gaming has led to the generation of data at the users' end in huge amounts, i.e., the term given is big data. Users are now enjoying streaming services in OTT platforms with anytime, anywhere connectivity, large industries and business houses now rely their work on internet services, be it a mobile broadband or home broadband usage. Such a revolution in availability of

DOI: 10.1201/9781003264477-8

services and connectivity is possible due to high-speed data networks at access areas, metropolitan areas, and core areas of connectivity.

8.1.1 Trends in Network Connectivity

Today almost all fields have generated data online. Further, the Covid-19 pandemic, where we faced conditions like lockdown, social distancing has boosted the usage of the internet by a large amount. It was only due to the availability of high-speed networks that the online education, other official and industrial work was possible on various platforms like zoom, Microsoft teams, google meet etc. Human beings and their work/learning survived because of good connectivity only. The recent launch of PM WANI CDoT 2020 in India as a public network makes sure the availability of connectivity within a city through wifi networks to the people for their daily requirements of data. It is estimated that by 2024, there will be 62 billion IoT devices that will be connected (King et al. 2019). In 2016, the open trial specification alliance (OTSA) was formed by Verizon 2016 for home broadband. In 2018, South Korea tested 5G speed in winter Olympics using the OSTA standard. 5G is a 3GPP mobile broadband standard for next generation. Verizon and AT & T started the service of fixed wireless 5G for home automation for multiple cities using 28 GHz and the OSTA standard. In 2018, AT&T and other players in the market accelerated the 3GPP 5G NSA Standard. In 2019, China started 5G mobility services using the 3GPP standard. In 2020, Japan, the US, and China initiated 5G deployment. China leads low frequency deployment in 5G for mobility at 3GHz and the US leads high frequency deployment at 28GHz for fixed wireless. By 2025, the number of 5th Generation equipment and base stations will reach 6.5 million that will cover broadband connectivity in areas such as augmented reality, virtual reality, gaming, smart city, industry automation, agriculture sector etc. enabled by IoT.5G is based on machine-to-machine communication that has low latency. 5G will require high quality fiber networks at backhaul to provide connectivity to small cells of 5G at the frontend. Figure 8.1 shows few domains where 5G will become a reality in the future. Software defined radio has now turned to a software defined network that makes generation and deployment of new applications in future networks.

A short comparison of 5G with 4G is given in Table 8.1. In 5G, fronthaul and backhaul will be merged to cross haul architecture in a single transport network that will combine different broadband services, both wired and wireless, packet transmission over optical fiber, and flexible services to both commercial and residential use.

8.1.2 Role of Optical Fiber in Broadband Technology

Due to its high bandwidth, optical fiber can accommodate big data generated by users at the end. It becomes crucial to analyze this data and understand the physical layer aspects from a networking view covering a large domain of connectivity.

There are many advantages of optical technologies as the ability to carry terabits of data in a second at long distance with less cost. Further, supporting DWDM (dense wavelength division multiplexing), amplification in optical domain, adding

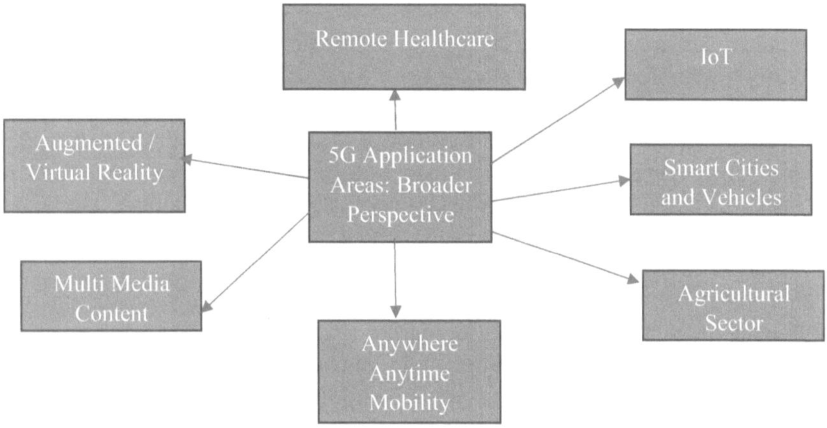

FIGURE 8.1 5G applications.

TABLE 8.1
4G with 5G Technology Comparison

Parameters	Comparison of 4G with 5G
Network latency	Reduced by the factor of 10
Users	100 times higher because of Internet of Things
Capacity	Per cell throughput is 10 times higher than 4G
Enhanced features	Radio access networks more complex, dual connectivity x2 and ex2 interface, enhanced coordinated multipoint
Speed	Higher in 5G
Latency	Low in 5G
Bandwidth	Higher in 5G
Applications	Automatize tasks

and dropping of wavelengths using optical multiplexing, coherent processing along with digital signal processing by mitigating the fiber impairments both linear and nonlinear like chromatic dispersion, attenuation, Kerr effects along with four wave mixing. SONET/SDH supports gigabits to terabits of data. Further, a tremendous increase in the number of users making cell sites denser and connecting cell towers through fibers and centralization of resources through cloud platforms through optical fiber in access networks in 5G has triggered research in future network connectivity.

8.1.3 The Current State of the Art of Optical Cabling in 5G

5G offers low latency approximately ten times less than 4G, high speed, high frequency, high capacity due to advancement in IoT devices to cover anywhere and anytime connectivity. These all are possible if the connectivity for the access, metropolitan, and core networks are highly reliable and scalable to cater for current and future requirements both in mobile broadband and fixed broadband networks. Network connectivity would be possible, which will be a mix of wired and wireless. Many manufacturers, like the Corning company Mullaley et al. 2018), have recently announced such connectivity options in the wired domain like optical fibers that cater for 5G network connectivity requirement issues due to its inbuilt features like tremendous capacity and large distance coverage without repeaters.

In 5G access networks (fronthaul):

Connectivity between indoor micro base stations and macro base stations need bend free deployment of cables without scarifying on its performance issues.

Many manufacturers like Corning, YOFC have announced bend improvement and bend insensitive fibers, which are in compliance to ITU G.657.A2/B2/B3 standards (Worton 2021) with reduced loss to cater for 5G deployment in a closed environment.

Further, having dense fiber deployment in a limited indoor environment poses stringent requirements on optical fiber manufacturers for its reduced diameters. Many manufacturers like Corning, Prysmian Group has introduced single mode fibres for such requirements by shaving off cable coating thickness and reducing it to few microns. Recent transmission at 25 km with multimode fiber by transferring petabits/sec has been demonstrated in European conference 2020 with increased complexity as it employed digital signal processing based on MIMO technology adjusting to more modes in MM fiber (Nokia bell labs 2020).

In 5G core networks (backhaul):

In data centers where the data is stored, backhaul is connected by fiber optical networks by using broadband multimode fiber in a short range. Recently OM5 is used as a multimode fiber for swift applications in data centers (CommScope 2017). As per the Telecommunications Industry Association, TIA-492AAAE standard published in June 2017 (Arlington 2017), a wide band multimode fiber (WBMMF) with core and cladding radius of 50/125 μm can support short wavelength division multiplexing (SWDM) transmission, hence OM5 fiber was designed to support a minimum of four low-cost wavelengths in the 850–950 nm range supporting SWDM. Due to its features to support high data rates up to 400Gbps in short distance and adopting higher modulation techniques and WDM technology, OM5 optical fiber becomes the choice to connect data centers under the 5G environment.

Long haul 5G networks without repeaters: By overcoming the nonlinear impairments and decreasing noise to optical signal ratio that occurs due to the use of higher modulation formats to support high data rate communications, many manufacturing companies like Corning and Infinera have manufactured ultra-low loss G.654.E fiber that has high data rate capabilities, supporting excellent long- distance transmission connectivity for deploying the 5G network.

For outdoor deployment fibers with anti-rodent and anti-bird features like GYKTS has also been manufactured by the YOFC company that consists of metallic strength, a stainless-steel tape armor and a polyethylene outer sheath to make optical fiber robust to deployment in sensitive environments (YOFC 2013).

8.2 ROLE OF MACHINE LEARNING IN OPTICAL NETWORKS

Optical networks' complexity is increasing in next generation connectivity, as networks are moving to elastic networks from static networks that cater for a variety of adjustable and dependent parameters like routing configurations, wavelength assignments, modulation formats, symbol rate, coding schemes etc. In order to automate the complex tasks in optical networking, an approach to machine learning will be a solution provider. Machine learning will help to analyze network data, automate network self-configuration, and manage faults in complex optical networks that are based on advance signal processing, compensation of fiber impairments, and coherent transmission and reception techniques (Musumeci et al. 2019).

Optical networks act as a backbone for the entire physical infrastructure providing coverage worldwide due to its inherent advantages like high capacity, low loss, greater distance coverage, fast transmission. So, it becomes crucial to analyze and understand the physical layer aspects from a networking view, covering a large domain of connectivity.

Optical networks are classified into three main broad categories as per the area covered as

- Access networks or shorthaul/fronthaul networks
- Metropolitan networks or mid-haul networks
- Core networks or long-haul networks

8.2.1 Machine Learning Approach to Optical Networks

Optical networks generate a lot of data that can be utilized to make networks more flexible as per the requirements through machine learning. Machine learning is a division of artificial intelligence where machines configure themselves in real time as per the data requirement for network design, configuration, and its operation. It is a move to make networks smart and flexible that feed the requirements without much human interference using intelligent algorithms. Machine learning techniques like artificial neural networks, clustering, and deep learning can be explored further in the area of optical networking.

An optical fiber system is itself complex because of its nature as non-linear, employability of coherent transmission/reception techniques, channel impairments, and its modeling. Further, employing flexibility to allocate network resources to such systems will increase the dimension of the problem to manifold. Machine learning is being considered as a pattern shift for designing such optical networks.

8.3 TECHNOLOGIES TO DESIGN MACHINE LEARNING BASED OPTICAL NETWORKS (SUPERVISED, UNSUPERVISED, REINFORCEMENT)

These techniques of machine learning can be applied both at the physical layer, network layer, and at cross layer platform. Broadly the classification of machine learning algorithms are as supervised learning, semi-supervised learning, unsupervised learning, and reinforcement learning.

Supervised learning: In this technique of learning, the algorithm is skilled on the information of the dataset that consists of some independent parameters like path, modulation rate, modulation type, wavelength, spacing between wavelength, BER and dependent parameters are predicted after training in correspondence to the new inputs. Prediction is based on independent input variables that results in the output variables dependent on input dataset. It is further classified as parametric models and non-parametric models. The techniques includeneural networks, deep learning with hidden layers, support vector machines, logical and linear regression, ADA boost, random forests, instance-based learning like k nearest neighbors, case-based reasoning.

Unsupervised learning: An agent learns patterns from the input without explicit output and identifies unusual patterns in data. Some techniques include principal component analysis, clustering K-means semi-supervised learning: In this type of learning, the concepts of both the approaches, i.e., supervised learning and unsupervised learning are taken to train and predict the output parameters.

Reinforcement learning: In this the algorithm feedback mechanism is strong that is required to learn on the effect of modifying some parameters by refining the behavior that is considered as a reward, e.g., Q learning, learning automata.

8.4 CHARACTERISTICS OF MACHINE LEARNING BASED OPTICAL NETWORKS

Optical networks must have the following characteristics in order to implement machine learning algorithms.

Reliable: Reliability is a concern in today's networks as users demand good connectivity at the time of day they desire. Users nowadays change the service provider if the quality of service in the area is unacceptable.

Flexible: People need connectivity at various devices due to advancements in IoT, machine leaning and hence flexibility in networks was desired to a large extent.

Less complex: Networks although complex, have resulted in better services as employing machine learning algorithms reduced the complexity of these networks.

High capacity: Due to the use of high bandwidth optical fiber as a backbone, these networks have inbuilt high capacity and hence are used as networks for today's and future networks.

8.5 RESEARCH OPPORTUNITIES OF MACHINE LEARNING BASED OPTICAL NETWORKS

Researchers around the globe are working in the area of optical transmission and network monitoring are considering machine learning approach a viable solution due to ongoing complexity of optical networks. (Musumeci et al. 2019) compiled a list of machine learning methodology based on the above classification for the physical layer as well as the network layer and put some light on current potential research directions. Amirabadi er al. (2019) investigates the ML algorithms for applications in optical communication as machine learning algorithms are well established and their applications and implementation in optical communication is at the beginning stage. Musumeci et al. (2019) classifies failure-management tasks and checks for the applications of machine learning and explains the same with the practical failure-management task. Main classifications discussed are bagging and random forests, Bayesian networks, artificial neural networks (ANN), deep learning, support vector machines, Gaussian processes, network kriging and cross-validation and statistical analysis of the results. Liu et al. (2020) discussed the problems for deploying AI-based methods in realistic system and listed the machine learning for quality of transmission (QoT) modeling techniques using modeling targets like BER, Q Factor, OSNR, SNR, Margins. Eira et al. (2015) quantifies traffic dynamics that justify the higher starting investment/cost in flexible-rate technology in dense wavelength division multiplexing (DWDM) networks. Morais et al. (2018) used a SDN-controlled optical network and proposed K-nearest neighbors, logistic regression, support vector machines, and artificial neural networks frameworks to acquire meaningful data for a knowledge database that can be used to train machine learning models for light path validation and suggested that machine learning is a valid alternative to estimate the QoT of a light path. Venkatesan et al. (2020) used regression technique and analyzed non-linearity problems of a sixty-four channel DWDM system. Rottondi et al. (2018) predicts the quality of transmission of unestablished light paths using machine learning based on the light path characteristics based on the parameters like total length of fiber, length of the longest/largest link and the number of light path links before actual deployment.

Some of the areas of application of artificial intelligence and machine learning techniques in the physical and network layer in optical networks are in transmitters, amplifiers, modulation formats, OSNR monitoring, quality of transmission estimation, recourse allocation, connection establishment, network reconfiguration and QoS, and dynamic bandwidth allocation in passive optical networks. Hence, the problems that come under optical transmission are classified as problems in the physical layer like nonlinear transmission systems, optical performance monitoring (OPM), and preventive fault detection and cross layer optimization in software defined radio and the problems that come under optical networking are classified as problems in the network layer like the quality of transmission estimation, recourse allocation etc. The following list compiles the research issues in these two categories as follows:

Transmitters: Zibar et al. (2015) characterizes laser amplitude and phase noise (Zibar et al. 2015) using Bayesian filtering and expectation-maximization technique. Brunton et al. (2014) used pattern methods and genetic algorithms to create a self-tuning method for mode-locked fiber lasers.

Optical amplifier: Huang et al. (2016) defined the regression model and used a kernel linear regression technique for studying power excursion in multi-hop EDFA networks. Huang et al. (2017) used a ridge regression model, which is a linear regression technique that will cope with the discrepancy between post-EDFA channel powers.

Linear impairments: Wu et al. (2009) used neural networks to identify CD, PMD and OSNR considering known modulation format and bit-rate. Tan et al. (2014) monitored CD, PMD, and OSNR using principal component analysis to identify modulation format/bit rate and monitors performance of optical signals.

Modulation format recognition: Tan et al. (2014) identifies modulation formats and bit rates from a recognized set using principal component analysis. Gonzalez et al. (2010) identifies modulation formats QPSK/8PSK/16QAM based on clusters and the number of levels in constellation diagram using clustering -=means in a cognitive digital receiver.

OSNR monitoring: Tanimura et al. (2016) monitored OSNR using adeep neural network trained with asynchronously sampled raw data and used five layers and 400,000 samples to get the accurate results, which resulted in elongated calculation timings. Thrane et al. (2017) used neural networks based on nonlinear regression for OSNR estimation and a modulation format classifier for systems up to 64QAM modulation formats and direct detection.

QoT estimation: Barletta et al. (2017) used random forests' classifier to forecast the likelihood of a candidate light path so that the BER will not exceed a given threshold. Oda et al. (2017) used a linear regression technique on BER information monitoring and a learning process to estimate the BER of each new service request. Panayiotou et al. (2017) used feed forward neural network to analyze the quality of transmission on a data driven dynamic multicast metro optical network using a machine learning supervised technique and suggested that the accuracy of such a technique is comparable to that of the QoT model and is around 92 to 95 %.

Resource allocation Petridou et al. (2008) used K-means clustering to propose a message scheduling algorithm that addresses sequencing of messages and assigning channel requirements for a WDM star network.

Reyes et al. (2017) modeled the resource allocation problem as Markov decision processes to optimize in dynamic optical networks.

Connection establishment Jia et al. (2016) used a back propagation neural network that predicted the arrival and holding time of potential connections by considering past experiences for routing and resource allocation in elastic optical networks

(a) Araujo et al. (2015) used neural networks and estimated the connection blocking probability using principal component analysis for optical networks.

Network reconfiguration Morales et al. (2017) used neural networks and performed reconfigurations based on the traffic amount and direction predicted in virtual network topology. Baldine et al. (2001) determined the frequency to reconfigure a broadcast WDM optical network based on Markov decision processes.To guarantee the QoS and bandwidth allocation dynamically in PONs Huang et al. (2008) and Ye et al. (2017) used neural networks. used Bayesian estimation and a prediction-based just-in-time active bandwidth allocation algorithm.

8.5.1 Data Analysis Using the Machine Learning Algorithm in Optical Transmission

A lot of research has been going on at two different layers of optical networks by applying various machine learning algorithms for data analysis so that the nonlinear domain can be modeled and adapted as per the data. The data can be historical data

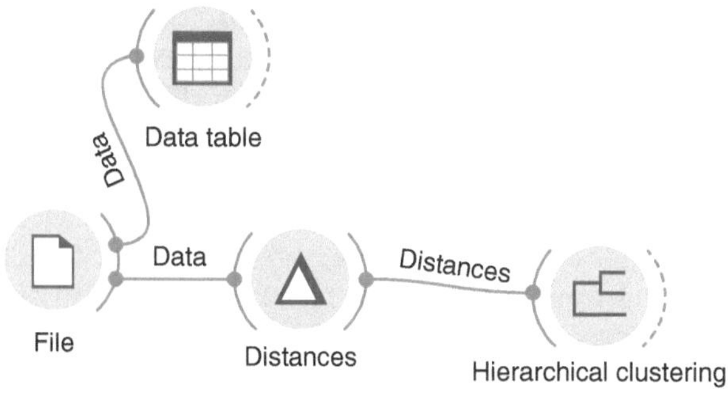

FIGURE 8.2 Analysis setup unsupervised machine learning (hierarchal clustering) in open-source tool.

collected from a real world scenario or simulated environment or it can be synthetically generated from the programs/algorithms written in software. In the literature both types of analysis have been carried out to make optical networks from static to dynamic and from uncast to multicast. A simulated work in this regard using regression modeling has been carried out by Venkatesan et al. (2020) employing 16, 32, 64 dense wavelength division system (DWDM) channels. The Germany backbone network (GBN) and continental long/ultralong-haul topology (CORONET) optical networks based on fixed-rate and flex-rate architecture were used to analyze networks in 2005 and 2018. There is an open-source digital tool "Orange" that can be used for data analysis under the unsupervised machine learning algorithm using the setup of Figure 8.2.

8.6 APPLICATIONS AND CHALLENGES OF GREEN MACHINE LEARNING BASED OPTICAL SYSTEMS AND NETWORKS AT DIFFERENT LAYERS

Application of the machine learning algorithm in optical communication has recently gained momentum due to its flexible and reliable approach. Today's users demand communication networks with high bandwidth, high data rate, high security and with reliable connections due to the sudden surge in internet applications seen in the last five years. Methodology or algorithm of machine leaning are well established and are in an advanced stage of implementation in various fields. These can be utilized to find, calculate, and omit defects by learning the system properties that will result in improving accuracy and performance of the system with less computation than otherwise needed. Machine learning trains itself through learning dataset and predicts the output that is not just reached.

As we are aware, due to the nonlinear nature of an optical channel, complex numerical calculations are needed and using machine learning for the optical channel can reduce this complexity to a large extent. Further complexity has risen exponentially due to a large number of adjustable parameters like modulation format, coding schemes, routing configuration, wavelength assignment, traffic grooming etc. due to the adoption of coherent transmission and reception techniques, advanced processing in digital domain for optical systems and networks. Further complexity of any system can be reduced by employing advanced DSP techniques. The highly complex optical communication and optical networks' complexity can be reduced by either DSP techniques or by machine learning. Recently Khan et al. (2019) had implemented DSP based ML for compensating the fiber nonlinearity without scarifying computation complexity in coherent optical orthogonal frequency division multiplexing (CO-OFDM) and also reviewed the existing ml approached and benchmark solutions for CO-OFDM.

Due to embedding internet activities in each domain of human life there is a race to increase bandwidth, data rate, security, and reliability of communication networks. As these characteristics are very important in judging the overall quality of communication links, high bandwidth transmission carrier link like fiber optic cable has shown great progress due to its low attenuation and ability to carry data at high speed. So, the systems that involve light as a carrier for transmission of data are considered as optical systems and the networks covering worldwide areas are called optical networks in various topologies like LAN, WAN, MAN using SDH/ SONET networks, employing various network configurations like tree, bus, mesh, star, and combinations of these in hybrid mode covering WDM, DWDM, TDMA, WDMA, OCDMA technologies. So, the classification is termed as optical systems at system level and optical networks at network level as follows:

(1) At system level: Optical channel modeling is becoming more complex (due to coherent technology catering different modulation formats, symbol rates, adaptive coding rates, and flexible channel spacing while catering already available nonlinear optical channel as the closed form expression for the optical channel is not defined.
(2) At network level: Due to advancements in both data plane and control plane for providing quality of service and transmission due to varied user requirements.

At data plane, elastic optical networks cater elasticity in allocating optical network resources like bandwidth allocation, transmission techniques (OFDM, Nyquist WDM), modulation formats (QPSK, QAM), and coding rates. So, this flexibility makes the resource allocation problem more challenging.

At control plane: Dynamic control in software defined networks (SDN) enables on demand reconfiguration and virtualization in terms of network reconfiguration, amplifier power settings, unexpected penalties due to nonlinearities that need integration between control elements like SDN controllers and optical performance monitoring at equipment level.

So, these flexible requirements pose severe challenges to system and network designers to decide the best system and network that can provide highly flexible and reliable communication for busy users' requirements.

So, machine learning, a branch of artificial intelligence, is considered an approach to make networks/systems more adaptive using complex mathematical and statistical tools.

ML applications can be useful in cross layer settings where the data available at the physical layer like BER in terms of signal quality, traffic samples can trigger changes at the network layer like routing, spectrum assignments and modulation, and coding format assignments.

Hence future optical networks are expected to run at much higher utilization than in the past, while providing strict guarantees on the standard quality of service (QoS).

Aggressive optimization and traffic engineering methodologies may suffer scalability issues and unacceptable computational complexity. Therefore, ML is regarded as an approach to automatized network self-configuration and fast decision making by relying on data.

Some applications of machine learning include fault prediction, intrusion detection, physical flow security, impairment aware routing, low margin design, traffic aware capacity reconfigurations. Here, both layers at physical and network layers find applications in the following domains.

Physical layer

QoT estimation (quality of transmission)
Optical performance monitoring (OPM)
Optical amplifier control
Modulation format recognition (MFR)
Nonlinearity mitigation
Network layer
Traffic prediction and virtual topology design
Failure detection
Flow classification
Path computation
Machine learning algorithms

There are various algorithms that are classified as follows based on the methodology in the artificial intelligence field. In this section, machine learning algorithms along with their implementation areas in optical systems and networks is discussed.

Machine learning algorithms specified in the literature are classified as follows:

(A) Supervised learning
(B) Unsupervised learning
(C) Reinforcement learning
(D) Semi-supervised learning

Supervised learning: Supervised learning tries to find the mapping between a pair of data sets. It is a kind of mentor approach where a student is in touch with a facilitator who provides continuous support and feedback on the accuracy of the results. It is used in the applications that include pattern recognition (classification) and regression (function approximation) type of problems.

Some examples of supervised machine learning algorithms are support vector machine (SVM), artificial neural networks (ANN), deep neural network (DNN), k nearest neighbor, random forest, regression.

SVM has lower complexity and is simpler and is used in the classification of data like the classification of received bit pattern (0, 1) as it is a binary classifier, differentiating between faulty fiber and the good fiber, i.e., where there exist two or more distinct groups of data. It is a widely used algorithm for the nullify effect of noises like nonlinear phase noise, laser phase noise, modulator nonlinearity, Kerr effect, amplified spontaneous noise, linear and nonlinear channel, and signal detection. However, extended SVM over standard SVM are needed to extract information from highly noisy overlapped constellations. Liang et al. (2018) demonstrated that SVM is also widely used for equalizing linear and nonlinear effects of optical fiber. In optical networks SVM too can be used to detect and predict optical link failures as well as quality of transmission estimation. As discussed by Mata et al. (2017).

Artificial neural networks: It is widely used in optical communication applications because of its results and can be applied to solve complex problems like nonlinear equalization given by Lyubomirsky (2015). It works on the principle of neurons present and connected with each other in the human brain. And it derives complex nonlinear relation between output and target. It can be further used for optical performance monitoring (OPM). And is a good choice for monitoring real time applications. Further, it can be used in adjusting optical amplification gain and monitoring and modeling of fiber nonlinearity in elastic optical networks (EON). Further, it can be used to predict network traffic as said by Ruan et al. (2018).

Deep neural network: Deep neural network works as given by Giacoumidis et al. (2018) on big data and employs many layers in between. Main application of DNN includes equalization, complex channel and system modeling as it can tackle deterministic and stochastic nonlinearity by exceeding the nonlinear Shannon-limit in coherent optical communications using 3D adaptive machine learning. Further Zibar et al. (2018) and Karanov et al (2019) proposed DNN can be used in both monitoring and identification. It can also be used in optimization of trans receiver parameters. It can be used for autoencoding, constellation reshaping, power adjustments, and modeling of system parts like optical amplifier, optical modulator etc.

Random forest: Random forest is based on a decision tree model and can be used in estimation of bit error rate BER of link and quality of transmission and also the estimation of a received optical power parameter as discussed by Barletta et al., Rottondi et al., and Toth et al. (2017).

Unsupervised learning: The unsupervised learning algorithm tries to find patterns and structure in unlabeled data and within it without giving reference from outside. There are various algorithms under unsupervised machine learning like c means, hierarchal clustering. It can combat stochastic parameter noise amplification, nonlinearity effects.

Some examples of unsupervised machine learning algorithms are k means clustering, expectation maximization, principal component analysis.

Reinforcement learning: These types of learning algorithms learn by rewards for its action.

Some examples of reinforcement machine learning algorithms are Q learning, deep reinforcement learning and are used to solve problems of optical burst switching (OBS).

Semi-supervised learning: Semi-supervised learning has not developed much in the area of optical communication till now, where the mixed approach of both the supervised learning and unsupervised learning is used.

8.7 ENERGY-EFFICIENT OPTICAL NETWORKS

As per the present scenario, as the number of devices is increasing day by day, and consuming a lot of energy, there is a need to develop and sustain optical networks that consume less power and hence become energy efficient. Researchers can devise methodology in this area too in order to meet the goals of sustainable living as optical networks will remain as a backbone for the years to come because of their unmatched performance and capabilities. Bathula et al. (2009) proposed an energy-efficient routing algorithm by trading off between quality of service and energy consumption for on off keying modulation. Chen et al. designed a mixed regenerator placement scheme for making optical networks energy efficient. They used optical networks in two different categories one is backbone optical networks and the other is optical access networks. Zhuet al. (2011) proposed a novel infrastructure in optical translucent networks that reduces the number of optical to electrical and then to optical regenerators to less than 50% and hence made optical networks energy efficient.

8.8 CONCLUSION

In this work, the trends in the network connectivity have been studied both in the wired and wireless domain considering 4G and 5G along with optical networks. Further, the scope of various machine learning algorithms for optical networks in the physical layer and the network layer are considered to make the networks dynamic and adapt to the requirements as per the data in real-time. Both the layers are interconnected and worked together, i.e., BER analyzed at the physical layer can trigger changes at the network layer like routing, spectrum, and modulation format and hence the term cross layer. Hence, it is observed that the optical networks are turning complex both at the physical layer and at the network layer and need attention so that the already developed machine learning algorithms in the literature can be applied to analyze and mine data in optical networks so that the cost involved in the actual deployment of fibers and their impairments can be reduced and networks can adapt to changes automatically and can provide quality of service.

REFERENCES

A. Eira, J. Pedro, J. Pires, 2015, "Optimal multi-period provisioning of fixed and flex-rate modular line interfaces in DWDM networks," in IEEE/OSA Journal of Optical Communications and Networking, vol. 7, no. 4, pp. 223–234, April, doi: 10.1364/JOCN.7.000223

A. Liang, C. Yang, C. Zhang, Y. Liu, F. Zhang, Z. Zhang, H. Li, 2018, "Experimental study of support vector machine based nonlinear equalizer for VCSEL based optical interconnect," in Optics Communications, vol. 427, pp. 641–647.

Arlington, Va. (August 9, 2017), https://tiaonline.org/press-release/tia-updates-data-center-cabling-standard-to-keep-pace-with-rapid-technology-advancements/

B. G. Bathula and J. M. H. Elmirghani, 2009, "Green networks: Energy efficient design for optical networks," in: 2009 IFIP International Conference on Wireless and Optical Communications Networks, pp. 1–5, doi: 10.1109/WOCN.2009.5010573

B. Karanov, M. Chagnon, F. Thouin, T. A. Eriksson, H. Bülow, D. Lavery, L. Schmalen, 2018, "End-toend deep learning of optical fiber communications," in Journal of Lightwave Technology, vol. 36, no. 20, pp. 4843–4855.

C. Rottondi, L. Barletta, A. Giusti, M. Tornatore, 2018, "Machine-learning method for quality of transmission prediction of unestablished lightpaths," in IEEE/OSA Journal of Optical Communications and Networking, vol. 10, no. 2, pp. A286-A297, doi: 10.1364/JOCN.10.00A286

C. Rottondi, L. Barletta, A. Giusti, M. Tornatore, 2018, Machine-learning method for quality of transmission prediction of unestablished lightpaths. in IEEE/OSA Journal of Optical Communications and Networking, vol. 10, no. 2, pp. A286-A297.

C. T. Huang, 2008, The study of balance traffic load with genetic algorithm for PON, in: 2008 International Conference on Intelligent Information Hiding and Multimedia Signal Processing, pp. 39–42.

C. Ye, X. Zhang, H. Huang, H. Feng, K. Zhang, 2017, Demonstration of 50Gbps IM/DD PAM4 PON over 10GHz Class Optics Using Neural Network Based Nonlinear Equalization. European Conference on Optical Communication (ECOC), pp. 1–3, doi: 10.1109/ECOC.2017.8346196

CommScope, 2017, Data Center High Speed Migration: Infrastructure issues, trends, drivers and recommendations, http://wirexpress.com/wp-content/uploads/Commscope_High-Speed-Migration.pdf

D. R. B. de Araujo, C. J. A. Bastos-filho, J. F. Martins-filho, 2015, "Methodology to obtain a fast and accurate estimator for blocking probability of optical networks," in IEEE/OSA Journal of Optical Communications and Networking, vol. 7, no. 5, pp. 380–391.

D. Zibar, A. Ferrari, V. Curri, A. Carena, 2019, Machine learning-based Raman amplifier design, in: Optical Fiber Communication Conference (pp. M1J-1). Optical Society of America

D. Zibar, L. H. H. de Carvalho, M. Piels, A. Doberstein, J. Diniz, B. Nebendahl, C. Franciscangelis, J. Estaran, H. Haisch, N. G.Gonzalez, et al., 2015, "Application of machine learning techniques for amplitude and phase noise characterization," in Journal of Lightwave Technology, vol. 33, no. 7, pp. 1333–1343.

E. Giacoumidis, J. Wei, I. Aldaya, L. P. Barry, 2018, Exceeding the nonlinear Shannon-limit in coherent optical communications using 3D adaptive machine learning. arXiv preprint arXiv:1802.0912

F. Morales, M. Ruiz, L. Gifre, L. M. Contreras, V. Lopez, L. Velasco, 2017, "Virtual network topology adaptability based on data analytics for traffic prediction," in Journal of Optical Communications and Networking, vol. 9, no. 1, pp. A35–A45.

F. Musumeci et al., 2019, "An overview on application of machine learning techniques in optical networks," in IEEE Communications Surveys & Tutorials, vol. 21, no. 2, pp. 1383–1408, Second quarter, doi: 10.1109/COMST.2018.2880039

F. Musumeci, C. Rottondi, G. Corani, S. Shahkarami, F. Cugini, M. Tornatore, 2019, "A tutorial on machine learning for failure management in optical networks," in Journal of Lightwave Technology, vol. 37, no. 16, pp. 4125–4139, 15 Aug.15, doi: 10.1109/JLT.2019.2922586

F. N. Khan, Q. Fan, C. Lu, A. P. T. Lau, 2019, "An Optical Communication's perspective on machine learning and its applications," in Journal of Lightwave Technology, vol. 37, no. 2, pp. 493–516.

I. Baldine, G. N. Rouskas, 2001, "Traffic adaptive WDM networks: a study of reconfiguration issues," in Journal of Lightwave Technology, vol. 19, no. 4, pp. 433–455.

I. Lyubomirsky, 2015, Machine learning equalization techniques for high speed PAM4 fiber optic communication systems. CS229 Final Project Report, Stanford University.

J. Mata, I. de Miguel, R. J. Durán, J. C. Aguado, N. Merayo, L. Ruiz, ... E. J. Abril, 2017, A SVM approach for lightpath QoT estimation in optical transport networks. In 2017 IEEE International Conference on Big Data (Big Data) (pp. 4795–4797). IEEE.

J. Thrane, J. Wass, M. Piels, J. C.Diniz, R. Jones, D. Zibar, 2017, "Machine learning techniques for optical performance monitoring from directly detected PDM-QAM signals," in Journal of Lightwave Technology, vol. 35, no. 4, pp. 868–875.

K. Mullaley, 2018, "5G networks impact on fiber-optic cabling requirements" Cabling Installation & Maintenance August. www.corning.com/in-building-networks/in/en/home/knowledge-center/5g-networks-impact-on-fiber-optic-cabling-requirements.html

K. Venkatesan, A. Chandrasekar, P.G.V. Ramesh, 2020, Machine Learning Based DWDM Design Using Regression Modelling. 10.21203/rs.3.rs-44204/v1.

King, 2019, www.corning.com/in-building-networks/in/en/home/knowledge-center/practical-iot.html

L. Barletta, A. Giusti, C. Rottondi, M. Tornatore, 2017, QoT estimation for unestablished lighpaths using machine learning, in: Optical Fiber Communications Conference and Exhibition (OFC), IEEE, 2017, pp. 1–3.

L. Barletta, A. Giusti, C. Rottondi, M. Tornatore, 2017, QoT estimation for unestablished lighpaths using machine learning, in Optical Fiber Communication Conference (pp. Th1J-1). Optical Society of America.

L. Ruan and Wong, E, 2018, Machine intelligence in allocating bandwidth to achieve low-latency performance, in: 2018 International Conference on Optical Network Design and Modeling (ONDM) (pp. 226–229). IEEE.

M. A. Amirabadi, 2019, "A Survey on Machine Learning for Optical Communication [Machine Learning View]" https://arxiv.org/pdf/1909.05148

M. C. Tan, F. N. Khan, W. H. Al-Arashi, Y. Zhou, A. P. T. Lau, 2014, "Simultaneous optical performance monitoring and modulation format/bitrate identification using principal component analysis," in Journal of Optical Communications and Networking, vol. 6, no. 5, pp. 441–448.

M. P. I. Dias, B. S. Karunaratne, E. Wong, 2014, "Bayesian estimation and prediction-based dynamic bandwidth allocation algorithm for sleep/doze-mode passive optical networks," in Journal of Lightwave Technology, vol. 32, no. 14, pp. 2560–2568.

N. G. Gonzalez, D. Zibar, I. T.Monroy, 2010, Cognitive digital receiver for burst mode phase modulated radio over fiber links, in: Optical Communication (ECOC), 2010 36th European Conference and Exhibition on, IEEE, pp. 1–3.

Nokia bell labs, www.nict.go.jp/en/press/2020/12/18-1.html

R. M. Morais and J. Pedro, 2018, "Machine learning models for estimating quality of transmission in DWDM networks," in IEEE/OSA Journal of Optical Communications and Networking, vol. 10, no. 10, pp. D84-D99, Oct, doi: 10.1364/JOCN.10.000D84

R. R. Reyes, T. Bauschert, 2017, "Adaptive and state-dependent online resource allocation in dynamic optical networks," in IEEE/OSA Journal of Optical Communications and Networking, vol. 9, no. 3, pp. B64–B77.

S. G. Petridou, P. G. Sarigiannidis, G. I. Papadimitriou, A. S. Pomportsis, 2008, "On the use of clustering algorithms for message scheduling in WDM star networks," in Journal of Lightwave Technology, vol. 26, no. 17, pp. 2999–3010.

S. L. Brunton, X. Fu, J. N. Kutz, 2014, "Self-Tuning Fiber Lasers," in IEEE Journal of Selected Topics in Quantum Electronics, vol. 20, no. 5, pp. 464–471.

S. Oda, M. Miyabe, S. Yoshida, T. Katagiri, Y. Aoki, T. Hoshida, J. C. Rasmussen, M. Birk, K. Tse, 2017, "A learning living network with open ROADMs," in Journal of Lightwave Technology, vol. 35, no. 8, pp. 1350–1356.

T. Panayiotou, S. Chatzis, G. Ellinas, Jan, 2017, "Performance analysis of a data-driven quality-of-transmission decision approach on a dynamic multicast-capable metro optical network," in IEEE/OSA Journal of Optical Communications and Networking, vol. 9, no. 1, pp. 98–108.

T. Tanimura, T. Hoshida, J. C. Rasmussen, M. Suzuki, H. Morikawa, 2016, OSNR monitoring by deep neural networks trained with asynchronously sampled data, in: 2016 21st Opto Electronics and Communications Conference (OECC) held jointly with International Conference on Photonics in Switching (PS), IEEE, 2016, pp. 1–3.

W. B. Jia, Z. Q. Xu, Z. Ding, K. Wang, 2016, An efficient routing and spectrum assignment algorithm using prediction for elastic optical networks, in: 2016 International Conference on Information System and Artificial Intelligence (ISAI), pp. 89–93.

Worton, 2021, https://community.fs.com/blog/itu-t-standards-for-various-optical-fibers.html

X. Chen, F. Ji, Z. Zuqing, 2013, "Energy-Efficient Resilience in Translucent Optical Networks with Mixed Regenerator Placement," in Journal of Optical Communication and Networking, vol 5, pp. 740–750.

X. Liu, H. Lun, M. Fu, Y. Fan, L. Yi, W. Hu, Q. Zhuge, 2020. AI-based modeling and monitoring techniques for future intelligent elastic optical networks. Applied Sciences, vol. 10, no. 1, p. 363. https://doi.org/10.3390/app10010363

X. Wu, J. A. Jargon, R. A. Skoog, L. Paraschis, A. E. Willner, 2009, "Applications of artificial neural networks in optical performance monitoring," in Journal of Lightwave Technology, vol. 27, no. 14, pp. 3580–3589.

Y. Huang, P. B. Cho, P. Samadi, K. Bergman, 2017, Dynamic power preadjustments with machine learning that mitigate EDFA excursions during defragmentation, in: Optical Fiber Communication Conference, Optical Society of America,, pp. Th1J–2.

Y. Huang, W. Samoud, C. L.Gutterman, C. Ware, M. Lourdiane, G. Zussman, P. Samadi, K. Bergman, 2016, A machine learning approach for dynamic optical channel add/drop strategies that minimize EDFA power excursions, in: Proceedings of ECOC; 42nd European Conference on Optical Communication, VDE, 2016, pp. 1–3.

YOFC, 2013, https://en.yofc.com/view/1722.html

Z. Zhu, July 2011, Mixed placement of 1R/2R/3R regenerators in translucent optical networks to achieve green and cost-effective design," in IEEE Communications Letters, vol. 15, no. 7, pp. 752–754, doi: 10.1109/LCOMM.2011.051011.110519.

9 Green Electronics for Future Communication Systems

Sakshi Mittal, Gurjit Kaur, and Manjeet Kumar

CONTENTS

9.1 INTRODUCTION

The increasing trends in buying and discarding electronic products have led to many environmental issues. With the advancement in technology, the rate at which electronic products are being disposed have increased tremendously. Discarding of these products has filled up the land area. Other methods of disposing, like burning, may release harmful chemicals that can affect the environment adversely. This leads to imposing the responsibility to use environment friendly and safe materials onto the shoulders of manufacturers, so that disposal has no harmful effect on the environment. This results in the need for green electronics. The main aim is to reduce the power consumption and increase the energy efficiency. Apart from this, some initiatives are to be made in order to reduce the CO_2 emission and reduce energy consumption.

DOI: 10.1201/9781003264477-9

Figure 9.1 shows the global waste electrical and electronic equipment (WEEE) production, disposal, their recovery estimates for 2010 (Bastiaan et al., 2010).

Figure 9.2 shows the global greenhouse gases emitted due to various human activities [IPCC, 2014]. It is evident from the graph that the majority emitted is carbon dioxide. The reason for CO_2 emission is the major use of fossil fuels.

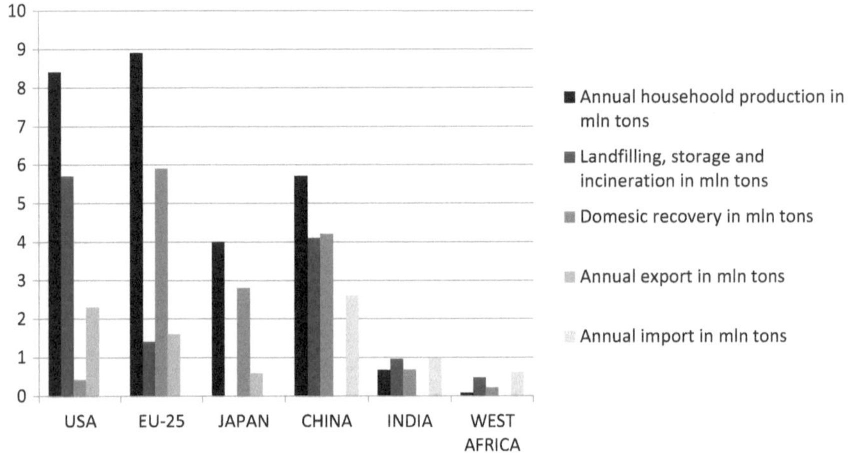

FIGURE 9.1 A global account of the waste electrical and electronic equipment (WEEE).

Source: Bastiaan et al., 2010.

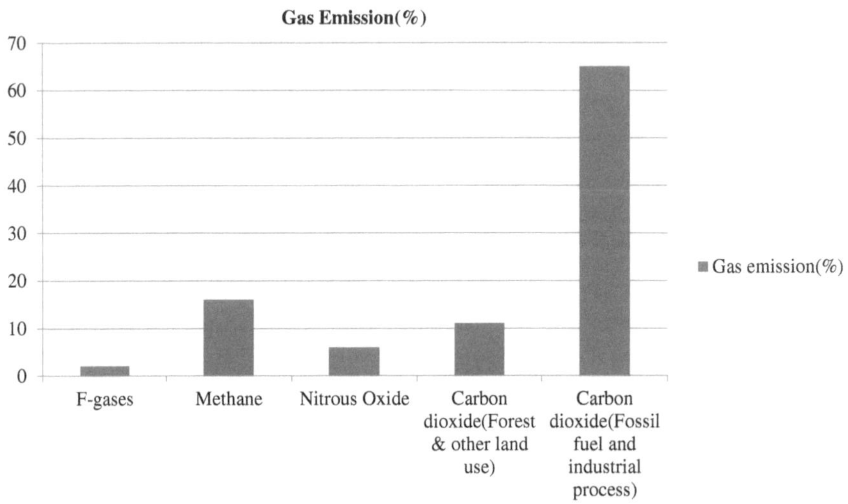

FIGURE 9.2 Global greenhouse gas emission.

Source: IPCC, 2014.

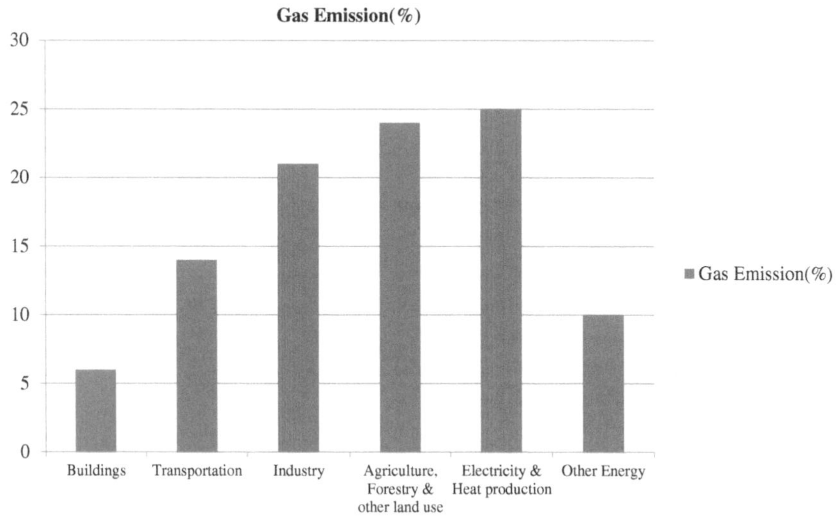

FIGURE 9.3 Greenhouse gas emission by economic sectors.

Source: IPCC, 2014.

Figure 9.3 shows the breakdown of emitted gases due to different economic sectors. It is evident from Figure 9.3 that electricity and heat production is the major phenomenon responsible for emission of greenhouse gases. Another major factor responsible is the deforestation and burning of waste at industrial sites. Hence, appropriate measures are needed in order to have an environment safe for survival.

9.2 ROLE OF GREEN ELECTRONICS FOR FUTURE COMMUNICATION SYSTEMS

The objective of proposing green electronics for a safe environment is summarized as follows:

(a) ***Energy saving and increase energy efficiency:*** With the rise in demand for higher data rates and devices with smart applications, there is a tremendous increase in the energy required to operate these devices. This will affect the environment significantly. Thus, it is required to have devices that consume less power, resulting in improving energy efficiency of the system (Alessio et al., 2016).

(b) ***Eliminating the use of materials that can prove to be harmful to nature:*** It refers to the material used in production of computers, cell phones, and other electronic products. With frequent launching of new products in the market, the discarding of electronic products has raised significantly, thus increasing the e-waste. This e-waste can prove to be hazardous to the environment if the material used for production is not chosen wisely. For example, plastic is often used on cable cords, which releases harmful chemicals when burned in the environment. Thus the responsibility lies with manufacturers to use materials that are environment friendly (NERC Blogs, 2015).

(c) ***Compliance with green electronics council:*** The green electronics council was formed in 2005 with the aim to provide certification to companies that offer green devices. The devices were assessed on various parameters and accordingly gold, silver, and bronze certificates are provided (NERC Blogs, 2015).

(d) ***Minimizing carbon dioxide emission and energy consumption:*** As evident from Figure 9.2, carbon dioxide is one the harmful gases, which is emitted in majority. Global warming, climatic changes, melting of polar ice caps, increase in sea level, and many more harmful effects are a result of carbon emissions trapped in the atmosphere. Thus the requirement for being environment safe for survival is the introduction of technologies to minimize carbon emission (Stephanie, 2020).

9.2.1 Green Communication Solutions

The role of green communication technique focuses on solutions like cognitive network, network coding, and a smart grid (Mustafa et al., 2020).

(a) ***Cognitive network:*** Cognitive network is a technology that makes use of some research areas like machine learning, computer network etc. to deal with the current problem in that domain.

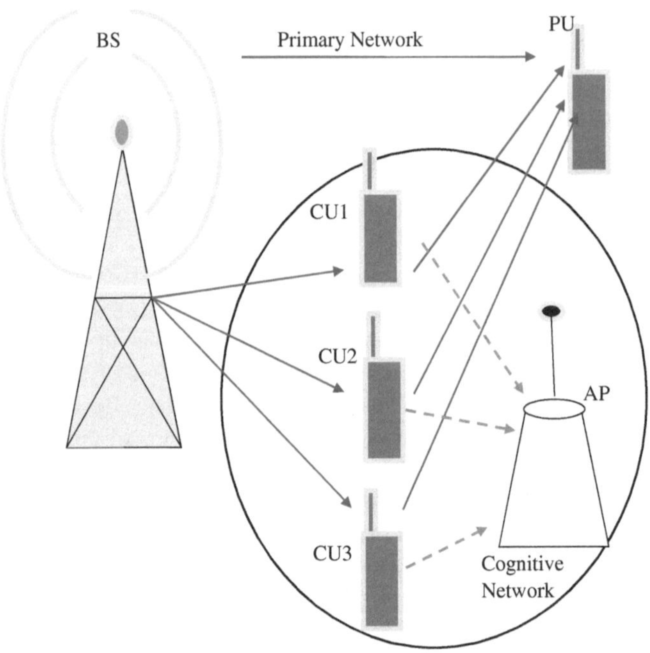

FIGURE 9.4 Resource allocation in a cognitive network.

This network learns from situations, plans, decisions, and actions taken in the past and utilizes it to improve future goals. It senses the requirement of active nodes and utilizes this information for spectrum allocation to secondary users. If the spectrum is found free, the spectrum can be utilized on a time sharing basis by other secondary nodes. In this manner, the spectrum can be utilized efficiently. Further the energy efficiency can be improved, thus leading to a green communication. Figure 9.4 shows the resource allocation to secondary users. Base station allocates the spectrum to a primary user. As and when cognitive or secondary users found any frequency band to be vacant, that part of the spectrum can be utilized for cognitive users. That part of the spectrum is allocated back to the primary user upon requirement.

(b) ***Network coding:*** Network coding enables intermediate nodes with the ability to code the information and transmit. Coding is generally done at end nodes. Like in a smart grid, each individual node is required to transmit information. Thus every individual node must have computational capabilities to recode the information and transmit it for security reasons. It can be said that this technique is a complement to the store and forward technique. It is called recode and forward. It works on dispensable paths to improve the performance of the network. Figure 9.5 demonstrates how the intermediate nodes are performing coding and forwarding to enhance the security system (Frank et al., 2021).

(c) ***Smart grid:*** Electric grid is a network of transformers, transmission lines, substations to deliver electricity from the utility to the consumer. Power generation is generally localized and is built around communities. This is a one way communication since 1890. These utilities provide electricity without knowing a consumer's individual requirement. This one way communication makes it difficult for the grid to determine ever changing and rising electricity requirements of consumers. This was replaced by the concept of a smart grid. A smart grid or network consists of computers, controls, automation, and equipment working together to respond to consumer's electric needs. Therefore, a smart grid provides a two way communication

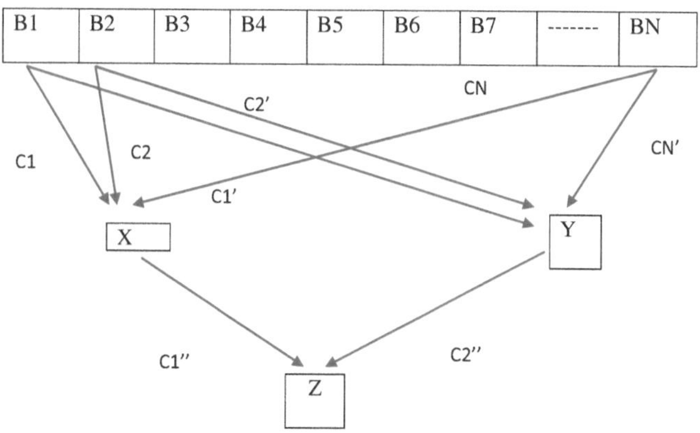

FIGURE 9.5 Network coding scheme.

where a consumer provides a feedback about their requirements for electricity (Gridtalk, 2019). This two way communication provides electricity and information to be exchanged between utility and consumer. This helps in making a smart grid help to manage our electric needs. It incorporates newer technologies such as solar and wind energy and plug-in electric vehicle charging. A smart grid helps in making the system more efficient, reliable, secure, and greener. Although there are some challenges that need to be taken care of.

(i) ***Customer participation:*** It requires customer participation to buy or sell electricity as per need from a power station. Handling customers' demand to shift electric load every time they need, can be a challenge.
(ii) ***Communication:*** A smart grid comprises of sensors, computers, renewable energy sources, and automation for feedback. Maintaining a continuous and effective communication between each node can be challenging.
(iii) ***Demand response applications:*** Utilities must be able to utilize the energy from demand mode for certain applications.
(iv) ***Security:*** Since communication is a foremost factor in a smart grid. Protection against theft, malicious activities, and abuse is a major concern.

Figure 9.6A shows the transfer of energy from power house to subscriber unit, whereas, Figure 9.6B shows the feedback path also from the subscriber unit to the power house. This feedback path contains information about a subscriber's need.

Methods that can be employed for advanced monitoring in smart grid are as follows:

(i) Efficient software algorithm and high speed computers: analyzing tool
(ii) Multi-agent system
(iii) Data acquisition applications like SCADA

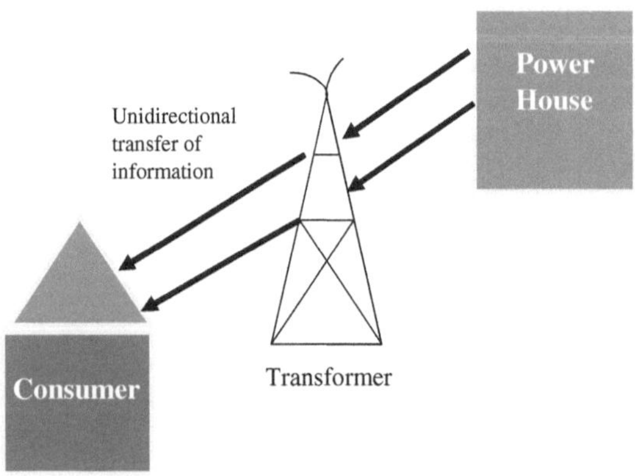

FIGURE 9.6A Electric grid.

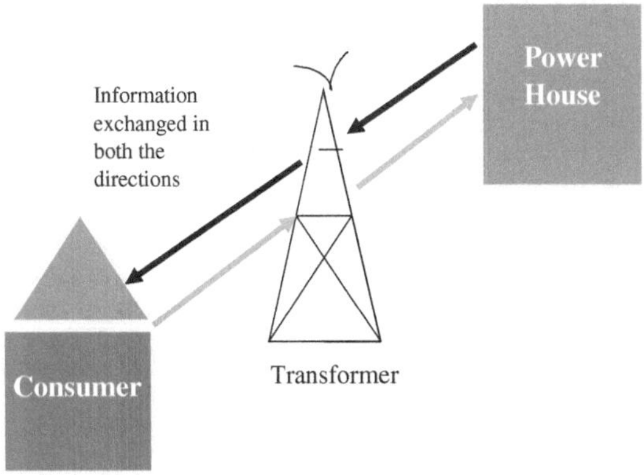

FIGURE 9.6B Smart grid.

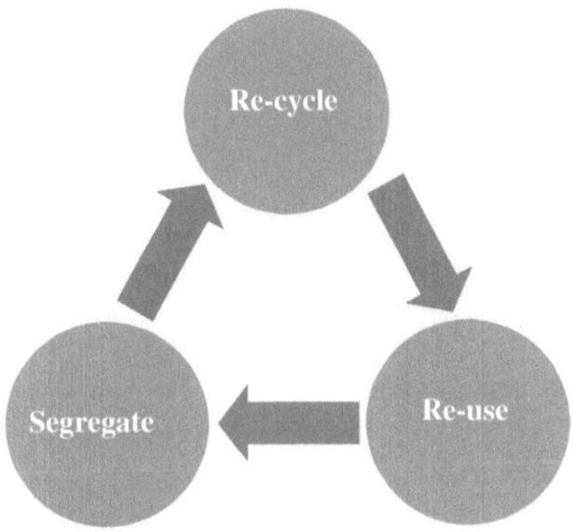

FIGURE 9.7 Green manufacturing.

9.3 GREEN MANUFACTURING

It refers to use of sustainable material in the production of electronic products like computer, television, mobile phones etc. It may be less expensive to use harmful products for manufacturing but results in harmful chemical effects on society as well as the environment. Manufactures play a vital role (Mustafa et al., 2020).

Figure 9.7 shows some key steps that manufactures or industrialists or any individual must follow to dispose the waste without actually harming the environment.

(a) ***Recycle:*** A large amount of waste around 75% is generally recyclable. The only thing required is the adequate knowledge about the materials that can be recycled. Separate containers for disposing metals, glass, toxins etc can help in recycling.
(b) ***Reuse:*** Try to find out ways to reuse metal, woods, glass etc.
(c) ***Segregate:*** Do not mix toxic waste with general waste that may contaminate rivers, oceans, and the environment.

9.4 ENERGY-EFFICIENT (GREEN) METRIC

Energy-efficient metric covers the parameters to assess the greenness of communication devices. Energy consumption at different levels such as component, equipment, and system level help in evaluating the energy efficiency (EE) in a wireless network (ETSI, 2009). Component level metric evaluates any particular part of a communication device, whereas, equipment level metric evaluates energy efficiency of the complete device. System level metric helps in evaluating the parameters viz. reduction in the energy consumption, its rating and area power consumption (Srivastav et al., 2020).

Energy efficiency: Defined as the ratio of system capacity to the total power consumed as denoted by eq. (9.1). Its unit is bits per joule (ETSI, 2009).

$$\eta_{EE} = \frac{Total\,data\,rate\left(bits\,per\,\sec\right)}{Total\,Power\,consumptionl\left(watts\right)} \tag{9.1}$$

EE can be increased by either increasing the system capacity or reducing the power consumption.

9.5 POWER ALLOCATION TECHNIQUES FOR GREEN ELECTRONICS

Power allocation enables source information to reach the destination efficiently, thus increasing the efficiency of the device.

There exist the following power allocation methods (Ding et al., 2014):

(a) ***Constant power allocation (CPA) or equal power allocation:*** Scheme is referred to as one of the optimal solutions. In this technique, an equal amount of power is shared among all subscribers. However the power allocation can be based on their requirement. This implies that a different amount of power can be allocated. But with CPA, equal power is distributed among all subscribers, irrespective of factors like how far the user is from the base station, whether a user is in active or sleep state, its signal to interference plus noise distribution etc. Hence it results in wastage of power.

(b) ***Channel inversion power allocation (CIPA):*** Unlike the CPA scheme where constant power is allocated, CIPA allocates considerable power to subscribers far away from the base station in order to achieve a stable transmission rate.

(c) ***Water-filling power allocation (WFPA):*** The subscriber with better channel conditions will be allocated power first thus optimizing its power. If any power is left, then that can be utilized to help the subscriber with poor channel conditions. Thus the outage probability of the worst subscriber will be minimized.

(d) ***Proportional fair power allocation (PFPA):*** The power is allocated as per the requirement of subscribers in such a way that relative throughput increment is identical for all subscribers. Thus this is a fair means of resource allocation.

9.6 ELECTRONIC ENERGY EFFICIENCY EFFORTS

Efforts are required to make produced waste back to the raw material, i.e., output is to be converted back into an input product. Therefore a feedback loop needs to be established to save energy. The main step towards minimizing the power consumption can be seen from the developments in the field of electronics such as light emitting diode, smart phones, laptops, other appliances etc.

9.6.1 LED

Light emitting diode, which generates heat in the form of light energy was first introduced in 1962. They work on the principle of electroluminescence. They offer a good picture quality but consume more power. LEDs were replaced by OLEDs in 1987. OLED or organic LED involves organic material in between conducting layers. Backlight is not required in the case of OLED as it works on the principle of self-luminescence (Sujith et al., 2020). They have comparatively better picture quality and consume less power. But they came at high cost and are less flexible as compared to LEDs.

9.6.2 Graphene

Graphene, a naturally occurring carbon allotrope was found in 1962. Various researches have been conducted to find the effect of graphene on human health. Negligible harmful effect of graphene has been observed (Ankit et al., 2018). Graphene is the thinnest material available that found its application in making strong and durable display screens for smart phones. Graphene can also be used for chip making, solar cells, pressure sensors etc. Graphene has a promising future due to its low production cost and environment safe nature.

9.6.3 CNF – Cellulose Nanofibers

A wood driven fiber is an ideal candidate for electronic applications due to less harmful effects on the environment. The following properties of CNF make it ideal for green communication: biodegradable, eco-friendly material and desirable electrical properties (Ansari et al., 2016).

9.6.4 Television

The transition from old, heavy, and bulky televisions to LCDs then to LED is a step towards advancement in technology. The six eras of television are shown in Figure 9.8.

The invention of black and white television started in 1927 and was then upgraded to color television in 1953 (Gilmore Blogs, 2017). Then LCD was discovered in 1968, which marked the beginning of a new generation of electronic devices. They used liquid crystal displays to generate images. These were replaced by an array of LEDs being used for image generation in 1977. HDTV was introduced in 1981 with

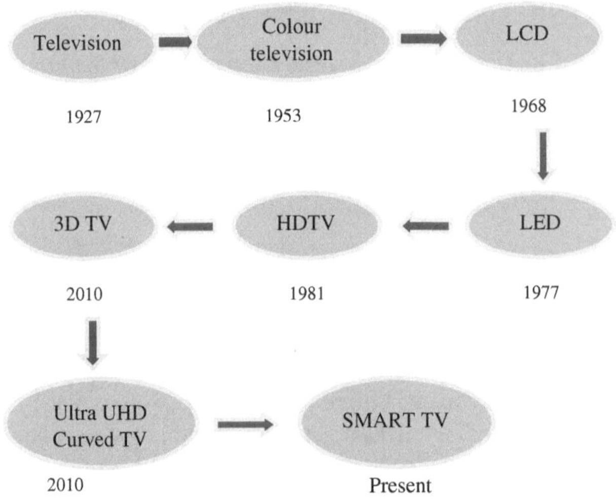

FIGURE 9.8 Evolution of television.

the aim of producing high definition pictures, which got replaced by smart television in 2007. The purpose behind this transition was reduced dimensions, flat or flexible screen, improved performance, high picture quality, and less energy consumption. The major technological changes include LCD and LED televisions. The difference among these two lies between the placement and the type of light used for picture generation. LEDs are more reliable and have a longer life span. Therefore LEDs are more efficient and consume less power. LEDs are a preferred choice over LCDs in terms of its application as well. This transition from a device consuming a lot of power to a device consuming less power is a set example of revolution towards green communication.

9.6.5 Mobile Phones

Mobile phones are battery operated. Hence their capacity is limited as the dimension of the device will increase with an increase in battery size. The requirement of the present generation is a device with all computer related processing in their pocket (Aaron et al., 2010). The evolution of mobile phone began with the first generation cell phone in 1980 and continues with deployment of 5G smart phones in 2019. These days' mobile phones are used for data processing, creating and sharing multimedia files, running different applications in addition to its basic feature of making a call. But running these applications in the background consumes a lot of power. Hence the device needs to be efficient in terms of energy saving. As per Mohammad et al. (2016), offloading is one of the main energy conserving techniques in mobile computing. Offloading means to transfer the execution of offloaded task from a mobile device to a computational infrastructure for example cloud. Cell phones in the late 80s used to be heavy and bulky. The discarding of these devices was a major issue. Burning of these devices releases harmful chemicals into the environment. This problem was solved with the invention of flexible displays for use in smart phones. These flexible displays are generally made up of OLED, which makes them environment friendly. Hence they also can be proved to be greener.

9.6.6 Computers

The first mechanical computer by Charles Babbage in 1822 was so heavy and the size of a room. These were based on vacuum tubes. With the invention of transistors and integrated circuits, the size of devices has reduced significantly. It has been a tremendous technological advancement from a large room size computer to a personal computer, a laptop, a tablet, and finally flexible displays. The big room size computer consumes a lot of power and a lot of space for its assembly and discarding was required. These days the devices are available in small size with improved performance but are harmful for the environment. Laptops are available with lead batteries and poly vinyl chloride for wire coating, which emit toxic material on burning. In spite of these limitations, laptops are much greener than desktop computers, which are large in size, consume more electricity, and other resources. Some steps at an individual level can help minimize energy and can prove to be a greener communication (Assim et al., 2006).

(a) Turn off computer when not in use
(b) Peripherals should be turned ON during use only
(c) Use power saving mode
(d) Use laptop or tablet instead of desktop as it consumes more electricity

9.6.7 Processors

Intel released their first processor 4004 in 1971. Starting from a 4-bit processor and 8-bit controller, Intel has launched a series of processors and controllers serving a wide range of applications. Processors require electrical energy for operation and dissipate this in the form of heat. Power dissipation varies from processor to processor. An initiative towards green communication was made by Intel with the launch of two quad core processors that require no more power to operate faster than previous processors. 45 nm manufacturing and reinvented transistor formula are the advancements by Intel that help to reduce power consumption and improve performance (Intel news, 2008).

9.6.8 Other Electronic Devices

Other home appliances such as geyser, street lights, home lightning system, power bank, batteries have also contributed in green communication by a smart move towards solar panel. Some solar devices are shown in Figure 9.9 (Solar Power Electronic Devices Blogs, 2016).

By utilizing solar panels for home appliances, electricity can be saved. Solar devices such as solar panels, street lights, water heaters, lanterns, solar home lightning systems etc. results in energy saving. Although the disadvantage associated was the unavailability of sunlight during night time. This was resolved with the invention

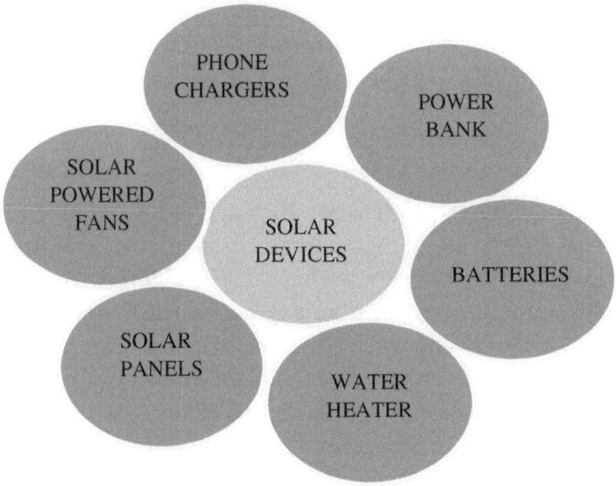

FIGURE 9.9 Solar devices.

of solar batteries and solar power bank. With the advancement in technology, the solar panels can be used to derive power for houses, thus lowering the electric energy consumed and reducing the amount of money required to pay electricity bills. Further advancements are required in this promising field. Thus, solar devices should be preferred for a green environment as it offers the following advantages:

(a) No emissions
(b) Renewable source of energy
(c) Requires minimum maintenance
(d) Long lasting

9.7 CHALLENGES OF GREEN ELECTRONICS FOR COMMUNICATION SYSTEMS

Everything comes at a cost. Electronic devices can be made greener but with the following challenges (Ahmed et al., 2017):

(a) ***Cost:*** The solution towards a green and safe environment increases the price of the device. Hence green electronic devices have a much higher cost than existing devices.
(b) ***Bandwidth:*** By utilizing more bandwidth, there can be reduction in energy consumption. As the transmission rate increases with increase in bandwidth (Anjuri et al., 2020). Green electronics aim at energy saving. Thus the electronics can be made greener at the expense of bandwidth.
(c) ***Spectrum efficiency:*** It can be defined as system throughput, i.e., the amount of data transmitted over a given spectrum. The transmission rate is directly proportional to the bandwidth and amount of available transmit power as per

Shannon's capacity theorem (Anjuri et al., 2020). Green communication aims towards minimizing power consumption. Hence the transmission rate will be impacted in order to make devices green.

9.8 RESEARCH OPPORTUNITIES OF GREEN ELECTRONICS SYSTEMS

With the advent of 5G, the aim is towards energy efficiency. With 5G, it is believed that energy consumption will be reduced to about one tenth of consumption in 2010. Further research persists in this field with the aim to further reduce the carbon dioxide emission in the near future. Figure 9.10 shows few emerging areas for future research work in the direction of green and sustainable communication.
5G technologies for green communication are

(a) ***Device-to-device communication:*** D2D communication is the direct communication path between two communicating devices without the signal going to base stations (Zihan et al., 2015). Although the communication is made under the control of base stations. The latency is also reduced as transmission is carried out in short range. Thus it results in an increase in the spectral efficiency of the network. With this technology, power consumption is reduced but at the expense of security (Jianting et al., 2013). Further research is required in this area to promote this technology for green communication. Figure 9.11 demonstrates the difference between cellular communication and D2D communication.

In the case of cellular communication, a direct communication link is established between the base station and user equipment. However, in the case of D2D communication, a connection is made between two devices in proximity without intervention of the base station.

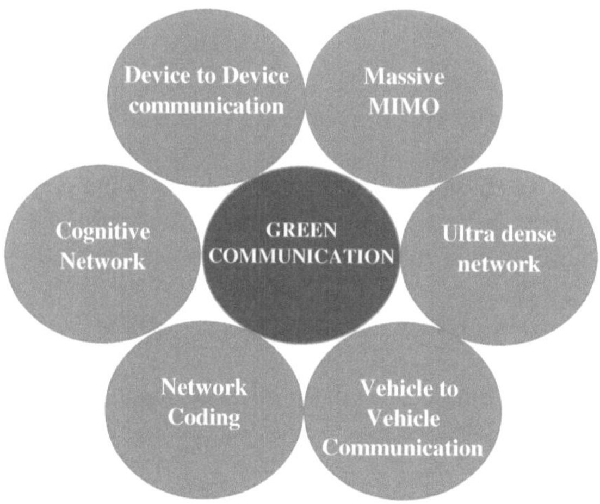

FIGURE 9.10 Emerging technologies for green communication.

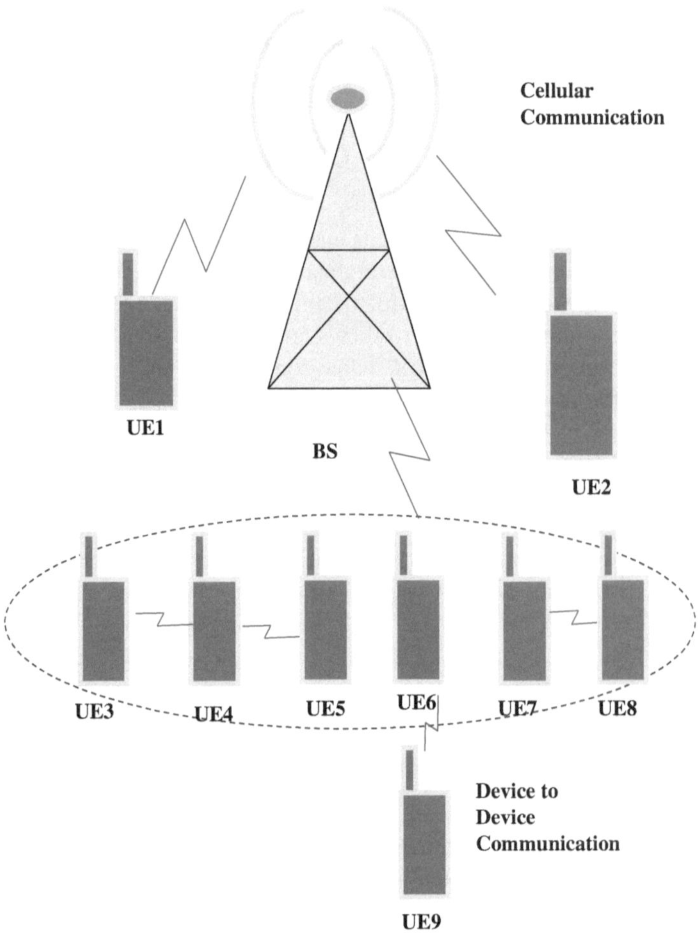

FIGURE 9.11 Device to device communication.

(b) ***Ultra dense networks:*** An ultra dense network means increasing the density in such a way that the number of base stations available is more than the number of users. It is based on small cell (pico cell and femto cell). As the numbers of base stations are more than the number of active users, the interference will be reduced significantly. With the introduction of this technology, it is possible for a device to communicate any other device within its proximity directly. The main advantages include reduced power emission and enhanced security due to the absence of intermediate nodes. Small cells, which are not serving any user equipment, will go into sleep mode thus conserving power. Figure 9.12 depicts the scenario for an ultra dense network where the number of base stations available for connection is much more than the number of users. It has been made possible with the concept of small cells. The size of a macro cell is greater than that of a pico cell, which can be deployed

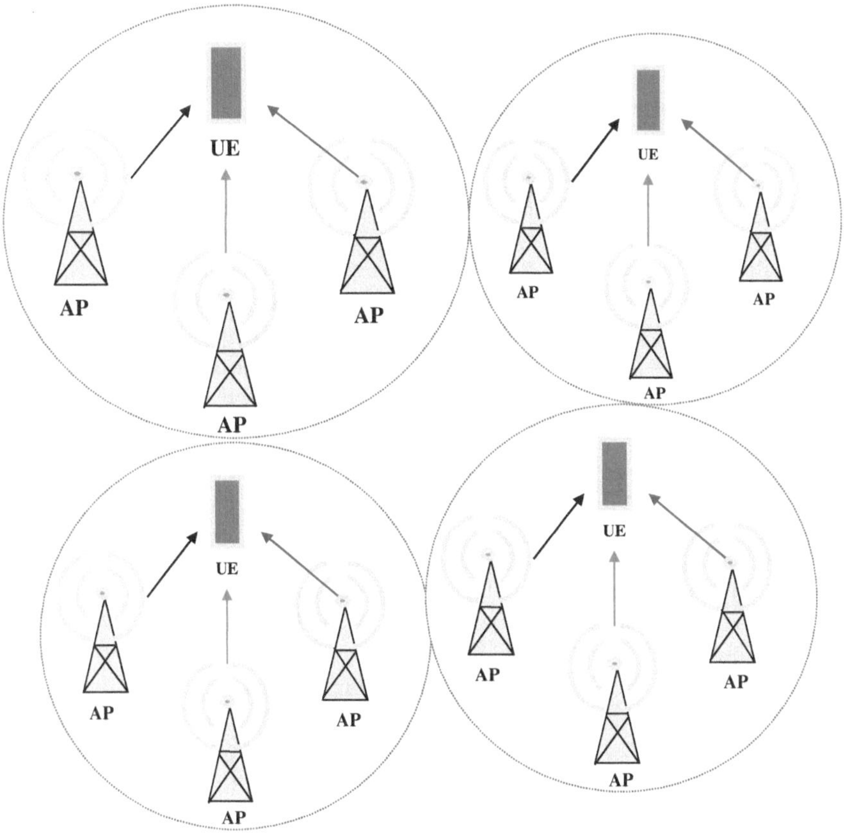

FIGURE 9.12 Ultra dense network.

for offices. However, the size of a pico cell is far greater than a femto cell, which can be deployed for a house or small business unit (Jialing et al., 2017). Some challenges need to be addressed for proper deployment of ultra dense networks:

(i) ***Propagation and system modeling:*** For D2D communication, new propagation models are needed as D2D communication will be a ground to ground link rather than tower to ground with a base station.

(ii) ***Mobility management and self-organization:*** Mobility is a major concern as a single subscriber is in the range of multiple small cells. Some standards need to be built to decide which base station a subscriber should connect with. With high mobility users, frequent handovers and allotment of user equipment is another area of concern. Rapid handovers can be avoided by use of a virtual cell. A group of small cells forming a large cell is considered a virtual cell.

(iii) **Additional functionality at the base station:** Future networks must include base stations to perform functions such as discovery, synchronization, authentication, power control, and selection of appropriate carriers (NIST Blogs, 2016). In the future, small cells can be deployed directly by customers without interference of service providers.

(c) ***Vehicle to vehicle (V2V) communication:*** Vehicle to vehicle communication is a communication between vehicles equipped with the same system in order to prevent traffic, avoiding accidents. This system will provide drivers with crucial information to avoid crashing when they need it. V2V technology provides drivers 360 degree awareness about vehicles equipped with a similar technology system within a short communication range. This offers information about vehicles in a range of about 300 m. Drivers will get warnings such as to stop when another vehicle is ahead. It can even warn about crucial vehicles that should not be passed. V2V uses a wireless protocol similar to Wi-Fi to allow short-range communication (Gupta et al., 2021). This short-range communication protocol is called dedicated short-range communication (DSRC). DSRC technology is combined with a global positioning system to form a low cost V2V communication system that provides information about similarly equipped vehicles within a communication range. It includes notifications about current position of GPS, speed acceleration, control information such as brake status, transmission state, steering wheel angle, vehicle's path history, and path prediction (NHTSA blogs, 2014). V2V includes the following application to avoid accidents (Mohamed El et al., 2020):

(i) The intersection movement assist application gives a warning message to driver when it in unsafe to enter intersection. At multiple pathways there can be fast moving vehicles that may result in collision if the vehicle does not stop even after the warning message.

(ii) Do not pass application gives a warning notification when it in unsafe to pass a slow moving application. This system works by detecting the other vehicles coming in the opposite direction. The driver may want to pass the slow moving vehicle without knowing that there could be a vehicle approaching at high speed in the opposite direction. This application can avoid collisions.

(iii) The emergency electronic brake light application will generate a signal to inform the driver that an out of sight car (ahead of many cars) has broken down. This will prevent sudden brakes and jerks.

(iv) Blind spot application allows a driver to virtually visualize what is happening in the area not visible to driver.

However, V2V does not include advanced technologies such as navigation, Internet access, assistant services, rear view cameras, and other advanced technologies (NHTSA blogs, 2014). The performance of vehicles and road safety can be enhanced with incorporation of V2V technology (Gupta et al., 2020). A lot of research is required in this direction to improve the road safety system and have a green future.

(d) ***Massive multiple input multiple output (MIMO):*** MIMO is used to enhance the capacity of a network by multiple transmissions and multiple reception, i.e., multiple input and multiple output. Massive MIMO is a replacement of conventional multiuser MIMO with the following advantages (Zhang, 2013):

(i) Conventional MIMO is designed to work with both types of duplexing (frequency and time) whereas massive MIMO is designed exclusively for time division duplexing to acquire knowledge about channel state.

(ii) Link quality varies with time and frequency. This variation may result from small scale fading and frequency selective fading. No such variations are observed in massive MIMO due to channel hardening.

(iii) Channel allocation has to be frequent to account for link quality. In massive MIMO, channel allocation planning can be done in advance as there are no variations in link quality with time.

(iv) Massive MIMO comprises an array of antenna at the transmitting and receiving side. It can be installed at the top of the base station as well as UE. 4G, LTE-Advanced are the applications of massive MIMO in which carriers are aggregated in a bundle. It will help to improve the coverage area, spectral efficiency, and data rate. It helps in green communication as well to reduce power radiations by efficient power allocation schemes.

(v) In conventional MIMO, third generation partnership project 3GPP specified the code book to transmit the feedback signal from user equipment UE to base stations. So that a transmitter can transmit the data efficiently by using appropriate precoding techniques. In massive MIMO, pilot signals are used for uplink purpose.

Figure 9.13 shows the concept of massive MIMO with a larger number of antennas placed to provide coverage in all directions, i.e., omni-directional antennas are generally used. Several hundreds of antennas are present to provide coverage to a few tens of users. It enhances the coverage area, increases the data rate, and helps reduce power radiations, thus leading towards green communication.

9.9 CONCLUSION

With a significant rise in demand for smart devices, the number of users and their demands have raised drastically. This in turn will increase the energy consumption. In order to save our environment from harmful electromagnetic (EM) radiations and reduce the amount of greenhouse gases, the concept of green communication is being proposed. An overview of green communication to minimize the consumption of resources, associated issues, and solutions are discussed in the chapter. Various

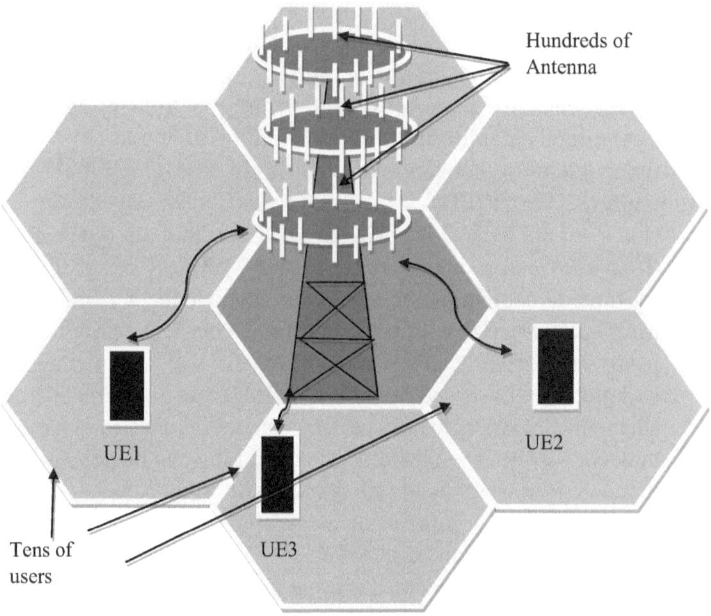

FIGURE 9.13 Massive MIMO with antenna array.

techniques to improve energy efficiency have been discussed. Future technologies dealing with green communication are also presented.

REFERENCES

A. Srivastava, M.S. Gupta, and G Kaur, 2020. Energy efficient transmission trends towards future green cognitive radio networks (5G): Progress, taxonomy and open challenges. Journal of Network and Computer Applications, Elsevier 168: 102760.

C. Aaron and H. Gernot, 2010. An Analysis of Power Consumption in a Smartphone. Proceedings of the 2010 USENIX conference on USENIX annual technical conference.

E. Zihan, K.W. Choi, and D.I. Kim, 2015. Distributed random access scheme for collision avoidance in cellular device-to-device communication. IEEE Transaction Wireless Communication 14(7): 3571–3585.

ETSI, 2009. Energy Efficiency of Wireless Access Network Equipment. ETSI Technical Committee Environmental Engineering (EE) 102. 706(1.1.1).

F. Ansari and L.A. Berglund, 2016. Tensile Properties of Wood Cellulose Nanopaper and Nanocomposite Films. Multifunctional Polymeric Nanocomposites Based on Cellulosic Reinforcements, chapter 4: 115–130.

Gridtalk, Podcast, 2019. SmartGrid.gov. U.S. Department of energy. www.smartgrid.gov/the_smart_grid/electric_vehicles.html

H.P.F. Frank, S. Eckehard, A.C. G Juan, L. Vincent, Z. Jiajing, L. Yun, S. Merve, S. Christian, S. René, T. Andreas, V. Andrés, F. Norman, and P.F. Gerhard, 2021. Tactile Internet-with Human-in-the-Loop. Elsevier. Communications and control Chapter 11: 249–275.

Intel News, 2008. www.computerweekly.com/news/2240085471/Intel-releases-green-processors

IPCC, 2014. Climate Change 2014: Mitigation of Climate Change.

K. Anjuri, G. Aarushi, K. Mayank, and S. Sachin, 2020. A Review on Energy Efficient Green Communication. International Journal of Innovative Research in Electronics and Communications 7(2): 8–11.

K. Ankit and K G Kapoor, 2018. Graphene in the horizon of green electronics. Proceedings 34th National Convention of Environmental Engineers & National Seminar on "Environmental Pollution and Climate Change.

L. Jialing, X. Weimin, L. Chih, Y. Chen, and S. Anthony, 2017. Ultra-Dense Networks (UDNs) for 5G. IEEE 5G Tech Focus 1(1) https://futurenetworks.ieee.org/tech-focus/march-2017/ultra-dense-networks-udns-for-5g?highlight=WyJ1ZG4iXQ==.

M. Ahmed, Nabeel, and U.R Masood, 2017. Green communications: Techniques and challenges. EAI Endorsed Transactions on Energy Web and Information Technologies 4(14): 1–6.

M. S. Gupta and K. Kumar, 2020. Application Aware Networks' Resource Selection Decision Making Technique using Group Mobility in Vehicular Cognitive Radio Networks. Vehicular Communications, Elsevier 26: 100263.

M.H. Mustafa, A. Lukman, A.R. Sami, A. Sameer, Mohammad A. Al-Mashhadani, and S.M. Ahmed, 2020. Green Communication Networks Challenges, Opportunities and Future Role. Journal of Communications 15(3): 256–262.

M.S. Gupta and K. Kumar, 2021. AFSOS: An auction framework and Stackelberg game oriented optimal network's resource selection technique in cognitive radio networks. IEEE Transactions on Network and Service Management 19(1): 61–72.

Marcus Gilmore Blogs, 2017. The Evolution of TV and How It Is consumed. www.theculturesupplier.com/the-big-screen/evolution-tv-consumed/

NERC Blogs, 2015. North east recycling council. https://nerc.org/news-and-updates/blog/nerc-blog/2015/09/01/what-does-green-electronics-mean

NHTSA Blogs, 2014. Vehicle to vehicle communication. www.nhtsa.gov/technology-innovation/vehicle-vehicle-communication

NIST Blogs, 2016. www.nist.gov/programs-projects/ultra-dense-networks

O. Stephanie Blogs, 2020. How Do Carbon Emissions Affect the Environment?.Green Matters. https://www.greenmatters.com/p/how-do-carbon-emissions-affect-environment.

R. Zhang and C.K. Ho, 2013. MIMO broadcasting for simultaneous wireless information and power transfer. IEEE Transaction Wireless Communication 12: 1989–2001.

S. Assim, 2006. Power Consumption in Handheld Computers. APCCAS 2006 - 2006 IEEE Asia Pacific Conference on Circuits and Systems: 1721–1724

S.S. Sujith, D. Deepak K., S. Shahnawaz, A.K.Y. Rohit, R.N Mangey, S. Aayushi, T. Fu-Ching, and J. Jwo-Huei, 2020. Approaches for Long Lifetime Organic Light Emitting Diodes. Advanced Science. DOI:10.1002/advs.202002254.

Solar Power Electronic Devices Blogs, 2016. Solar Energy technologies Office, Office of energy and renewable energy, energy.Gov, U.S. Department of Energy. www.energy.gov/eere/solar/solar-power-electronic-devices

T. Mohammad, E. Alan, and T. Loai, 2016. Studying the Energy Consumption in Mobile Devices. Proceeding of 13th International Conference on Mobile Systems and Pervasive Computing (MobiSPC 2016), Procedia Computer Science 94: 183–189.

Y. Jianting, 2013. Secrecy-based access control for device-to-device communication underlaying cellular networks. IEEE Communication Letter 17(11): 2068–2071.

Z. Alessio, S. Luca, B. Giacomo, J. Eduard, and D. Merouane, 2016. Energy-efficient power control: a look at 5G wireless technologies. IEEE Transaction Signal Processing 64(7): 1668–1683.

Z. Bastiaan C. J., H. R. Krikke, and J. Venselaar, 2010. Handling WEEE Waste Flows: On the Effectiveness of Producer Responsibility in a Globalizing World. International Journal of Advanced Manufacturing Technology 47(5–8): 415–436.

Z. Ding, S. M. Perlaza, I. Esnaola, and H. V. Poor, 2014. Power Allocation Strategies in Energy Harvesting Wireless Cooperative Networks. IEEE Transactions on Wireless Communications 13(2): 846–860.

Z. Mohamed El, Y. Ahmed, and I.G. Ahmed, 2020. Vehicle To Vehicle "V2V" Communication: Scope, Importance, Challenges, Research Directions and Future. The open transportation journal 14(1): 86–98.

10 Green Internet of Things (IoT) Technologies for Communication Systems

Puja Priya and Gurjit Kaur

CONTENTS

DOI: 10.1201/9781003264477-10

10.1 INTRODUCTION TO GREEN IoT

As per the present growth in digital technology, almost every section of society has been transformed. This has enhanced the greater use of smart and dedicated application devices that use the Internet of Things (IoT) and which has changed the way of living. IoT technology is the international network topology that contains several electronic gadgets such as smart wireless sensors, actuators, and machines, can be implanted as the electronics devices and can perform sensing of the signal, processing, and transmission of signal over the IoT network (Lee & Lee, 2015).

To develop the IoT infrastructure a huge number of sensors, various communications tools/protocols, and signal handling components are required. In IoT infrastructure around billions of sensing gadgets are connected and a large amount of information transmission takes place over this network. The information is processed using different types of processors (Singh et al., 2019). All the sensing devices and information processing components are fitted with a battery so that they can provide better performance. The communication devices and network devices are used to offer uninterrupted network connections for a better quality of service (QoS) for the IoT infrastructure. To manage high QoS, devices used in IoT infrastructure utilize more energy. It becomes critical to pay consideration to the energy competent devices that form an IoT infrastructure (Madakam et al., 2015).

The main focus of the present research on enhancing energy utilization in IoT infrastructures is based on energy utilization in wireless communication, designing, and signal processing. Thus, improving numerous prevailing methods (Parasher et al., 2018), which create an IoT infrastructure that is more energy competent by focusing on the devices at various levels of IoT infrastructure.

The main objective of this chapter is to bring the present work in the energy competent IoT infrastructure by investigating the previous works, recognizing their current stand, challenges, applications, future of IoT energy competent infrastructure toward green IoT (Albreem et al., 2018).

10.2 LAYERED INFRASTRUCTURE OF IoT

The layered infrastructure of IoT reflects the power greedy devices at each level. The main focus of the layered IoT infrastructure is to gain its presence and uniqueness (Burhan et al., 2018) by sensing, monitoring, transmitting, and processing huge information. It is a combination of hardware and software tools, communication procedures, and various management technologies. The present infrastructures differ with the type of usage, project, QoS factors, interoperability, etc.; this infrastructure also highlights the key power starving nodes, various protocols, middleware devices, and applications. The layered IoT infrastructure is classified into five layers: perception, transport, processing, network, and application as represented in Figure 10.1. Elements related to every layer are present in Figure 10.2. These elements can be sub-divided into power-starving and non-power-starving elements.

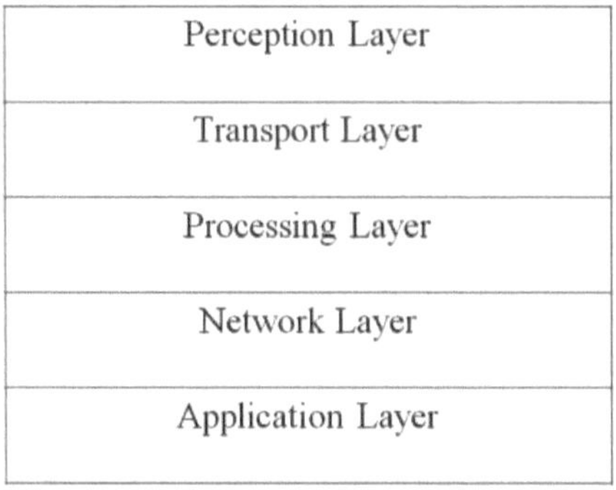

FIGURE 10.1 Layered infrastructure of IoT.

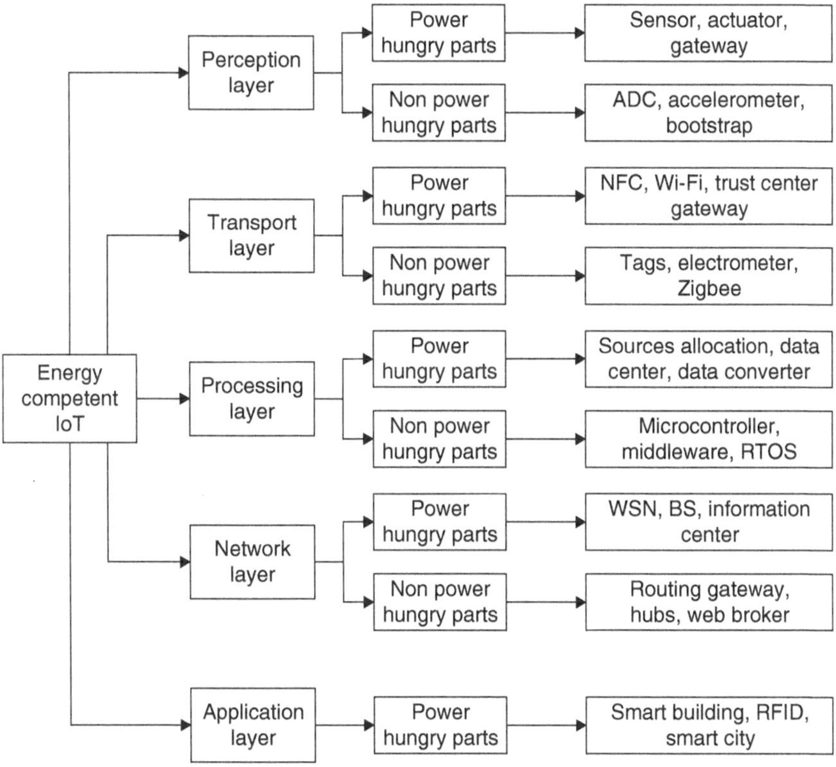

FIGURE 10.2 Organization of IoT factors.

10.2.1 Perception Layer

The use of the perception layer is to sense information, planning signal processing, and forming communication links to form a network using several other wireless sensor nodes. The main goals of perception layer are auto-arranged sensing and weight management (Bansal et al., 2021). The elements of this layer consist of sensors, RFID (Madakam et al., 2015), sink nodes, actuators, gateway nodes (Jedda et al., 2012), and software to resolve an issue (applications like bootstrap software) of which sink nodes, gateway nodes, actuators, and inactive RFID are power starving. There are various applications of sensors such as manufacturing, logistics, industry, ADC, accelerometers, vehicles, GPS (Michael et al., 2006), etc.

10.2.2 Transport Layer

The transport layer is utilized as a channel sandwiched between the perception and process layer, which helps in the allocation of information from the perception to the processing layer (Polese et al., 2019). Various elements in the transport layer consist of Bluetooth, RFID (Weinstein, 2005), WiFi (Gu et al., 2019), etc. Transport layer examples are readers, tags, electrometer, trust center, etc.

10.2.3 Process or Middleware Layer

In the processing layer, a huge quantity of information is examined, development, and accumulated. Process layer helps in the examination of the bottom layers, allotment of capital for competent storing space in practical and physical devices. This layer transforms the information into the needed form (Dagher et al., 2014). The various elements comprise a processing unit, middle wares, and information bases. Among these components, the processing unit utilizes most of the power.

10.2.4 Network Layer

The network layer ensures a connection link among various devices of another layer as a total structure model. Various elements of these layers are wireless sensor nodes (WSNs), cloud servers, and information gateway nodes. Among this WSNs (Yu et al., 2008), base stations (Ullah et al., 2019), information centers, and directing gateway nodes are the most power-starving elements (Salman & Jain, 2017).

10.2.5 Application Layer

This layer provides the service benefits of the subordinate layers to the users. It gives the imagination of the process information by performing intelligent tasks. There are various examples of this layer such as smart vehicles, health, smart home, smart cities, and so on. Controlling units in this layer are power-hungry devices (Karagiannis et al., 2015).

10.3 ENERGY OPTIMIZATION METHODS FOR GREEN IoT

Sensors, transmission, execution, and actuation are the main source of energy utilization among each layer and across all the layers. In the IoT infrastructure when devices interact with each other, they utilize high energy.

Based on several factors for which many of the energy competent methods are suggested are sleep off scheduling method and selective identification, information barrier lowering, lowering of information path distance, wireless communication, and implementation on tradeoff for communication, new communication methods, and allocation of mission proficient nodes and supplies on multi-core. Energy improvement methods are functional below various coatings of the IoT organization (Srivastava et al., 2020).

10.3.1 Methods for Perception and Processing Layers

Power-starving elements present beneath perception and processing layers, are sensor nodes and entry nodes that execute information detection, analysis, filtering, guesswork (intelligent task), and compactness along with implementations. The sensor's functionality is also reliant upon the battery life. Gateway or relay nodes function as a storing barrier, managing sensor services, and permitting information transportation. Gateway nodes measure sleep time duration of sensor nodes based on factors like battery life, contradictory elements, quality of service (QoS), the occurrence of an incident and alarm indications to manage the communication network. Diverse executing and supply selector that come beneath the processing layer can manage the entry nodes, select resources, and pre-implement the information. Thus, the parts beneath these layers have huge energy utilization issues.

The energy competent job scheduling algorithm is suggested here on the foundation of the sleep time duration of sensor nodes. The algorithm can also lower energy utilization of sensor nodes. The dynamic voltage and frequency scaling (DVFS) (Le Sueur & Heiser, 2010) method was also utilized along with the job scheduling algorithm to lower energy utilization of BIG LITTLE processor infrastructure. In the same way, it is suggested to employ a task scheduling method on heterogeneous cores' network by using time division multiple excess (TDMA). Selective and compressive sensing methods are likewise suggested for lowering energy utilization. Various other methods like the quality of service (QoS) depend on the sensors for job duty cycle allotment. Selective and compressive sensing methods also lower the energy utilization of sensor and gateway nodes. Small energy competent methods of processing layer allot server and workload sharing for information centers, various energy utilizing devices, assistance between the nodes at the same time of executing.

10.3.2 Methods beneath Transport and Network Layers

Transport and network layers manage sensing and networking features like directing the analysis of load, and information for communication devices. They also consist of skills like communication of information, local communication, re-communication, and evading of package collision and handshaking of devices. These communication devices utilize a large amount of energy. Further, network communication, locality evidence in routing, inter-functionality of heterogeneous items, and enormous files' association utilize a large amount of power. With the massive increase in consumers and product-based service providers, there is an expansion in network program, which as a consequence has huge energy utilization. In wireless sensor networks (WSN), sensor nodes and communication devices convey data for a big duration of time, subsequent in huge energy utilization.

Energy optimization methods comprise multi-hop and intelligent route in WSN to base stations (BS). All these methods track changes in documents as per hop-to-hop point thus lowering the energy utilized in a larger route of transmission. Two methods are used to enhance the communication path by making groups along with the network devices and executing layers. These route enhancement methods result in lowering energy utilization. An additional method with similar communication link between sensors and clipping of network demands utilizing pruned adaptive routing practice. Pruned adaptive routing delivers an enhanced path to an objective node thus actively dropping the energy utilization. There are limited optimization methods, which operate on hubs and changes to curtail route extent in files communication. System on chip (SoC) is used to handle network load and sinking energy utilization. Vigorous transferring of packages using contact points, narrow/bottomless sleep mode, check-pointing organization state mechanism, and data fusion are energy competent methods present in transport and network layers. Mixed integer linear programming (MILP) (Floudas & Lin, 2005) and Graph algorithms (van LEEUWEN, 1990) are used in scheduling the communication, which decreases energy utilization in the network.

10.3.3 Methods beneath the Application Layer

Application layer activates the procedure and shows yields are extremely battery-dependent. Every grid application, portable sensors, activators, controller like HVAC, consumes large calculation time since they execute the jobs like information organization, gridding, and manual handling, etc. Information development on the foundation of HTTP utilizes more power due to huge analyzing overhead. Such a large time of calculation and huge operation applications utilizes a large amount of energy at the application layer. There are several energy enhancement methods present at the application layer. Machine learning algorithms are used to forecast energy utilization by identifying several events in the application layer.

Diverse energy utilization methods mentioned in the application layer consist of MQTT (Akbar et al., 2018) (message queue telemetry transport) and CoAP (Chen, 2014) (constrained application protocol), CoAP can substitute HTTP, stalking tool, and MQTT-S/MQTT-SN. For IoT arrangements, an application of smart city utilizes methods like dispersed management (Kuorilehto et al., 2005) and fog computing (Yi et al., 2015) that generates an active and dispersed energy management outline. Also, methods like bulk transportation increases additional costs in terms of huge volume, sophisticated delivery competence, little energy utilization, and fewer budgets.

TABLE 10.1
Energy Saving Methods

Elements	Reasons for Energy Utilizations	Energy-Saving Techniques	Variables Improved
Sensors	Information regular sensing	Selective sensors, sleep scheduling (Haimour & Abu-Sharkh, 2019)	Sensor deployment, QoS, mobility, additional resources
RFID	Detection of sensed information	Passive sensor (Pereira et al., 2017)	Communication delay, performance
Sink nodes	Storage and monitoring	Tasks to core mapping (Birhanu et al., 2017)	Budget, communication delays, complexity
Datacenter	Processing	Distribution of load (Ferreira et al., 2016)	Complexity, QoS, accuracy
Gateways nodes	Communication connection among sensor nodes	Handle storing and scheduling events (Motlagh et al., 2020)	Quality of data, power handling, battery level of sensor node
Resource allocation	Allotment of information	Tuning and allocating sensing and processing elements (Li & Da Xu, 2020)	Implementation scheduling, effectiveness, output
Wireless sensor nodes	Communication and networking	Multiple hop routing (Salameh et al., 2020), cluster head communications (Lee et al., 2009)	Coverage, transmission power level, movement, network life duration
eNode/base station	Processing and information fetching	Optimal communications path routing between sensors and cloud sources, effective load balancing algorithm (Liu et al., 2016)	Load allotment, topology placement
Processor	Fast computing, allocation of a specific task	Fog computing and edge computing, dynamic tasks allocation and scheduling (Huu & Tham, 2013)	QoS cost, difficulty, resources, output
Actuator	Showing response	Estimate and following of various types of energy utilization (Bansal et al., 2021)	Security, pattern applications. privacy

10.4 CHALLENGES OF GREEN IoT

Presently a lot of studies on IoT networks are existing, scientists are still looking for recognizing the power-starving elements/features in IoT networks and are looking for optimization clarifications. A few of the researches are explained in this section. There are still present scientific disputes some are explained below:

1. Proposing energy competent executing components at the processing layer is needed to tackle as much of these executing components that utilize most

TABLE 10.2
Energy Competent Methods

Layer Approach	Issue Focused	Methods	Factors	Disadvantages
Perception (Kumar et al., 2020)	Sensing and data accumulations	Sleep on and off schedule, local edge computing, and intelligence service schemes	Latency and bandwidth	Regular sensing by the gateway, not efficient in real-time
Transport (Zemrane et al., 2018)	Layer communication	Zigbee, low power Bluetooth and IPv6	Connection time	Remote nodes and execution time security
Processing (Saqlain et al., 2019; Eghbali & Lighvan, 2021)	Control and processing	Resources, cooperative relaying, compressive sensing, data fusion, storage management	Execution scheduling, topology positioning, spectrum load distribution	Internal OS and CPU consumption enhances, heterogeneous duty scheduling enhances the difficulty
Network (Cotrim & Kleinschmidt, 2020; Pirozmand et al., 2020)	Communication	Multi-hop communication, game theory	Transmission reliability, network life span	Cluster placement, mobility, flexibility
Application (Sanaa et al., 2020)	Visualization and communication	MQTT, WCCFS	Location awareness, throughput	No disadvantages

of the energy in storage, organization, and executing of huge quantities of information.

2. Sensors used in the perception layer should be ended with energy competent features because of the huge power utilization required for the uninterrupted sensing device.
3. Regular sensing, management by the gateway, execution by keeping the information in focus and processors, network programmer, lengthy applications are major challenges in these gadgets as due to this, additional energy is utilized by these gadgets.
4. Similarly, vast quantities of features relate to these layers like a huge quantity of real-time information, interactivity, and smart procedures that are essential to make an energy competent organization.

10.5 GREEN MACHINE TO MACHINE TECHNOLOGY (M2M)

Green machine-to-machine communications technology (M2M) is an innovative technology that has attracted both academia and industry. M2M communication technology also shows new trends in industrial uses. A new study has proven that the estimated digit of linked machine-type communication devices in IoT infrastructure will be roughly 50 billion by 2020 (Alam, 2018). And these digits will exceed human-to-human (H2H) in the coming future. M2M communication is used to connect the computer with other electronic devices and store data as shown in Figure 10.3, where the machine represents an object that shows electrical, and mechanical, environmental, and electronic properties.

M2M devices contain a limited number of resources like battery and computation devices for maintaining long working time. To maintain energy utilization in green M2M applications it becomes very essential to rather increase output, as MTCDs are required to send a few pieces of information with a bounded energy. In the current scenario, various researches have been carried out to enhance the performance (Parasher et al., 2019) of energy utilization in M2M technologies. Information computation is also considered an additional essential problem in M2M communications technologies. Various resources limited MTCDs cannot depend only on their bounded reserves to execute their computational requirement. Conventionally, mobile cloud computing (MCC) has been largely accepted to permit suitable contact to an allocated pool of computing sources (Fernando et al., 2013). The space amid the cloud server and the MTCD are normally huge, cloud computing examination may not offer assurance to fewer latency functions and quick transmission of information from the MTCD to the cloud may not be that realistic or comes under budget. Further, for all MTCDs to form a direct connection with the cloud server has shown unrealistic and expensive results. To handle these problems, fog computing, which is also known as mobile edge computing (MEC) has been researched and placed executing sources

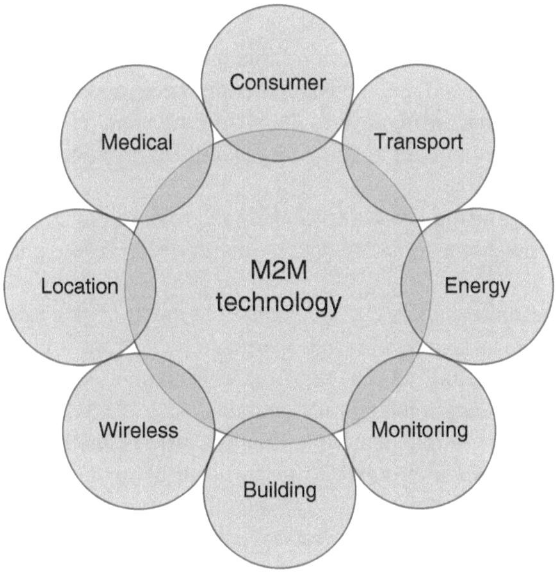

FIGURE 10.3 Showing M2M communication.

near to the last machines (Abbas et al., 2018). For high performance and low-cost utilization in M2M communication technology, MEC can permit MTCDs to operate computation relived off their computational job to the MEC server by using wireless cellular networks. Much work has been done on energy utilization and computation in M2M communication technologies, but both the parameters are discussed separately. For best performance in green M2M communication technology, it becomes essential to consider these two parameters jointly.

10.5.1 Key Elements and Technologies for M2M Communication Technologies

In this section key elements and technologies are presented for M2M communications such as MEC and WNV.

10.5.1.1 Mobile Edge Computing (MEC) in M2M Communications

MEC is a new growing technology; this technology is identified as a vital technology for future systems and also for both information technology (IT) and cloud computing-based servers at the system edge. The typical elements of MEC are listed below:

On-grounds: MEC technology can be executed in separation from another part of the system, and can only operate local sources. It is permitted to be essential for a few allocations like security and flexibility.

Closeness: MEC poses the capability to straightly perform the computing task this can be attracted by a few particular applications. This happens because of closeness to data resources.

Lesser computing: MEC servers are set near to confined customers, this could help in lowering the computation time, and also show many advantages like fast processing speed, good customer experience, and lower congestion.

Network framework information: Real-time network information utilized by various services and applications. By using MEC mobile service can distinguish and monetize.

Locality understanding: MEC servers help in providing local area information to influence low down signaling data to identify each linked gadget's position.

M2M communication technology is an essential element of IoT that mostly utilizes a huge amount of MTCDs and generates a large amount of information. This process needed a difficult computing process to extort information. As shown in Figure 10.4 IoT or M2M communication technology contains MEC. In M2M technology, information computing and implementation require various methods to perform the required task. It becomes impossible to perform all computations with low energy utilization and more efficiency especially when the computing task is intense (Kaur et al., 2019). Luckily computing operations can be implemented on either MEC servers or restricted gadgets for optimization methods.

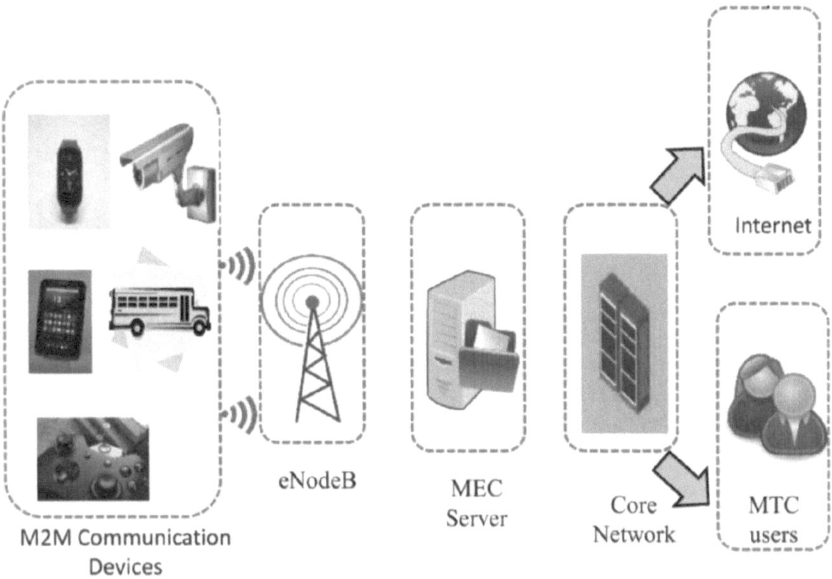

FIGURE 10.4 MEC in IoT and M2M communication.

10.5.1.2 Wireless Network Virtualization (WNV) and e-SIM in M2M Communications

For network infrastructure management and utilization of network resources, WNV is believed to be an efficient and reliable method. The hypervisor is an essential component in WNV, which is maintained by the main operating method, analyzing the network and associating main network sources to various QoS, examining applications or network needs. MNV can help in enhancing new technologies and shows more flexibility in designing the communication methods such as M2M communication technology. In current technology, various methods are used with M2M communications technology. Nevertheless, in realistic telecommunication networks, arbitrary MTCD grouping is undesirable as large and several MTCDs have distinguished QoS needs (Tomar et al., 2018).

In today's M2M communication technology, various methods are used widely. Though arbitrary MTCD cluster formation is unacceptable for practical implementation of the cellular network because large and several MTCD have the distinguished quality of service (QoS) needs. To execute WNV network M2M technology shows another issue. A new technology eSIM is welcomed in M2M communication. Figure 10.5 presents the selection step among virtual networks using eSIM and complete collection and surrender steps among virtual networks. All these steps are explained below.

(a) Suppose one particular MTCD has connected to virtual network 1, whose details are saved and allowed in its eSIM.

(b) With this, MTCD will connect to another virtual network 2, and share the need for network 2 is transmitted to the main subscription manager.
(c) After receiving the positive response as per the network connects' need, the subscription manager will send the outline details of virtual network 2 to the eSIM.
(d) Then the eSIM triggers the details of virtual network 2 and at the same time end virtual network 1.
(e) MTCD finally enables the alarm to the contribution manager that the whole button is ended during virtual network 2.

eSIM shows many benefits in MTCDs in two ways. Firstly, eliminating SIM cards from MTCD is a very difficult task, particularly if MTCD is placed in any distant site. It is infeasible to start MTCD to shift to a further virtual network through the real SIM. Also, as MTCDs are increasing in number, the typical SIM needs to be modified in various regions, which makes it hard to continue the service among networks. As compared to typical SIMs, eSIM can change virtual network paths more suitably when compared with SIM. Also, this technology will minimize the hurdles for the development of WNV in M2M technology.

10.5.1.3 Software Dependent Networks (SDN) in M2M Technology

A network with active and rapid designs can support both WNV and MEC in M2M technology; this will also make energy-efficient resources between virtual

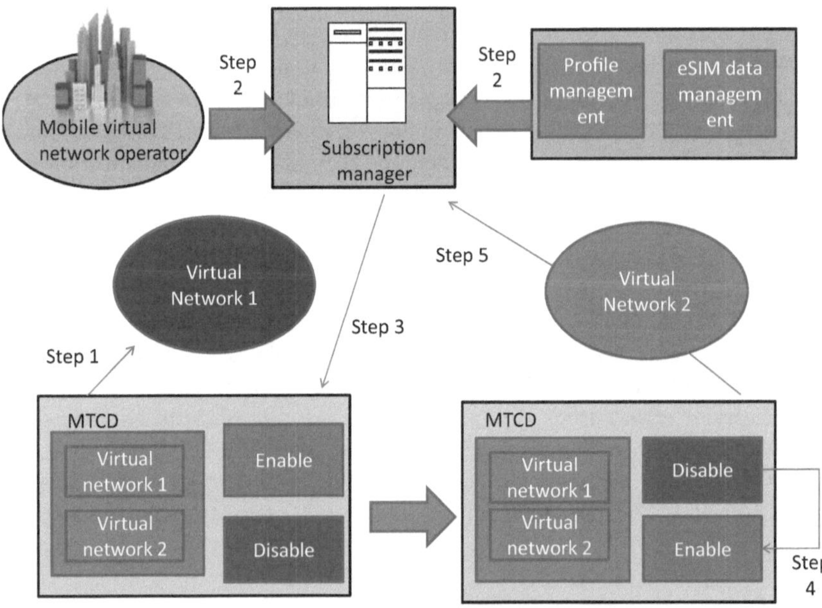

FIGURE 10.5 Switching procedure among virtual networks through via eSIM.

networks. Although the present network infrastructure and procedure were not made to compel with the new changes in technology. So, with a huge amount of various network parameters, procedures, and interactions it becomes infeasible for the network components to adjust and implement as per a large number of various procedures and standards. SDN can offer a capable solution to design MEC and WNV with M2M technology (Tomar et al., 2018). However, the whole network is managed by the software and does not depend upon the hardware for changes in methods or planning. As per the research, SDN manager could provide a whole view of network usages and breaks the perpendicular connections of the networks. Hence, SDN managers can be distinguished from information frameworks and can be centralized in SDN managers. As we compare the present M2M infrastructure, various network resources and procedures can be rapidly implemented, organized, and function.

10.6 NETWORK OUTLINE WITH GREEN M2M COMMUNICATION

In this part, how network outline design is implemented with MEC, WNV, and SDN are designed for green M2M technology is introduced. Also, implementation aspects of green M2M technology are presented.

10.6.1 Outline for Green M2M Technology

M2M network with MEC, WNV, and SDN is presented in Figure 10.6 showing a virtual and software M2M network.

Physical radio sources: There are whole M infrastructure providers (InPs) giving wireless approaches for proving services. For now, the entire quantity of MTCDs is N, and the mth InP positions the mth cellular system, which acquires Nm (1. Nm . N) MTCDs and one eNodeB.

Wireless network virtualization: As shown in Figure 10.6, the hypervisor will split and assign the virtual sources to g mobile virtual network operators (MVNOs). Also, every MVNO will provide associated services and applications for its subscribers.

Virtual network (VN) source: As presented in Figure 10.7 the gth VN contains a virtual eNodeB with a virtual MEC server and Ng (1. Ng. N) MTCDs. While system sources can be kept and resources block (RB) overcrowding can be neglected by collection supported transmission. A collection-supported method is used to manage huge admission with quality of service (QoS) guarantee promise. In the represented outline, a single MTCD has the highest arithmetic mean off-channel benefit over additional MTCDs, which are designated as the planner in the collection. The highest procedure is believed with the same sized time enclosures and every time enclosure comprises of K time slots. Every MTCD will identify the RB and plan to admission the eNodeB or the planner at the starting off every time slot. When MTCD is in use it shows different communication capacities Cnx,r(k). Cncoor,r´(k) represents the present communication rate in the connection between the planner and the virtual eNodeB.

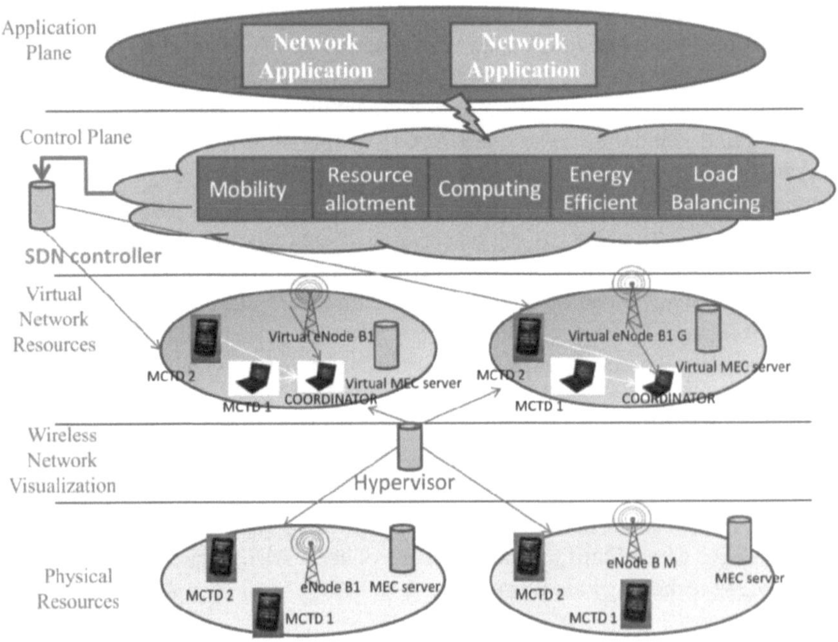

FIGURE 10.6 An infrastructure of virtualized and software-defined M2M communication system with MEC server.

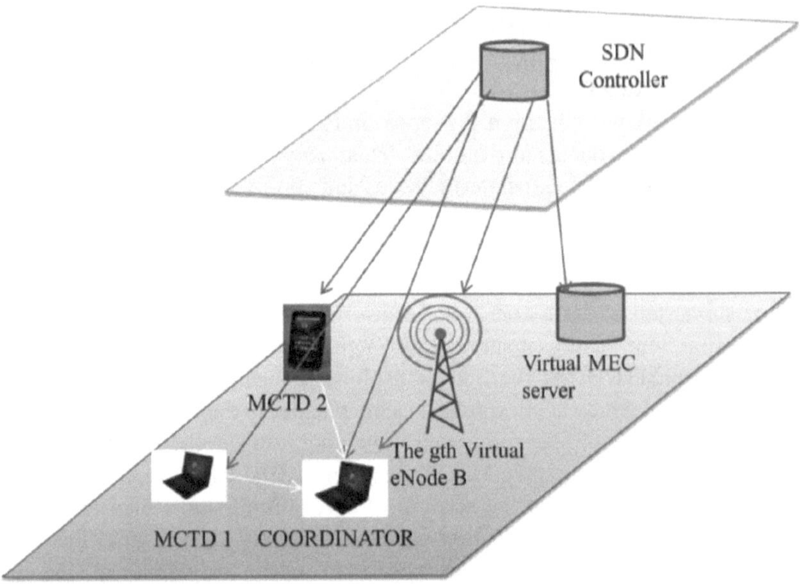

FIGURE 10.7 gth virtual network.

Control mode: Using a SDN controller, various plans and standards related to MEC and MNV can be incorporated. This best admission method as well as the good calculating steps will be chosen by the SDn controller. This will enhance the performance and probability.

Application mode: Mainly it contains unlike network functions in M2M technology networks. The focus of this network is to design and implement those network functions.

10.7 OPERATION FACTS OF GREEN M2M TECHNOLOGY

The calculating method and energy utilization method are shown below.

Computing method: The ability to tackle the computing jobs present in MTCDs, planner, and MEC servers. Hence, previously the MTCD enters the network, requires choosing where to manage the calculating jobs. Suppose the nth MTCD wants to perform computing jobs, which are represented as Inx (anx, bnx). Here anx means the size of input information used and bnx means the sum of CPU cycles needed for computing jobs.

Confined computing: The implementation time determines the computing ability (CPU cycles per second) of the nth MTCD. And this is shown as tln x(k). However, if the MTCD selects to enter the planner, computing jobs will be implemented on the planner. So, during this case, computing jobs devolved firstly and after this are forwarded to the planner via wireless connections. After devolving the planner will implement the computing jobs for benefit of the nth MTCD. The whole implementation time comes out to be tl′nx (k).

Mobile frame computing: This computing is dissimilar from confined computing. In this, the computing job requires to be devolving to the MEC server. Specifically, the MEC server will separate a total computing job into two. First computing jobs devolve and second input information transmission. After doing this information devolving the MEC server will implement computing jobs. The overhead of the MEC method is presented as tcnx(k) in terms of execution time.

Energy utilization method: As MTCDs are energized by batteries with a bounded energy. Energy utilization cannot be overlooked. For long MCTD operation life, regarding energy efficiency is essential by two factors: first is the energy utilization in execution time and the second is by computing jobs. Energy utilization in the execution process requires one MTCD that tries to execute system requirements to identify the RB initially. After this, it inhabited the RB to transfer signals if they are offered. Hence energy utilization in the execution step should be classified into two parts and they are RB execution sensing and data communication. However, in typical networks, the energy utilization consumption for RB sensing is mostly ignored but in this, it is considered. Also, for information computing jobs energy utilization has three distinguished factors, which are computing in the confined MTCD or on the planner or devolving to the MEC server. For local components' calculating jobs, energy utilization is given as Eln x(k). In the case of computing jobs' implementation on the planner, input information requirements are devolved and transfer is shown as El´nx(k). Also, for computing jobs worked on the MEC server, energy utilization consumed for devolving and sending input information is given as Enx(k).

10.8 ENERGY AWARENESS IN IoT

Energy awareness in IoT is a way of communicating the ideas or developing more awareness among the customers and encourages them towards energy savings through organizing campaigns. Energy awareness planning involves correct planning, a set of tasks that will help in creating energy awareness. By using sensor devices, one can record and demonstrate awareness and guide and forecast future energy requirements. This will help in creating an understandable image among the observers and encourage them to preserve energy and lower its consumption.

Energy awareness at the school level is one of the greatest platforms to generate awareness in children. By using posters, videos, games, plays, teachers can be aware of students regarding the idea of green IoT (Garlisi, 2019; Amaxilatis et al., 2017). For this, the great steps involved are planning, accumulating real-time information, creating awareness material by creatively using the information so that students can grab the theme easily, distributing books or templates that have proper instruction about the green IoT, and so on.

Industries or any business will show long-term advantages by organizing energy awareness programs. Green IoT awareness programs can help in generating an energy-saving environment in offices. Awareness can be planned as per the specific form of business and their kinds of energy utilization. A few of the methods involved are guiding employees to accept an energy-saving environment, cheering employees to reorganize the parts where they can minimize energy consumption, putting encouraging slogans on the notice boards, walls, canteen area, etc of the office. By using emails, meetings, office newsletters, offices can encourage employees to be energy aware. Awareness programs depend on the nature of the business. Also, by collecting feedback, further improvements to energy awareness programs can be made in the future.

Preparation for the energy awareness program is a very essential step. The right preparation is essential for a successful energy awareness program. The next step involves selecting the necessary recipient that will show a wide impact on the program. The next step involves collecting real-time information to support the program. The information may contain present energy consumption examples, entire energy budget, estimated energy utilization, etc., after information collection the next step is to make the program objects. Program objects should be creative, easy to understand, show the positive impact of green IoT, and compare it with the current scenario. Posters, videos, slogans, messages, and demos are the best way to convey the benefits of the green IoT. When the objects are ready, next is to identify the best way to deliver the message. After all the above steps, the final part is to carry out the awareness program. Awareness programs should be arranged in a way that conveys a message with the greatest awareness. After, the program's feedback form is a must for enhancement of future awareness plans. Figure 10.8 shows the energy awareness programs steps below.

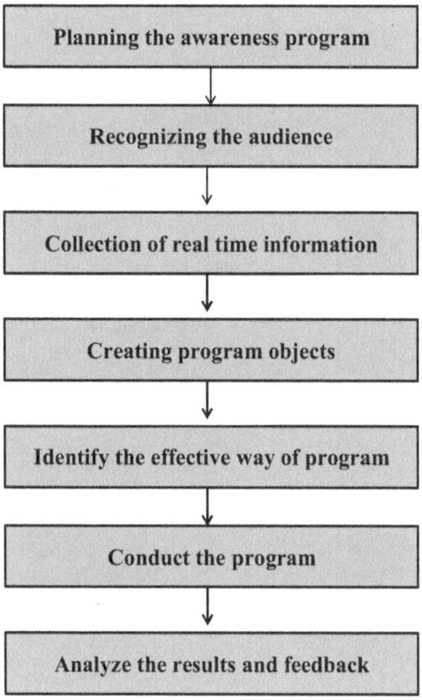

FIGURE 10.8 Energy awareness programs steps.

10.9 APPLICATIONS OF GREEN IoT

The tremendous changes that happen in our environment have brought the development of IoT technology. Even though due to over generation of e-waste, harmful emissions and unnecessary energy utilization have enhanced the developmental cost. Green IoT is predicted to provide sustainable development for the future and can help us to maintain a green environment. In the coming days, we will see many devices, gadgets, smart sensors, and many more that will operate and form a communication link among each other to gain their work smartly and intelligently for a green environment. Hence, green IoT applications show more interest in energy saving, lower CO_2 emission, and lowering pollution levels. Green IoT will help industries to lower the greenhouse gas effect and help the environment. Adopting green IoT can help IoT in analyzing various energy resources, which are eco-friendly and lower the harmful effects of IoT on the environment (Tomar et al., 2017) (Kaur, Tomar, & Singh, 2018). Therefore, green IoT is having several applications, which are environment friendly, economical, significant, sustainable, and provide benefit to human health as shown in Figure 10.9. A few are explained below.

(a) **Smart Healthcare:** In these various smart sensors or actuators are used to monitor, analyze, and capture patient health on a real-time basis. These sensors remain connected to the internet for the collection and processing of information. This application is an evolution in healthcare infrastructure

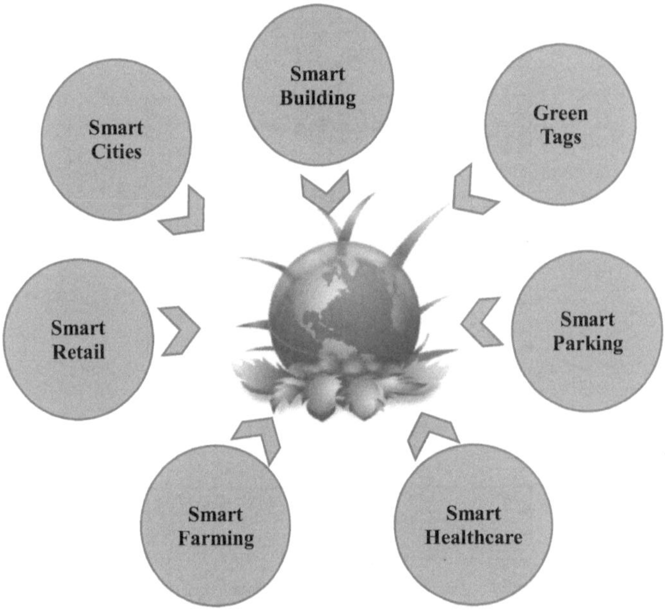

FIGURE 10.9 Green IoT applications.

and helps the doctors to monitor the patient remotely, lower the cost of treatment, improve the service, and enhance accessibility (Chanak & Banerjee, 2020).

(b) **Smart Home:** Green IoT allows home devices like heating, lighting, and electronics machines to be handled distantly by a laptop/computer or smart phone. The mobile or computer listens to voice instructions to perform their task at home. These instructions can be differentiated among occupants for personal talk and reactions. Smart home TV, computer, phone, and other smart devices can combine into a single device. For green IoT, there are a few new things to be considered like green design, green operation, green manufacturing, and green reuse will lower harm on the environment (Mokhtari et al., 2017).

(c) **Manufacturing Mechanization:** Now the manufacturing sector has been mechanized with smart machines these can automatically do work with very little human intervention or no human intervention using the internet. This mechanization is based on green IoT technology (Krasniqi & Hajrizi, 2016).

(d) **Smart Network:** An efficient smart network is all about equality. It means a network that can change, correct, or recorrect to transmit efficient, good quality, and economical energy optimally. A smart network can provide customers the capability to contribute to the resolution. The future smart network will utilize economical remote memory, less energy consumption, reliable and secure network for green IoT-based technology (Li et al., 2018) (Bekara, 2014).

(e) **Smart City:** A new trend in developing more energy-efficient green IoT-based technology. A smart city is the most hopeful application of green IoT solutions. Using IoT in smart city applications can contribute towards a sustainable smart world. So, for smart cities, devices should contain smart sensors so that they can communicate with each other wirelessly and smartly. Smart devices in smart cities can sense the objects around them and process this sensed information and communicate with each other (Prasad et al., 2005) (Kaur et al., 2018). These can be used in smart city applications such as life smart traffic organization, smart automobiles, smart buildings, and smart parking, and so on. This will enhance the quality of life and improve unhealthy environmental conditions (Bansal et al., 2021; Dagher et al., 2014).

(f) **Smart Farming:** This will help farmers to compete with various issues they face during farming. This will resolve and help find methods to lower the water shortage, cost management, and land utilization. Green IoT in smart farming will help to provide smart and sustainable farming and food production (Prathibha et al., 2017).

(g) **Green WSN:** This network depends on the green energy conservation method, radio optimization method, and green routing methods that help in lowering energy utilization. WSN is made up of lower needed sensors that consume restricted power and have limited storage ability. Additionally, lowering storage ability, lower information capacity is performed by smart information algorithms and this will trigger sensors when needed.

(h) **Smart Environment:** As the number of vehicles is increasing, more migration of people and various other modern practices have knowingly harmed the air quality (AQ) in recent years. Analyzing the damage air quality index is one of the very difficult and complex tasks. Using IoT technology will help in improving the air quality index and lower the climate change effects. For lowering of natural outcomes like waste removal, several associations and municipal organizations come together with a target to upgrade operational waste management outlines. Using RFID sensors tags on dustbins makes the management of waste smarter. This will help farmers to handle the complexity they face.

10.10 FUTURE OF GREEN IoT

For the future development in green IoT, the following research needs to be carried out for the best and competent results.

(a) More development and research on unmanned aerial vehicles (UAV) to substitute the large quantity of IoT devices mostly in large area analysis or monitoring like agriculture, traffic, military, logistic, etc. This will minimize the number of devices, reduce power utilization, lower network complexity, and lower pollution. UAV is future competent technology and can help in the implementation of green IoT with a low budget and more efficiency.

(b) Integrating sensor cloud technology in the wireless sensor network and mobile cloud computing application is the new research area. This will enable green IoT development. This will help in a more energy-efficient network and provide services in WSN and mobile cloud management.
(c) Machine computerizations can have a very critical role in green IoT. M2M technology can minimize energy utilization and harmful emissions. This can be used in an emergency like traffic control or to take faster action when accidents occur.
(d) More research is required towards IoT-based device designs for lower CO_2 discharge and energy utilization. This will enable a greener environment.
(e) For more energy efficiency for green IoT technology, radiofrequency device's energy utilization should be given importance. As they are prime blocks in IoT systems.
(f) More research is required on improving QoS by focusing on more constraints like bandwidth, output, delay, etc. This will help in developing more efficient and effective green IoT.
(g) For sustainable development of a green environment, more research is required for less energy utilization IoT networks, developing more energy-efficient resources, lowering the IoT harmful effect on human health, and reducing the environmental impact.

10.11 CONCLUSION

Green IoT is more sustainable and it can be utilized for better future growth. It can allow our future environment to be more sustainable, low polluting, healthier, economical, and can provide enhanced quality of services. Green IoT can be used for green networks, green communication, green design, green applications, green sensor networks, RFIDs, smart cities, smart buildings, smart industries, smart healthcare, green supply chain management, plastic recycling, water management, smart parking, and so on. This chapter has focused on the present work and possible technologies for future green IoT.

ACKNOWLEDGEMENT

This work is supported by Delhi Technological University, Department of Electronics and Communications, New Delhi-110042

REFERENCES

Abbas, N., Zhang, Y., Taherkordi, A., & Skeie, T. (2018). Mobile Edge Computing: A Survey.IEEE Internet of Things Journal, 5(1), 450–465. https://doi.org/10.1109/JIOT.2017.2750180

Akbar, S. R., Amron, K., Mulya, H., & Hanifah, S. (2018). Message queue telemetry transport protocols implementation for wireless sensor networks communication – A performance review. Proceedings – 2017 International Conference on Sustainable Information Engineering and Technology, SIET 2017, 2018-Janua, 107–112. https://doi.org/10.1109/SIET.2017.8304118

Alam, T. (2018). A Reliable Communication Framework and Its Use in Internet of Things (IoT). International Journal of Scientific Research in Computer Science, Engineering and Information Technology, 5(May). https://doi.org/10.36227/techrxiv.12657158

Albreem, M. A. M., El-Saleh, A. A., Isa, M., Salah, W., Jusoh, M., Azizan, M. M., & Ali, A. (2018). Green internet of things (IoT): An overview. 2017 IEEE International Conference on Smart Instrumentation, Measurement and Applications, ICSIMA 2017, 2017-Novem(November), 1–6. https://doi.org/10.1109/ICSIMA.2017.8312021

Bansal, M., Kumar, A., & Virmani, A. (2020). Green IoT: Current Scenario & Future Prospects. Journal of Trends in Computer Science and Smart Technology, 2(4), 173–180. https://doi.org/10.36548/jtcsst.2020.4.001

Bekara, C. (2014). Security issues and challenges for the IoT-based smart grid. Procedia Computer Science, 34, 532–537. https://doi.org/10.1016/j.procs.2014.07.064

Birhanu, T. M., Li, Z., Sekiya, H., Komuro, N., & Choi, Y. J. (2017). Efficient Thread Mapping for Heterogeneous Multicore IoT Systems. Mobile Information Systems. https://doi.org/10.1155/2017/3021565

Burhan, M., Rehman, R. A., Khan, B., & Kim, B. S. (2018). IoT elements, layered architectures and security issues: A comprehensive survey. Sensors (Switzerland), 18(9), 1–37. https://doi.org/10.3390/s18092796

Chanak, P. & Banerjee, I. (2020). Congestion Free Routing Mechanism for IoT-Enabled Wireless Sensor Networks for Smart Healthcare Applications. IEEE Transactions on Consumer Electronics, 66(3), 223–232. https://doi.org/10.1109/TCE.2020.2987433

Chen, X. (2014). Constrained Application Protocol for Internet of Things Abstract: Table of Contents: Wireless and Mobile Networking, 857, 1–12. www.cse.wustl.edu/~jain/cse 574-14/index.html Back

Cotrim, J. R., & Kleinschmidt, J. H. (2020). LoRaWAN mesh networks: A review and classification of multihop communication. Sensors, 20(15), 4273.

Dagher, R., Mitton, N., & Amadou, I. (2014). Towards WSN-Aided navigation for vehicles in smart cities: An application case study. 2014 IEEE International Conference on Pervasive Computing and Communication Workshops, PERCOM WORKSHOPS 2014, 129–134. https://doi.org/10.1109/PerComW.2014.6815178

Eghbali, Z., & Lighvan, M. Z. (2021). A hierarchical approach for accelerating IoT data management process based on SDN principles. Journal of Network and Computer Applications, 181, 103027.

Fernando, N., Loke, S. W., & Rahayu, W. (2013). Mobile cloud computing: A survey. Future Generation Computer Systems, 29(1), 84–106. https://doi.org/10.1016/j.future.2012.05.023

Ferreira, J., Dantas, J., Araujo, J., Mendonca, D., Maciel, P., & Callou, G. (2016). An Algorithm to Optimize Electrical Flows of Private Cloud Infrastructures. Proceedings – 2015 IEEE International Conference on Systems, Man, and Cybernetics, SMC 2015, 771–776. https://doi.org/10.1109/SMC.2015.144

Floudas, C. A. & Lin, X. (2005). Mixed integer linear programming in process scheduling: Modeling, algorithms, and applications. Annals of Operations Research, 139(1), 131–162. https://doi.org/10.1007/s10479-005-3446-x

Gu, Y., Zhang, X., Liu, Z., & Ren, F. (2019). WiFi-based real-time breathing and heart rate monitoring during sleep. 2019 IEEE Global Communications Conference, GLOBECOM 2019 – Proceedings. https://doi.org/10.1109/GLOBECOM38437.2019.9014297

Haimour, J. & Abu-Sharkh, O. (2019). Energy efficient sleep/wake-up techniques for IOT: A survey. 2019 IEEE Jordan International Joint Conference on Electrical Engineering andInformation Technology, JEEIT 2019-Proceedings, 459–464. https://doi.org/10.1109/JEEIT.2019.8717372

Huu, T. T. & Tham, C. K. (2013). An auction-based resource allocation model for green cloud computing. Proceedings of the IEEE International Conference on Cloud Engineering, IC2E 2013, 269–278. https://doi.org/10.1109/IC2E.2013.21

Jedda, A., Khair, M., & Mouftah, H. T. (2012). Connected coverage for RFID and wireless sensor networks. Procedia Computer Science, 10, 1046–1051. https://doi.org/10.1016/j.procs.2012.06.146.

Karagiannis, V., Chatzimisios, P., Vazquez-Gallego, F., & Alonso-Zarate, J. (2015). A Survey on Application Layer Protocols for the Internet of Things. Transaction on IoT and Cloud Computing, 3(1), 11–17. http://icas-pub.org/ojs/index.php/ticc/article/view/47

Kaur, G. & Tomar, P. (2018). "Handbook of Research on Big Data and the IoT, IGI-Global, International Publisher of Progressive Information Science and Technology Research USA. Total Pages, 568, ISBN13: 9781522574323|ISBN10: 1522574328|DOI: 10.4018/978-1-5225-743, EISBN13: 9781522574330

Kaur, G., Tomar, P., & Singh, P. (2018). Design of cloud-based green IoT architecture for smart cities. In Internet of Things and Big Data Analytics Toward Next-Generation Intelligence (pp. 315–333). Springer, Cham.

Kaur, G., Tomar, P., & Tanque, M. (2019). "Integrating Artificial Intelligence to Solve Pervasive IoT Issues, Elsevier, Cambridge, MA, USA.

Krasniqi, X. & Hajrizi, E. (2016). Use of IoT Technology to Drive the Automotive Industry from Connected to Full Autonomous Vehicles. IFAC-PapersOnLine, 49(29), 269–274. https://doi.org/10.1016/j.ifacol.2016.11.078

Kumar, A., Saha, R., Alazab, M., & Kumar, G. (2020). A lightweight signcryption method for perception layer in Internet-of-Things. Journal of Information Security and Applications, 55, 102662.

Kuorilehto, M., MarkoHannikainen, M., & Hamalainen, T. D. (2005). A Survey of Application Distribution. EURASIP Journal on Wireless Communications and Networking, 5, 774–788.

Le Sueur, E. & Heiser, G. (2010). Dynamic voltage and frequency scaling: The laws of diminishing returns. HotPower 2010–2010 Workshop on Power Aware Computing and Systems.

Lee, I. & Lee, K. (2015). The Internet of Things (IoT): Applications, investments, and challenges for enterprises. Business Horizons, 58(4), 431–440. https://doi.org/10.1016/j.bushor.2015.03.008

Lee, S. H., Lee, S., Song, H., & Lee, H. S. (2009). Wireless sensor network design for tactical military applications: Remote large-scale environments. Proceedings – IEEE Military Communications Conference MILCOM. https://doi.org/10.1109/MILCOM.2009.5379900

Li, Y., Cheng, X., Cao, Y., Wang, D., & Yang, L. (2018). Smart choice for the smart grid: Narrowband internet of things (NB-IoT). IEEE Internet of Things Journal, 5(3), 1505–1515. https://doi.org/10.1109/JIOT.2017.2781251

Li, X., & Da Xu, L. (2020). A review of Internet of Things—resource allocation. IEEE Internet of Things Journal, 8(11), 8657–8666.

Liu, Y., Shu, W., & Zhang, C. (2016). A parallel task scheduling optimization algorithm based on clonal operator in green cloud computing. Journal of

Madakam, S., Ramaswamy, R., & Tripathi, S. (2015). Internet of Things (IoT): A Literature Review. Journal of Computer and Communications, 03(05), 164–173. https://doi.org/10.4236/jcc.2015.35021

Michael, K., McNamee, A., & Michael, M. G. (2006). The emerging ethics of humancentric GPS tracking and monitoring. International Conference on Mobile Business, ICMB 2006, 1–10. https://doi.org/10.1109/ICMB.2006.43

Mokhtari, G., Zhang, Q., Hargrave, C., & Ralston, J. C. (2017). Non-Wearable UWB Sensor for Human Identification in Smart Home. IEEE Sensors Journal, 17(11), 3332–3340. https://doi.org/10.1109/JSEN.2017.2694555

Motlagh, N. H., Mohammadrezaei, M., Hunt, J., & Zakeri, B. (2020). Internet of things (IoT) and the energy sector. Energies, 13(2), 1–27. https://doi.org/10.3390/en13020494

Parasher, Y., Kedia, D., & Singh, P. (2018). Examining Current Standards for Cloud Computing and IoT. In Examining Cloud Computing Technologies Through the Internet of Things (pp. 116–124). IGI Global.

Parasher, Y., Singh, P., & Kaur, G. (December 5, 2019). Green Smart Town Planning. Green and Smart Technologies for Smart Cities, CRC Press, 19–41 379, ISBN 9781138318090.

Pereira, F., Correia, R., & Carvalho, N. B. (2017). Passive sensors for long duration internet of things networks. Sensors (Switzerland), 17(10). https://doi.org/10.3390/s17102268

Pirozmand, P., Ghafary, M. A., Siadat, S., & Ren, J. (2020). Intrusion detection into cloud-fog-based iot networks using game theory. Wireless Communications and Mobile Computing, 2020.

Prasad, R. S. & Kaur, G. (2005). "Optical Communication", Galgotia Publications, ISBN: 9788175154858, 8175154853

Prathibha, S. R., Hongal, A., & Jyothi, M. P. (2017). IOT Based Monitoring System in Smart Agriculture. Proceedings – 2017 International Conference on Recent Advances in Electronics and Communication Technology, ICRAECT 2017, 81–84. https://doi.org/10.1109/ICRAECT.2017.52

Salameh, H. B., Otoum, S., Aloqaily, M., Derbas, R., Ridhawi, I. Al, & Jararweh, Y. (2020). Intelligent jamming-aware routing in multi-hop IoT-based opportunistic cognitive radio networks. Ad Hoc Networks, 98, 102035. https://doi.org/10.1016/j.adhoc.2019.102035

Salman, T. & Jain, R. (2017). Networking protocols and standards for internet of things. Internet of Things and Data Analytics Handbook, 215–238. https://doi.org/10.1002/9781119173601.ch13

Sanaa, E. L., Bajit, A., Barodi, A., Chaoui, H., & Tamtaoui, A. (2020, December). An optimized security vehicular Internet of Things-IoT-application layer protocols MQTT and COAP based on cryptographic elliptic-curve. In 2020 IEEE 2nd International Conference on Electronics, Control, Optimization and Computer Science (ICECOCS) (pp. 1–6). IEEE.

Saqlain, M., Piao, M., Shim, Y., & Lee, J. Y. (2019). Framework of an IoT-based industrial data management for smart manufacturing. Journal of Sensor and Actuator Networks, 8(2), 25.

Singh, P., Dixit, V., & Kaur, J. (2019). Green Healthcare for Smart Cities. Green and Smart Technologies for Smart Cities, 91–130.

Srivastava A., Gupta M. S., & Kaur G. (2020). Energy efficient transmission trends towards future green cognitive radio networks (5G): Progress, taxonomy and open challenges. Journal of Network and Computer Applications, Elsevier, 168, 102760.

Tomar, P. & Kaur, G. (2017). Examining Cloud Computing Technologies through the Internet of Things, IGI-Global, International Publisher of Progressive Information Science and Technology Research Books and Journals, USA, DOI: 10.4018/978-1-5225-3445-7, ISBN13: 9781522534457, Pages 311.

Tomar, P. & Kaur, G. (2018). "Green and Smart Technologies for Smart Cities, CRC Press, Taylor & Francis Group, Dec 2019, Boca Raton: CRC Press, Ist Edition, ISBN 9781138318090 – CAT# K391489, https://doi.org/10.1201/9780429454837.

Tomar, P. & Kaur, G. (2019). Artificial Intelligence and IoT-Based Technologies for Sustainable Farming and Smart Agriculture, IGI-Global, International Publisher of Progressive Information Science and Technology Research Books and Journals, USA.

Ullah, H., Gopalakrishnan Nair, N., Moore, A., Nugent, C., Muschamp, P., & Cuevas, M. (2019). 5G communication: An overview of vehicle-to-everything, drones, and healthcare use-cases. IEEE Access, 7(c), 37251–37268. https://doi.org/10.1109/ACCESS.2019.2905347

van Leeuwen, J. (1990). Graph Algorithms. In Algorithms and Complexity. ELSEVIER B.V. https://doi.org/10.1016/b978-0-444-88071-0.50015-1

Weinstein, R. (2005). RFID: A technical overview and its application to the enterprise. IT Professional, 7(3), 27–33. https://doi.org/10.1109/MITP.2005.69

Yi, S., Li, C., & Li, Q. (2015). A survey of fog computing: Concepts, applications and issues. Proceedings of the International Symposium on Mobile Ad Hoc Networking and Computing (MobiHoc), 2015-June, 37–42. https://doi.org/10.1145/2757384.2757397

Yu, T., Bo, G., & Ling, T. (2008). Designand application of wireless sensor networks for ground verification ofremote sensing. 2008 International Conference on Apperceiving Computing and Intelligence Analysis, ICACIA 2008, 381–384. https://doi.org/10.1109/ICACIA.2008.4770048

Zemrane, H., Baddi, Y., & Hasbi, A. (2018, October). Comparison between IOT protocols: ZigBee and WiFi using the OPNET simulator. In Proceedings of the 12th International Conference on Intelligent Systems: Theories and Applications (pp. 1–6).

11 Green Cloud Computing Based Future Communication Systems

Puja Priya and Gurjit Kaur

CONTENTS

11.1 INTRODUCTION

With the growing demand for "cloud computing" in various fields, it has become essential to initiate the "green computing" concept for lower power consumption, minimization of carbon dioxide emissions, and re-cycling of power and energy for more efficient and valuable usage. Energy over consumption has introduced a new challenge as it leads to the addition of more greenhouse gases and increases the present-day global temperature. So, it has become essential to move towards "green technologies" (Garg et al., 2012). As the internet has become a part of daily life, cloud computing has appeared as the main technology that could work wirelessly or in wired systems too. To manage sustainability in the technical world, it has become essential in the last few decades to lower the extreme usage of energy resources. Therefore, researchers are now showing interest in the green and sustainable development of information and communication technologies (ICTs), which will help in

DOI: 10.1201/9781003264477-11

the development of a sustainable environment by eliminating all the negative effects that have taken place in the last few decades. The government is now also putting stress on manufacturers to minimize negative impacts by following environmental law and regulations. In terms of ICTs, the green parameters are characterized as green ICT's ecological reliability, environmental reliability, reasonable computing, and green computing. As studied, the relationship between sustainability and ICTs should visualize the opposite effects of ICTs on the environment based on current and future scenarios. But the development of new and appealing technologies has made the industries and customers ignore the environmental problems (Conley & Greene, 2006). Due to the present environmental changes, and compulsion from international environmental organizations, ICTs are now working with environmental law and regulations (Singh et al., 2019). This has generated a focus towards monitoring and protection of the ecosystem too (Parasher et al., 2018). The implementation and design of various methods for ICTs still poses some limitations, like budget, less time and focus towards overcoming challenges, profit-making industries, etc.

Cloud computing is the sub-branch of ICTs. Many issues have been created against these technologies. There is a significant push from government departments to minimize its negative impact on the environment.

The origin of a green cloud computing system is highly linked to the "green" data centers as these data centers play a key role in cloud computing technology. As per the survey, the energy usage by the data centers is 1% of the year 2018 (GeSI, 2020). As per the GeSI report, carbon dioxide discharge has increased from 1.3% in 2010 to 2.3% in 2020 due to ICTs (Data center knowledge, 2020). If data centers are created on the green computing concepts it will help to protect the environment. The main objective of the chapter is to introduce present data usage on the green cloud computing concept and to identify the major challenges and their remedies. This chapter will introduce readers to an overview of cloud computing, new research areas on a green cloud computing system, the future progress of green computing, and issues related to its implementation (Srivastava et al., 2020).

Cloud computing has appeared as a versatile technology over the last few decades because of its capabilities for handling high data, complex applications, and the storage of large amounts of data, without even the need for additional hardware (Atrey et al., 2013). It contains cloud-based data centers that can analyze the data using technology and can adjust themselves as per the computer environment. Cloud computing ideology has gained interest from various organizations mainly because it offers less investment in hardware and software installation. Cloud computing is a mature concept that has emerged from service-oriented architecture (SOA), micro-service architecture, parallel computing, grid and allocated computing, virtualization, and containerization (Parasher et al., 2019). Some ideas like parallel computing, distributed computing, and virtualization are also very old, whereas ideas like SOA, micro-service architecture, grid computing, and containerization are new ideas. Due to more research and development, cloud computing technology is evolving itself from a hardware and software viewpoint. The following standards are the new ICT solutions' model provided by cloud computing; these are service-centric, universal, expandable, expenditure-based, and self-service. Cloud computing is identified by its standards.

Clouds in cloud computing technology are virtualized data centers and the applications provided by this depend on the service subscription by the user as shown in Figure 11.1. Clouds need a high amount of energy to work, for example, in the present data centers with approx. 1000 racks, 10 Megawatt of power is required for it to work. This results in large working costs. Therefore, for a data center, power cost is a very important parameter for its overall working costs (Parasher et al., 2019). As per the European Union report, it has become essential to reduce carbon dioxide discharge from 15% to 30% by 2020, to maintain a global world temperature increase below 2 degrees. This makes energy usage and carbon discharge by cloud systems a major environmental issue (Radu, 2017).

The various literature presents that cloud computing shows a real shift in conventional data centers into highly energy-efficient technologies by utilizing concepts like virtualization and load consolidation. In most of the mobile applications high traffic of users leads to less use of resources and high energy waste of data centers (Tomar et al., 2018). Cloud data centers can minimize the energy wastage through server merging and sharing of a similar physical host of non-identical loads using the virtualization method while keeping unused off. Now various researches are coming to light with proof that a cloud can minimize the carbon footprint of organizations. Using a cloud infrastructure, small organizations can reduce carbon discharge by 90%

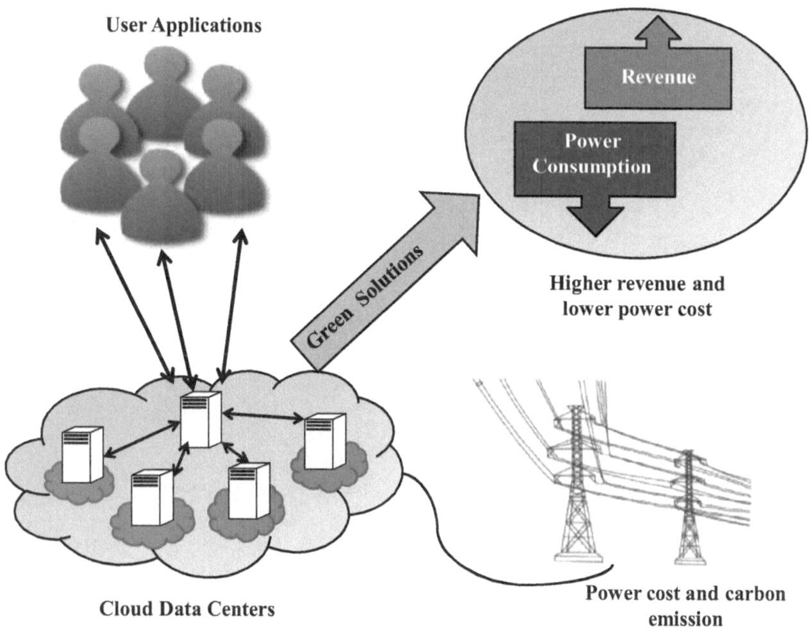

FIGURE 11.1 Cloud Computing architecture.

whereas big organizations can reduce this by 30 to 60%. Due to this, it has become essential to move towards the "green cloud computing" concept (Kaur et al., 2018).

11.2 FEATURES OF CLOUD COMPUTING

The main features of cloud computing are given in Figure 11.2 and are explained below:

1. **Computer Oriented:** Features like computation, storage, and networkability are computer oriented and this can be obtained at several stages that include virtual machine (VM) and outlined stages.
2. **Service Oriented:** To maintain cloud computing services' infrastructure, a service-oriented approach is used. It contains various devices to provide such services. It may be a software platform or complete infrastructure to provide the services.
3. **Versatile:** Operations like computation, storage, and networking are the necessary attributes needed for flexible cloud computing applications, depending upon the runtime for a better quality of services (QoS).
4. **Active and Distributed:** Cloud computing resources are computerized and they are distributed to provide extraordinary performance and make the cloud

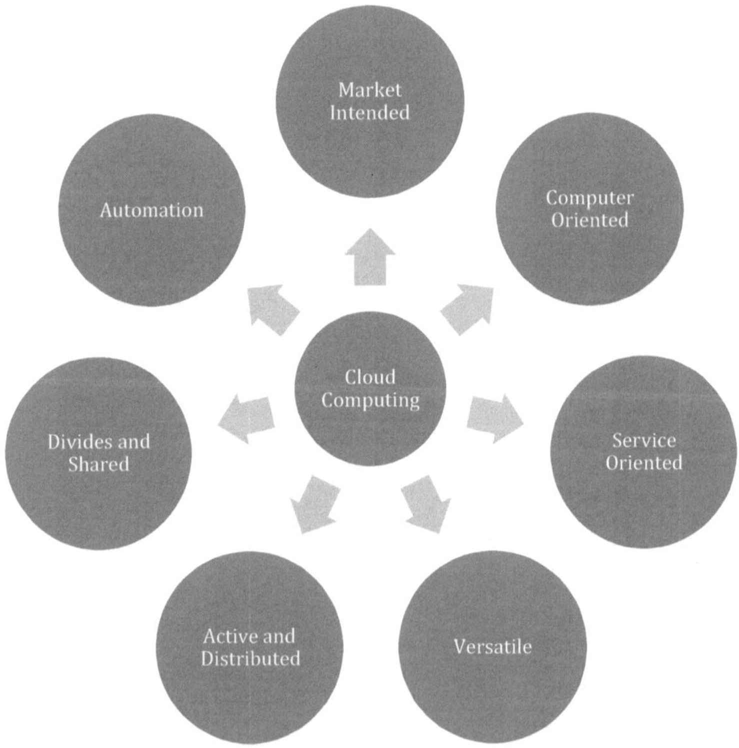

FIGURE 11.2 Features of cloud computing.

services more dependable. To make this happen, these resources are active and can be modified as per user needs, like software, network orientation, and so on.

5. **Divided and Shared:** In cloud computing, the infrastructure is divided and shared. Here resources assist various users with active allotment as per the needs of their specific applications. This infrastructure is also known as the "multi-tenant infrastructure." Practically users have no idea about how to manage the physical resources, the placement of the resources, and various other users with whom resources are shared.
6. **Automation:** Cloud shows an automation nature so that it can give largely independent services. This can be done by handling cloud infrastructure on its own to manage errors, failures, or performance lowering.
7. **Market Intended:** To use cloud computing services, users pay as per the "pay-per-use" or "pay-as-you-go" basis. The budget model changes as per the quality-of-service requirement of the applications. This will determine the efficacy aspect of cloud computing. This model will enhance the management and development of resources' treatment.

11.3 GREEN CLOUD COMPUTING

Green cloud computing is the extension of the cloud computing concept, which designs, generates, and employs digital data centers to minimize effects on the environment. This not only minimizes energy consumption but also lowers organization operating time.

Green cloud computing allows users to use the benefits of cloud data centers while lowering their harmful outcomes on the environment, which indirectly harm human well-being. It contains systems such as the following:

1. **Green design:** In this cloud, infrastructure design contains cloud data centers that provide energy-efficient assistance, hardware and software applications, and other various devices that utilize minimized energy than its equivalent.
2. **Green generation:** In this cloud, infrastructure will produce lower waste during reuse, which helps in a more sustainable environment.
3. **Green operation:** It lowers energy generation while using cloud infrastructure.

11.3.1 Green Cloud Computing Infrastructure

From the various researches on cloud computing energy efficiency, power, and performance management it has been concluded that there are many loopholes. Many attempts have been made to make cloud computing a more sustainable network. If the document dimensions are large enough then the network will utilize more energy. So, it becomes necessary to execute applications locally than in the cloud for green solutions (Parasher et al., 2019). Many types of research have been carried out with a focus on one particular parameter of cloud computing and the other parameters have been neglected, which lower the overall energy competence management. Various

TABLE 11.1
Green Parameter Power Calculation

Parameters	Overview	Formula
Power usage effectiveness (PUE) (Horner & Azevedo, 2016)	It is the fragment of complete energy utilized by the facilities of a data center to the complete energy utilized by IT devices	$PUE = \frac{Complete\ energy\ used\ by\ facilities}{Whole\ energy\ use\ by\ devices}$
Carbon usage effectiveness (CUE) (Azevedo et al., 2010)	Measurement of greenhouse gases discharged in the environment by data centers	$CUE = \frac{Total\ greenhouse\ gases\ discharged\ by\ data\ centers}{Total\ energy\ usage\ by\ devices}$
Water usage effectiveness (WUE) (Zoie et al., 2017)	Measurement of water usage by data centers every year	$WUE = \frac{Yearly\ water\ utilization}{Whole\ energy\ use\ by\ devices}$
Energy reuse effectiveness (ERE) (Tschudi et al., 2010)	Measurement of the benefits of energy reuse by the data centers	$ERE = \frac{Complete\ energy - reuse\ energy}{Whole\ energy\ use\ by\ devices}$
Data center infrastructure competence (DCiE) (Alnazir et al., 2017)	Measure energy competence of data centers	$DCiE = \frac{Complete\ device\ power}{Whole\ service\ power} x100\%$
Data center productivity (DCP) (Brill, 2007)	Measure quantity of necessary work done by data centers	$DCP = \frac{Complete\ required\ work}{Sum\ of\ resources\ to\ execute\ work}$
Compute power competence (CPE) (Dubois et al., 2016)	Measure total quantity of energy usage during computing	$CPE = \frac{Complete\ energy\ used\ by\ IT\ devices}{PUE}$
Green energy coefficient (GEC) (Zhang et al., 2014)	Quantity of green energy usage by data center	$GEC = \frac{Utilized\ green\ energy}{Whole\ energy\ used}$
Space, wattage, and performance (SWaP) (Jain et al. 2013)	Measure space and energy needed by data centers	$SWaP = \frac{Quality}{Space\ x\ power}$
Data center energy productivity (DCeP) (Sego et al., 2012)	The measure of operation performed by the data center to amount of energy utilized to fulfill the work	$DCeP = \frac{Complete\ required\ operation\ performed}{Whole\ energy\ used\ to\ execute\ work}$
Energy reuse factor (ERF) (Tschudi et al., 2010)	Measure reusable energy	$ERF = \frac{Using\ reused\ energy}{Whole\ energy\ used\ by\ devices}$

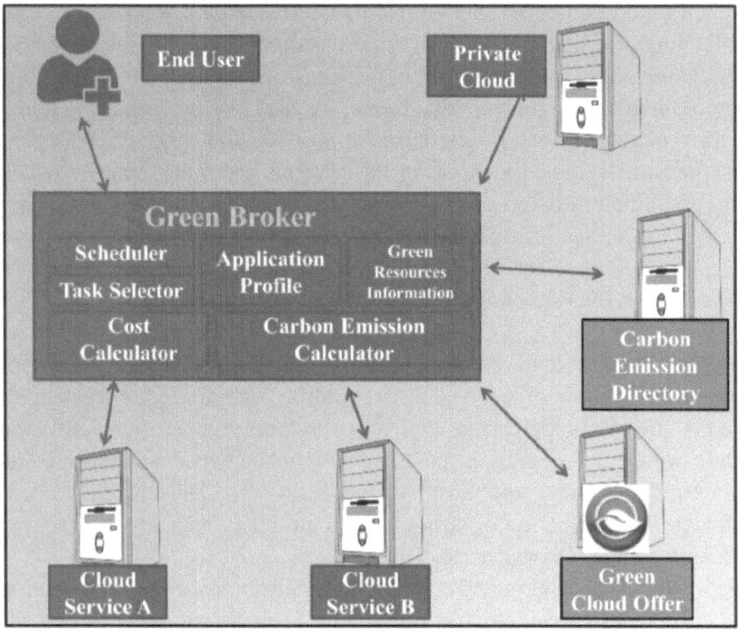

FIGURE 11.3 Green cloud infrastructures.

other operations are reallocated to distribute the workload, which will help in energy-efficient working without affecting the issues of virtualization (Tomar et al., 2019). Cloud computing infrastructure is looking for an energy-efficient solution that can minimize energy consumption and carbon dioxide discharges without affecting the cloud computing market. Figure 11.3 represents the green cloud computing architecture intended to minimize energy consumption. The task of this green cloud computing structure is to build a green cloud for both consumers and suppliers (Kalan, 2014).

In green cloud infrastructure, customers request for their cloud service using a new middleware green broker that handles the green cloud supplier to transfer the customer's request. There are three types of customer requests such as software, programmers, or infrastructure (Atrey et al., 2013). The cloud suppliers can record their facilities in the form of "green offers" in the public record, which is read by a green broker. The green offers contain green facilities, rates, and time for its reading to reduce carbon discharge (Tomar et al., 2017). Green brokers get information on the latest condition of energy specifications for utilizing numerous cloud facilities from carbon discharge records. The carbon discharge record manages all the information linked to energy-efficient cloud facilities (Garg et al., 2011). This information may contain PUE and cooling competence of cloud data centers, which is delivering facilities, infrastructure budget, and the rate of carbon discharge. Carbon discharge is measured by the green broker for all the cloud suppliers who are providing the demand for cloud facilities. Then it chooses the group of facilities that will provide the lowest carbon discharge and can purchase these facilities on the behalf of customers.

The green cloud computing sub-structure is created in a way that it can follow the overall energy utilization of the supplier-customer request. This depends on the carbon discharge record and green cloud facilities of two major elements. It records the energy competence of every cloud supplier and also encourages suppliers by making their facility green. From the customer window, the green broker takes vital steps like analyzing and choosing the (Prasad and Kaur 2005) cloud facility depending upon the customer quality of service (QoS) needs while maintaining the lowest carbon discharge during their service. Usually, a customer can access any of the three cloud facilities SaaS, PaaS, and IaaS. Executing these facilities should also be energy-efficient. Each layer of the cloud supplier should provide "green" facilities.

- **SaaS level:** SaaS majorly provides software updates of their data centers or assets from IaaS suppliers. The SaaS suppliers are required to create and calculate the energy competence of their software design, placement, and execution. While performing operation, the SaaS supplier selects the data centers that are only energy-efficient and closer to the customers. The lowest set of similar confidential information of customers should be managed by utilizing energy-efficient storage (Chavan et al., 2013).
- **PaaS level:** This level provides a path for application design. This level ensures and encourages energy-efficient application deployment. This can be completed by including numerous energies' outlined software like Joule Sort. This software is used to measure the energy needed to handle an external design. Further, this path itself can be outlined for copious code level developments, which can include present complier through the energy-efficient operation of the various applications. Like application design, cloud infrastructure also permits the placement of customer applications on a hybrid cloud. So, to gain the highest energy efficiency, this path designs the application and selects which part of the application information should be executed in the cloud infrastructure (Cloud, 2011) .
- **IaaS level:** The IaaS layer plays a vital role in the complete green cloud computing architecture, as it promotes individual services and also promotes various other services provided by clouds. They utilize the newest techniques for ICT and cooling packages for the least energy-consuming infrastructure. By using techniques like virtualization and consolidation, energy competence is further enhanced by turning off unused servers. Numerous energy meters and sensors are deployed to measure the latest energy of every IaaS service provider. This data is shown on a regular basis by cloud suppliers in the carbon discharge record. Several green planning and resource-proving rules and regulations ensure the lowest utilization of energy (Li et al., 2010). Also, cloud providers develop numerous green plans and costing plans for providing benefit to the customers to use their amenities during less traffic or the highest energy-efficient hours.

11.3.2 Features of Green Cloud Computing

There is major distress among the various organizations about cloud computing energy wastage by the data centers. Due to these various technologies, new ideas are emerging from cloud providers for improvements and more efficient energy usage than typical computing. Hence this new energy-efficient cloud infrastructure helps in lowering carbon discharge and also lowers the ICTs' infrastructure themselves by multi-occupancy. The main technology for efficient energy usage clouds is "virtualization" (Xu et al., 2014). This technique helps in enhancing energy competence for cloud service providers by reducing the economic dimensions linked with various organizations using the same cloud infrastructure. Virtualization is a procedure in which reasoning groups or a sub-part of computing materials are shown, which can be used in a way that helps the real arrangement. By consolidation of the unused servers in the configuration of numerous virtual machines (Liu et al., 2016) sharing similar physical servers at large use, organizations or industries can benefit from a large amount of savings for capacity, good management, and energy efficiency.

As per the survey, the subsequent four are the main features that have authorized cloud computing to minimize energy consumption and carbon discharges from ICT. From these cloud features, companies can minimize carbon discharges by almost 30% per customer by placing their demands on the cloud. This makes the cloud data centers highly efficient. Some of the main features are shown in Figure 11.4.

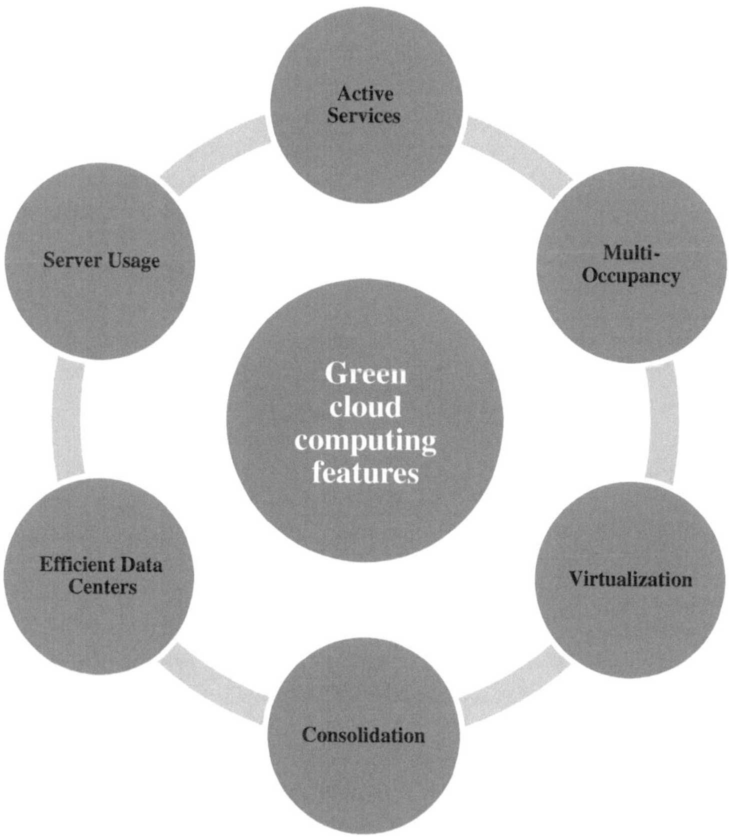

FIGURE 11.4 Characteristics of green cloud computing.

1. **Dynamic Services:** In a typical arrangement, data centers and personal infrastructure helped to support the worst-case command. This makes the ICT companies place additional infrastructure than required. The purpose for having such active services is (a) it is very challenging to identify the sudden command at a particular occasion like in network applications; (b) to provide good quality of services, like during the FIFA World Cup, when its website gains a huge amount of traffic during the matches. To manage such a spike during a particular time in a year, operating large numbers of servers during the whole year is not energy efficient. Thus, active infrastructure service is required to conserve the energy usage with the minimum usage of the resources. Such situations can easily be handled by cloud infrastructure. To add more resources to the cloud infrastructure during heavy traffic conditions, virtual machines can be used by another host. As the demand is generated, cloud service providers analyze and project the condition and assign the resources accordingly. Applications with less traffic can be merged on a single server. This makes the data centers always remain in an active state to serve the demand. And this leads to lower energy consumption.
2. **Multi-Occupancy:** Using this feature, cloud infrastructure can minimize overall energy utilization and related carbon discharges. Like SaaS, the level provides numerous organizations with the same cloud infrastructure. This method is more energy-efficient than any other method because it does not install various copies of software on numerous infrastructures. Also, organizations show fluctuating demands, so multi-occupancy on the single server lowers the overall spike in demand, which will indirectly reduce the requirement of additional cloud infrastructure. Hence, lower fluctuation solutions are better for quality and more energy savings.
3. **Server Usage:** Sometimes the server usage goes below 5 to 10% of its average use. So, by using the virtualization technique, several applications' management and execution can be achieved on a single server separately. This will increase usage by 70% and help in minimizing the number of vigorous servers. Elevated usage of server results in high power consumption, the high server running time to perform more execution with the same power consumption.
4. **Efficient Data Centers:** Power utilization of data centers plays a very critical role in the overall energy consumption of cloud infrastructure. Using the right technology can minimize the overall energy competence and help in improving "power usage effectiveness" (PUE) of the cloud data centers. In today's time, the best infrastructure for cloud solution providers can gain PUE limits as a minimum as 1.1 to 1.2. This is approx. 40% more than a typical data center's power efficiency. The server infrastructure in the shape of modular containers, water- or air-based cooling, or advanced power supervision using power supply optimization that can enhance PUE in data centers. Also, cloud computing permits services to select data centers that are operating at improved PUE values. This can be accomplished by a high-speed network, virtualized services, and calculating, analyzing, and reporting data centers.

5. **Virtualization:** This idea is implemented to execute various virtual systems on a single hardware device using separate steps. Virtualization permits the formation of numerous VM (virtual machines) to perform various jobs at a particular time. Virtualization is not a new idea; it was used earlier too, for example, in 2nd generation devices.

   Withdrawing high performance from devices, lowering the usual expenditure on infrastructure, and efficient use of assets are the major benefits of virtualization. High-speed processing, less power consumption, the use of high-end assets, and money savings are the major requirements of virtualization, this helps in the green cloud's design. Active workload management with VMs, resource sharing across VMs, design of secured VMs, and energy optimization techniques for virtualization is the latest and most used activities of green clouds.
6. **Consolidation:** In green cloud computing, the idea of consolidation is defined as "the process of placing various data centers' common information execution applications on a single server using virtualization technique." This is the major sub-job obtained from virtualization and it is engaged to execute the process level load management, proper use of VMs, and lowering the power use.

11.4 ALGORITHMS FOR GREEN CLOUD COMPUTING

Green cloud computing infrastructure aims at efficient computation with stabilized usage of resources for computing to lower the energy usage or carbon discharge. Various algorithms have been designed based on energy efficiency, supply management, task scheduling, carbon dioxide discharge, etc. (Larumbe et al., 2013; Ferriera et al., 2015; Huang et al., 2014) amongst many are discussed below.

11.4.1 GREEN CLOUD COMPUTING TASK SCHEDULING ALGORITHM

Green cloud computing algorithm is used to maintain energy competence depending upon the priority task scheduling by selecting virtual machines (VMs) for the task execution (Assiraj et al., 2019) as shown in Figure 11.5. This algorithm has the following features:

(a) Tasks are suggested to the VM manager by the customer in terms of a small and large quantity of resources in MHz frequency that it needs across with the SLA occurrence.

$S_i = \{R_{min}, R_{max}\}$ determine the minimum and maximum need of the server.

$V_i = \{F_{min}, F_{max}\}$ determines the minimum and maximum operating frequency of VM.

(b) VM manager accepts the task and measures the pressure on the several servers, which is given by

$W_i = \{P_i * R_i\}$,

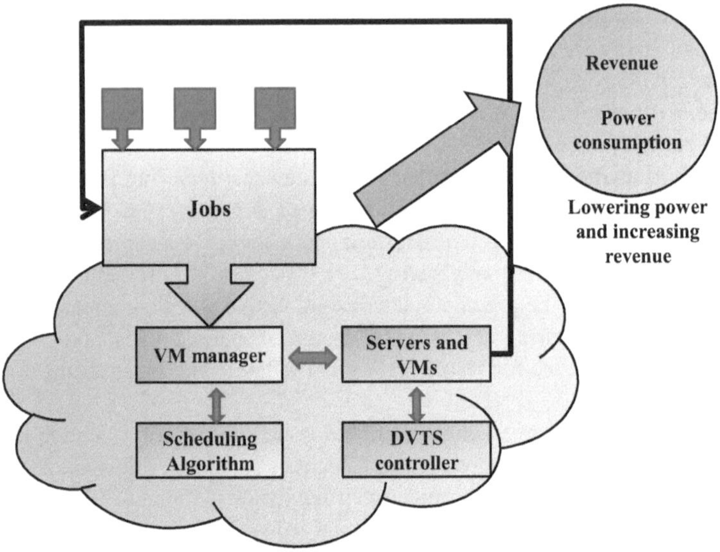

FIGURE 11.5 Green cloud computing task scheduling algorithm.

where P_i represents the power value of the server and R_i, represents resources acquired by the server.

(c) VM manager suggests the tasks to scheduling the algorithm for examining the finest as per the task need.
(d) VM managers by using a scheduling algorithm can select the VMs as per the influence in descending direction.
(e) VM is assigned to perform a task by the scheduling algorithm keeping energy consumption as the main focus. Servers that utilize minimum energy are selected first and sent to the VM manager.
(f) If no suitable server can assure the need, then step (e) should be repeated.
(g) If the task cannot be completed because of an unreachable VM, a message is sent to the VM manager by the DVFS controller.
(h) A complete list of hang servers is received and if there is no hang server then the selection of VM for completing a task cannot be achieved.
(i) And if there are hang servers then add them to the present server and go to step (c).

11.4.2 Power-efficient Virtual Machine Assignment Algorithm

Virtual machines (VM) are assigned to physical devices and predict that physical devices consume power that enhances linearly power consumption as the load increases. For this issue, two algorithms are present below and both the algorithms can resolve this issue.

The major issue of virtual assignment of algorithms (Lee and Lee, 2015) depends on the decision-making power as the request approaches the system without having any information about future requests early. Hence, no prediction can be made to give superior and minor logic on its ruthless ratio, i.e., no prediction is available for the output analysis.

For the first algorithm, analyze the ß-VMA (virtual machine algorithm), the main feature of this algorithm is that the load of the physical device or machine should remain as close as possible to x^*, to reduce the physical device power consumption. ß-VMA holds the physical device load within $\left\{\frac{\xi^*}{\beta}, \frac{2x^*}{\beta}\right\}^3$. As it arrives ß-VMA set VM d_i to a new physical device when it has a larger load than $\frac{x^*}{\beta}$ or to a present physical device s_j whose load $l(A_i) \leq \min\{x^*/ß, C\}$ if any. Also, the coming request has to be suitable in the chosen physical device, i.e., $l(A_i) + l(d_i) \leq C$. If no physical device can meet this request, then VM d_i is assigned to the new physical device. In the case where $C = \infty$ then here only one or zero physical devices with $l(A_i) \leq \min\{x^*/ß, C\}$.

ß, make the algorithm more tuned. Suppose if ß = 2, it shows better outputs than lower ß values. Theoretical outputs do not necessarily display good outputs in real positions.

In the second algorithm, α-VMA, where α takes "rate" as the parameter for the physical device as a reference. Parameter α is a convex function where the minimum value is getting when loading $x = x^*$. This algorithm α-VMA assigns input load d_i to a physical device only if three criteria are met.

(a) Physical device s_k load is below $\min\{x^*, C\}$ and the new VM d_i fit in the physical device or machine. That is $l(A_k) + l(d_i) < \min\{x^*, C\}$.
(b) When adding the input load to the physical device its rate decreases. That is $l(A_k) + l(d_i) < \alpha(l(A_j))$.
(c) The chosen physical device is the one that has minimal out of all physical devices meeting at (a) and (b). That is $\alpha(l(A_j) + l(d_i))$ is minimal.

If no such VM is present or $l(d_i) > \min\{x^*, C\}$, d_i is given to a new physical device. α-VMA permits that a physical device is given more than x^* load if that lowers its rate. Permit more packed partitions than ß-VMA.

11.5 HIGH-PERFORMANCE COMPUTING (HPC) FOR GREEN COMPUTING

High-performance computing (HPC) is defined as the capability to execute information and achieve difficult computation at high rates (Cousik, 2019). For example, our computing device such as a laptop or computer with high-speed processors (3 GHz) can execute approx. 3 billion computations per second. This is far quicker than any human can get. The leading example of HPC solutions is the "supercomputer."

A supercomputer consists of thousands of computing branches that operate simultaneously to execute and end one or more jobs. This type of computing is known as "parallel processing." It is like handling thousands of PCs linked together. It combines computing power to execute the jobs faster.

For more advanced researches, more information is needed. With the advance in HPC more development is taking place for economic development. So HPC is fundamental for scientific, manufacturing, and economic progress.

As new technologies arrive, such as the Internet of Things (IoT) (Patil, 2019), artificial intelligence (AI), and 3D imaging, improvements are being made in the area and the quantity of information that industries can use for their expanding needs. Processing the data in real-time is a challenging task like in live streaming, forecasting, product testing analysis, etc. To achieve this, industries need to depend upon a more accurate IT infrastructure for execution, storage, and monitoring of a large amount of information.

11.5.1 HPC Architecture

HPC solutions have three major elements

1. Evaluate
2. Network
3. Memory

To create HPC architecture, evaluating servers are linked to form a network, which together creates a cluster. Software programs and algorithms are executed together in a cluster. An HPC cluster contains hundreds to thousands of computing servers that form a network. Each computing server is known as a "node." These nodes work in parallel with each other, which will enhance processing speed and result in high-speed execution of the job. The cluster is a linked network that is connected to memory to catch the output. Simultaneously, these elements work to finish multiple jobs.

To perform a maximum number of jobs, each element must be in synchronization with the other elements. A similar memory element must be able to read and run information to and from computing servers as fast as possible. In the same way, the network elements should support high-speed movement of information among computing servers and memory. If one element cannot match with the rest of the elements, the execution of the whole architecture degrades.

11.5.2 Applications of HPC

HPC solutions can be used in various applications such as the following.

1. **Research:** HPC helps researchers to find various renewable energy sources, the understanding universe, weather forecast, etc.
2. **Technology:** AI and machine learning (ML) can be used in HPC online fraud, automated vehicles, advancement in the health sector, etc. (Parasher et al., 2019).
3. **Manufacturing:** It can be used in the design of new manufacturing products, supply chain management, product testing, etc.
4. **Capital Management:** It can be used to analyze real-time stock analysis and trading.

5. **Resource Management:** It can be used to enhance the reliability in natural resource analysis and management.
6. **Entertainment Industry:** HPC can be used to add special effects, live telecast, and movie editing.

11.6 GREEN DATA CENTERS IN CLOUD COMPUTING

The increasing energy requirement, money management, and a need to gain extra from the present expenditure are enabling present cloud service providers to acquire the finest methods to produce data centers that work greener. To develop energy-efficient data centers, numerous finest methods have been introduced to accelerate the competence of each component from the electrical system level to the processor level.

The first level: This level includes the intelligent establishment of the data centers and selecting their place. There exist two very essential parameters such as energy supply and energy competence of the components. Therefore, the data centers are built in such a way so that the generation of electricity would require renewable sources of energy like the sun or wind energy. Presently, the data centers' positions are to be determined based on their topographical properties such as weather conditions, fiber optics' compatibility, and they reach an abundant supply of economical energy levels. A major issue of cloud suppliers is that business and energy sources are also realized as budget use, not as carbon discharge.

The second level: This level includes issues within the data center's cooling system that expose mostly one-third of the total energy usage. Various studies have preset those temperature fluctuations within data centers, resulting in the lowering of the authenticity of ICT systems. For managing data centers' temperature, two methods are used to cool down the temperature such as air- and water-based temperature management systems. Both of these methods straightly lower the temperature of the device rather than cooling the whole region. Therefore, new methods for temperature management are introduced based on liquid cooling, nano-liquid cooling methods, in the server, in the rack, and row temperature reduction by the organizations. Apart from this, the external temperature/weather condition influences the energy need of the cooling mode. In some organizations, outside cool air is utilized to detach heat from the data center.

Third level: In this level data center power consumption is raised on the placement of current power competence servers and processors. Minimum energy consumption of the processors can lower the energy utilization of the IT infrastructures to a large extent. Various methods are present for new energy management servers in the market from various dealers like Intel, Qualcomm, etc. where these are proving to be better operation per watt systems. These new server models have introduced new concepts like clock gating and power gating for lowering CPU clock run time and switching off the block that is idle, respectively. Also increasing energy conservation and computing speed can be attained by the utilization of multi-core processors. This also needs software that can operate on multi-core CPU conditions for a power-efficient multi-core system. This led to the usage of virtualization and consolidation technologies to minimize the energy needs of IT systems.

Likewise, storage area networks (SAN) permit the development of an efficient storage network that accumulates all storage. Utilizing energy-efficient disks like tiered storage (Solid-State, SATA and SAS) permits improved energy efficiency.

Fourth level: Power supply is also an essential element that requires its implementation in an energy-efficient way. Its goal is to provide the server requirements by converting large voltage ac current to low voltage dc current, which is needed by all of the electrical components. These electrical circuits lose some energy in the form of heat, this happens due to dissipation by other fans inside power supply units (PSUs). The complexity of a PSU determines the energy competence of the circuit. Some researchers have proved the incompetence of the PSUs. Even today data center servers are manufactured with reduced quality of efficient PSUs. This problem can be solved by using energy star certified PSUs, which are 80% efficient.

11.7 RESEARCH AREA FOR GREEN CLOUD COMPUTING

Various methods are proposed for future research to lower the power utilization in the cloud, a few are presented below.

1. **By lowering CPU power dissipation**
   The processor utilizes electrical power for charging (dc supply) for its working, for the switching activities for the devices present in it, for transistor cooling, and various chips. Processor extra heating will be dispersed in the environment. The heat produced by the processor varies from device to device, for example, mobile phones utilize less power than the processors of laptops, desktops, notepads, etc. So, by using cooling in such devices, power dissipation can be lowered.
2. **By adding new technology for clock gating**
   Clock gating is used in the hardware as the switch whose function is to activate and de-activate the clock. The clock will get activated only when the hardware block is performing some work and the clock will be off when the block is not performing any work. This method is very useful and is used in various circuits. This method can also help in lowering the power dissipation.
3. **By utilizing divided level power**
   Dividing at the processor level is defined as the division along the horizontal plane. In this method, the processor does not share the power level with the microchips to form a communication link among a CPU and motherboard. This motherboard will maintain the diving level power resources and provide a distinct power supply to the processor and the microchip. This can also perform nested division. This means dividing the power level into any number of blocks. Figure 11.6 shows the division level power and combined level power.
4. **By utilizing energy competent processors:**
   With a high level of voltage scaling and frequency scaling, the requirement of processor, clock cycle of a processor, or maybe both voltage and clock cycle of a processor can be altered to minimize the power consumption of CPU.

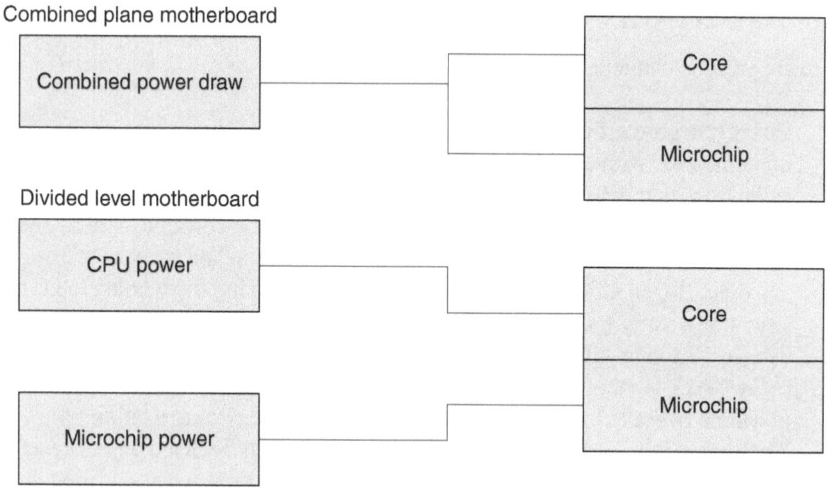

FIGURE 11.6 Division level power and combined level power.

5. **By utilizing renewable energy sources:**
   In cloud computing, all the data centers needed diesel producers to give standby power, and with this, data centers dissipate carbon dioxide as well as greenhouse gases (GHGs). By using renewable sources of energy like hydro, wind, solar energies to produce electricity to maintain the power needs and cooling needs of such data centers in clouds.
6. **By utilizing energy competent storage:**
   By inventing new and modern energy competent storage it becomes essential to replace the present cloud storage. The average life cycle of a data center is around 9 years. So, for replacing present data centers, engineers can use energy competent memory like solid-state memories. Solid-state memories need low cooling as compared to hard disk drives because they do not contain any moving mechanical parts.
7. **By lowering cooling needs:**
   Earlier cooling was performed by using mechanical refrigerators that use provision of compact interior the data centers or by using cooled water, which will be externally provided to man air supervisor for chilling of the parts. Nowadays, free cooling is initiated for cooling needs as compared to mechanical cooling. A free cooling system is made to lower the need for cooling. As per this, if the environmental temperature is lower or at a dangerous point, then a mechanical refrigerator can give uninterrupted or unplanned cooling on its own. A free cooling system does not limit the needed fan energy required for cooling purposes, it only eradicates the requirement of mechanical cooling.

11.8 CHALLENGES OF GREEN CLOUD COMPUTING

There are various challenges of green cloud computing.

1. **Energy organization:** In today's time, most of the clouds are designed and implemented with multi-core CPUs. This enhances the demand for design techniques that help in energy improvement and good management with multi-core functionality of CPUs. In clouds, data centers also consume a lot of energy, where data centers are used to accumulate information and manage it using the software. So, an efficient energy organization, analyzing the whole cloud infrastructure, energy handling system, and smart energy system that have accurate decision-making infrastructure are the new challenges in this domain. As per today's ICT technology development, we need global and smart organizations to handle overall cloud infrastructure for proper energy organization.
2. **Virtualization:** Various previous studies were largely focused on the design of the high energy efficient cloud virtualization operation, but the virtualization process is still facing various optimization-related issues. So, the design of the methodology with the best techniques to develop the whole cycle of virtualization operation is an essential challenge for research. Another challenge for virtualization (Overview, 2011) is to create the best automated VM (virtual machine) with feasible and active assets' allotment and splitting services without harming the cloud execution.
3. **Multi-occupancy:** This is an essential parameter for a green cloud, but presently it is facing privacy and security issues. So, designing the highly secure multi-occupancy infrastructures while keeping privacy protection is highly recommended and faces further challenges for future study.
4. **Consolidation:** Design of smart backup in VM's consolidation, multi-featured threshold value measurement, supporting the vital resources, and server recess management have become other challenges in this domain.
5. **Sustainable:** This domain is majorly focused on environmental focused software design, i.e., carbon discharge measurement software to measure the collision of the cloud on the environment. The requirement to design a global platform to verify the clouds with series based on multiple factors of green cloud computing.

11.9 CONCLUSION

As the requirement for cloud computing is increasing every day, demand for high power utilization has also increased and this comes as one of the main challenges in future cloud computing infrastructures. Due to high power consumption, operation cost increases, and this leads to an increase in carbon footprint. So, it becomes very significant to lower the power utilization in cloud computing for environmentally friendly solutions. Hence, power coherent remedies are essential to lower the effects of cloud computing on the environment. To solve this problem a new concept called "green cloud computing" is the new technology. But this technique also faces many challenges like sustainability, multi-occupancy, energy organization, virtualization, and consolidation. Various research is going on in this domain and is showing a positive impact on future development. This chapter has discussed several features, algorithms, and challenges of green cloud computing.

ACKNOWLEDGEMENT

This work is supported by Delhi Technological University, Department of Electronics and Communications, New Delhi-110042.

REFERENCES

Alnazir, M. K. A. M., Mustafa, A. B. A. N., Ali, H. A., & Yousif, A. A. O. (2017, January). Performance analysis of Cloud Computing for distributed data center using cloud-sim. In 2017 International conference on communication, control, computing and electronics engineering (ICCCCEE) (pp. 1–6). IEEE.

Arjona Aroca, J., and Fernández Anta, A. (2016). Empirical comparison of power-efficient virtual machine assignment algorithms. Comput. Commun., vol. 96, pp. 86–98, DOI: 10.1016/j.comcom.2016.05.005

Assiroj, P., Hananto, A. L., Fauzi, A., and Hendric Spits Warnars, H. L. (2019). High-Performance Computing (HPC) Implementation: A Survey. 1st 2018 Indones. Assoc. Pattern Recognit. Int. Conf. Ina. 2018 – Proc., no. September, pp. 213–217, DOI: 10.1109/INAPR.2018.8627040

Atrey, A., Jain, N., & Iyengar, N. (2013). A study on green cloud computing. International Journal of Grid and Distributed Computing, vol. 6, no. 6, pp. 93–10.

Atrey, N. Jain, & N. C. S. Iyengar. (2013). A Study on Green Cloud Computing, Int. J. Grid Distrib. Comput., vol. 6, no. 6, pp. 93–102, DOI: 10.14257/ijgdc.2013.6.6.08

Azevedo, D., Patterson, M., Pouchet, J., & Tipley, R. (2010). Carbon usage effectiveness (CUE): A green grid data center sustainability metric. The green grid, 32.

Banerjee, S. & Jain, S. (2014). A survey on Software as a service (SaaS) using quality model in cloud computing. International Journal of Engineering and Computer Science, vol. 3, no. 1, pp. 3598–36020.

Brill, K. G. (2007). Data center energy efficiency and productivity. The Uptime Institute-White Paper.

Chavan, P., Patil, P., Kulkarni, G., Sutar, R., & Belsare, S. (2013). "IaaS cloud security," Proc. – 2013 Int. Conf. Mach. Intell. Res. Adv. ICMIRA 2013, no. December, pp. 549–553, 2014, DOI: 10.1109/ICMIRA.2013.115

Cloud, P. (2011). "PaaS – Cloud Computing," no. October 2013, [Online]. Available: http://deletemalware.blogspot.com/2011/07/paas-cloud-computing.html

Cousik, T., Shafin, R., Zhou, Z., Kleine, K., Reed, J., and Liu, L. (2019). "CogRF: A New Frontier for Machine Learning and Artificial Intelligence for 6G RF Systems," no. Ml, pp. 1–8.

Ferreira, J., Dantas, J., Araujo, J., Mendonca, D., Maciel, P., and Callou, G. (2016). An Algorithm to Optimize Electrical Flows of Private Cloud Infrastructures. Proc. – 2015 IEEE Int. Conf. Syst. Man, Cybern. SMC 2015, pp. 771–776, DOI: 10.1109/SMC.2015.144

Garg, S. K. & Buyya, R. (2012). Green cloud computing and environmental sustainability. Harnessing Green IT: Principles and Practices, 2012, 315–340.

Garg, S. K., Yeo, C. S., & Buyya, R. (2011, August). Green cloud framework for improving carbon efficiency of clouds. In European Conference on Parallel Processing (pp. 491–502). Springer, Berlin, Heidelberg.

Conley, J. G., & Greene, J. (2006). Environmentalism contained: A history of corporate responses to the new environmentalism (Doctoral dissertation, Princeton University).

Dubois, D., Rucker, D. D., & Galinsky, A. D. (2016). Dynamics of communicator and audience power: The persuasiveness of competence versus warmth. Journal of Consumer Research, 43(1), 68–85.

Horner, N., & Azevedo, I. (2016). Power usage effectiveness in data centers: overloaded and underachieving. The Electricity Journal, vol. 29, no. 4, pp. 61–69.

Huang, J., Wu, K., and Moh, M. (2014). Dynamic Virtual Machine migration algorithms using enhanced energy consumption model for green cloud data centers. Proc. 2014 Int. Conf. High Perform. Comput. Simulation, HPCS 2014, pp. 902–910, DOI: 10.1109/HPCSim.2014.6903785

Jain, A., Mishra, M., Peddoju, S. K., & Jain, N. (2013, April). Energy efficient computing-green cloud computing. In 2013 international conference on energy efficient technologies for sustainability (pp. 978–982). IEEE.

Kaur, G., Tomar, P., & Singh, P. (2018). Design of cloud-based green IoT architecture for smart cities. In Internet of Things and Big Data Analytics Toward Next-Generation Intelligence (pp. 315–333). Springer, Cham.

Kaur, G., Tomar, P., and Tanque, M. (2019). "Integrating Artificial Intelligence to Solve Pervasive IoT Issues, Elsevier, Cambridge, MA, USA.

Larumbe, F. & Sansò, B. (2013). "A tabu search algorithm for the location of data centers and software components in green cloud computing networks," IEEE Trans. Cloud Comput., vol. 1, no. 1, pp. 22–35, DOI: 10.1109/TCC.2013.2

Lee and K. Lee. (2015). The Internet of Things (IoT): Applications, investments, and challenges for enterprises. Bus. Horiz., vol. 58, no. 4, pp. 431–440, DOI: 10.1016/j.bushor.2015.03.008

Li, Y., Li, Y., & Jiang, C. (2010). A survey of virtual machine system: Current technology and future trends. 3rd Int. Symp. Electron. Commer. Secur. ISECS 2010, no. September 2017, pp. 332–336, DOI: 10.1109/ISECS.2010.80

Liu, Y., Shu, W., & Zhang, C. (2016). A parallel task scheduling optimization algorithm based on clonal operator in green cloud computing. J. Commun., vol. 11, no. 2, pp. 185–191, DOI: 10.12720/jcm.11.2.185-191

Overview, E. (2011) Virtualization: A Review and Future. Am. J. Inf. Technol., vol. 1, no. 1, pp. 1–37.

Parasher, Y., Kedia, D., & Singh, P. (2018). Examining current standards for cloud computing and IoT. In Examining Cloud Computing Technologies Through the Internet of Things (pp. 116–124). IGI Global.

Parasher, Y., Singh, P., & Kaur, G. (2019). Green Smart Security System. In Green and Smart Technologies for Smart Cities (pp. 165–184). CRC Press.

Parasher, Y., Singh, P., & Kaur, G. (2019). Green Smart Town Planning. In Green and Smart Technologies for Smart Cities (pp. 19–41). CRC Press.

Patil, A. and Patil, R. An analysis report on green cloud computing current trends and future research challenges (March 19, 2019). Proceedings of International Conference on Sustainable Computing in Science, Technology and Management (SUSCOM), Amity University Rajasthan, Jaipur, India, February 26-28, 2019, pp. 813–820.

Prasad, R. S. & Kaur, G. (2005). "Optical Communication," Galgotia Publications, Greater Noida, Uttarpradesh, ISBN: 9788175154858, 8175154853

Radu, L. D. (2017). Green cloud computing: A literature survey," Symmetry (Basel), vol. 9, no. 12, DOI: 10.3390/sym9120295

Sego, L. H., Marquez, A., Rawson, A., Cader, T., Fox, K., Gustafson Jr, W. I., & Mundy, C. J. (2012). Implementing the data center energy productivity metric. ACM Journal on Emerging Technologies in Computing Systems (JETC), 8(4), 1–22.

Singh, P., Dixit, V., & Kaur, J. (2019). *Green Healthcare for Smart Cities*. In Green and Smart Technologies for Smart Cities (pp. 91–130). CRC Press.

Srivastava, A., Gupta, M. S., & Kaur, G. (2020). Energy efficient transmission trends towards future green cognitive radio networks (5G): Progress, taxonomy and open challenges. Journal of Network and Computer Applications, 168, 102760.

Tomar, P. & Kaur, G. (2017). Examining Cloud Computing Technologies through the Internet of Things, IGI-Global, International Publisher of Progressive Information Science and Technology Research Books and Journals, USA, DOI: 10.4018/978-1-5225-3445-7, ISBN13: 9781522534457, Pages 311.

Tomar, P. & Kaur, G. (2018). "Green and Smart Technologies for Smart Cities, CRC Press,Taylor & Francis Group, Dec 2019, Boca Raton: CRC Press, Ist Edition, ISBN 9781138318090 – CAT# K391489, https://doi.org/10.1201/9780429454837

Tschudi, B. I. L. L., Vangeet, O. T. T. O., Cooley, J., & Azevedo, D. (2010). ERE: A metric for measuring the benefit of reuse energy from a data center. White Paper, 29, 82–83.

Tomar, P. and Kaur, G. (2019). "Artificial Intelligence and IoT-Based Technologies for Sustainable Farming and Smart Agriculture, IGI-Global, International Publisher of Progressive Information Science and Technology Research Books and Journals, USA.

www.datacenterknowledge.com/energy/study-data-centers-responsible-1-percent-all-electricity-consumed-worldwide

www.gesi.org/research/smart-2020-enabling-the-low-carbon-economy-in-the information-age

Xu, L., Wang, K., Ouyang, Z., & Qi, X. (2014, August). An improved binary PSO-based task scheduling algorithm in green cloud computing. In 9th International Conference on Communications and Networking in China Key Lab of Broadband Wireless Communication and Sensor Network Technology, Lili Xu (pp. 126–131). IEEE.

Zhang, Q., Sun, Y., Xu, W., & Zhu, D. (2014). Organic thermoelectric materials: emerging green energy materials converting heat to electricity directly and efficiently. Advanced Materials, 26(40), 6829–6851.

Zoie, R. C., Mihaela, R. D., & Alexandru, S. (2017, October). An analysis of the power usage effectiveness metric in data centers. In 2017 5th International Symposium on Electrical and Electronics Engineering (ISEEE) (pp. 1–6). IEEE.

12 Green Artificial-Intelligence-Based Communication System

Asbah Masih and Gurjit Kaur

CONTENTS

12.1 INTRODUCTION TO GREEN AI-BASED COMMUNICATION SYSTEMS

Advancement in the area of communication systems have led to the need for faster, robust, and more reliable systems. AI-based communication systems are used to fulfill the diverse requirements of the increasing communication systems (Kaur et al. 2019). AI-based technologies help in resource management and naturally handle the issues of large-scale complex communication systems. These technologies provide self-configuration, self-healing, and also self-optimization to the communication system thus making it more reliable and robust.

With the present 5G, there is an exponential rise in the network infrastructure along with the massive increase in the energy cost and thus the need to manage it is

DOI: 10.1201/9781003264477-12

more than ever before. To deliver high throughput services, 5G has been launched in many countries and the energy cost of 5G base stations is much more when compared to 4G base stations. An increase in power amplification (PA) in multiple input multiple output (MIMO) antennas and the processing of growing data are the two major parts that contribute to the consumption of energy. It has been shown that information and communication technology (ICT) accounts for more than 20% of the total electricity consumed. This increase in power requirements is resulting in an increase in carbon dioxide content in the atmosphere. The increase in carbon dioxide deteriorates the environment and also affects human health (Gandotra and Jha, 2017).

To solve this increasing energy crisis and reduce carbon footprint green communication systems is the need of the hour. The energy consumption of the network components can be reduced by both software and hardware techniques. For green communication both reduces the operating expenditure cost and carbon emissions by introducing a more user-friendly environment.

Two major components of green communication systems are energy-efficient network design and energy harvesting is practiced (Chien and Yang, 2015). Artificial intelligence is the best method to shift automatic network management towards the green era. Green artificial intelligence is based on the idea of progress while ensuring to protect the environment.

12.2 GREEN AI-BASED COMMUNICATION SYSTEMS

The need for better and more energy-efficient systems has led to various green AI-based communication systems.

12.2.1 Green AI-based Wireless Communication System

The increase in the volume of traffic with high user data rates has led to the development of AI-based wireless communication systems. However, a significant part of the total energy consumption of information communication technology is utilized by wireless communication systems. Thus, there is a need for green AI-based communication systems (Patil and Goti, 2013). These systems ensure efficient use of energy. Various energy-efficient strategies used in wireless systems are shown below. Figure 12.1 shows the energy-efficient strategies used in AI-based wireless systems.

12.2.2 Green AI-based Wired Communication System

The growth and demand for optical communication have increased significantly over the years. Optical fibers have ultra-high bandwidth, their power consumption is low, and they have high transmission gain. These systems however undergo various losses because of the transmission of light through the fiber. Optical communication includes non-deterministic events. AI-based optical communication systems that overcome the problem of uncertainty and are used to solve the problem of decision making. Deep reinforcement learning (DRL) is used in optical communication networks for resource allocation, to automize routing resolutions, configuration etc. (Chen et al. 2019a, Andreoletti et al. 2020, Wang et al. 2021). To achieve a holistic

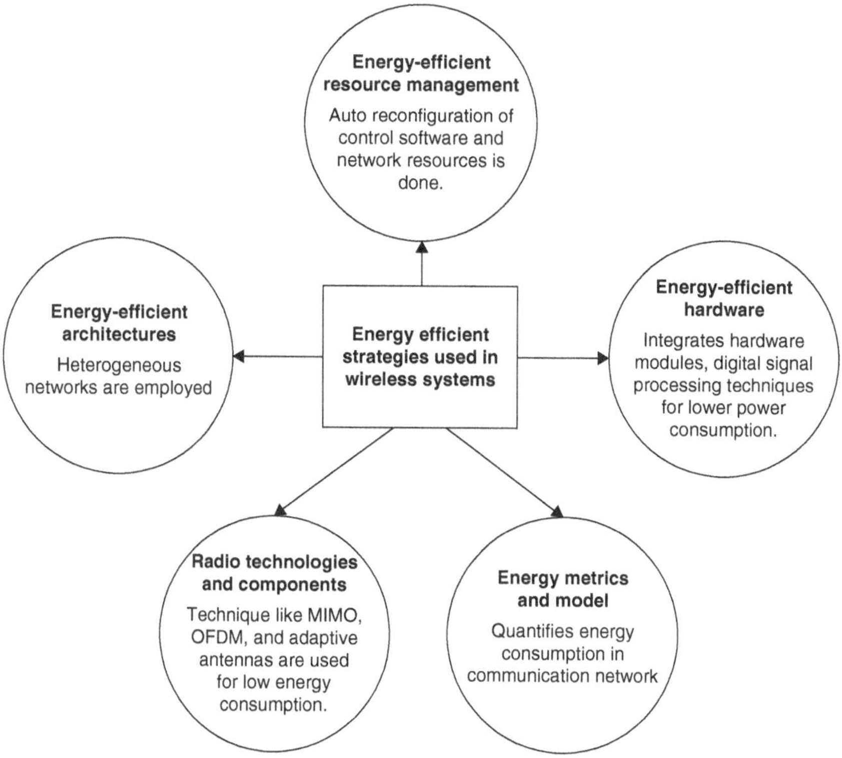

FIGURE 12.1 Energy-efficient strategies used in AI-based wireless systems.

approach green AI-based optical communication systems are used. Energy consumed by the optical networks is reduced by optimization of space between the repeaters. It typically has ordered a few thousand kilometers. For these systems inefficiencies and overheads are reduced to reduce energy consumption.

12.3 ENERGY HARVESTING METHODS BASED ON AI-BASED TECHNIQUE

To reduce the CO_2 emissions and move towards a cleaner and safer environment one of the important aspects of green communication is the energy harvesting method (Gandotra and Jha, 2017). In this method, renewable energy sources like wind energy, solar energy are used to charge the device batteries.

Since energy harvesting uses renewable energy sources and it complicates the working as well as management of cellular networks (Srivastava et al. 2020). To track the dynamic energy harvesting source and also optimize operations, AI techniques are used. Depending on whether it is controllable and predictable, energy harvesting techniques are grouped into various groups. AI stands in the uncontrollable but predictable energy group. AI models here are used to get the relationship between the future harvesting power and related sources so that it can use the results to reconfigure the network in advance. AI models can also be used to directly make changes to network management policy from the harvesting-related factors. In the case of partially controllable RF energy harvesting technique, base stations' power control and transmission scheduling are optimized by AI. AI can reduce the consumption of energy for UAV enabled base stations by optimizing the trajectory. Figure 12.2 shows the energy harvesting sources for ICT based on AI technique.

AI solves the issues of various energy harvesting sources (Srivastava et al. 2021):

- The major issue with solar energy is its inconsistency as the incoming solar energy is variable, thus there is a need for correct prediction for the amount of solar radiation received. This amount determines how much energy can be utilized at the sensor nodes. AI techniques help to correct predictions about the amount of solar radiation that will be received, thus solving the major drawback. Figure 12.3 shows the AI-based solar energy harvesting method.

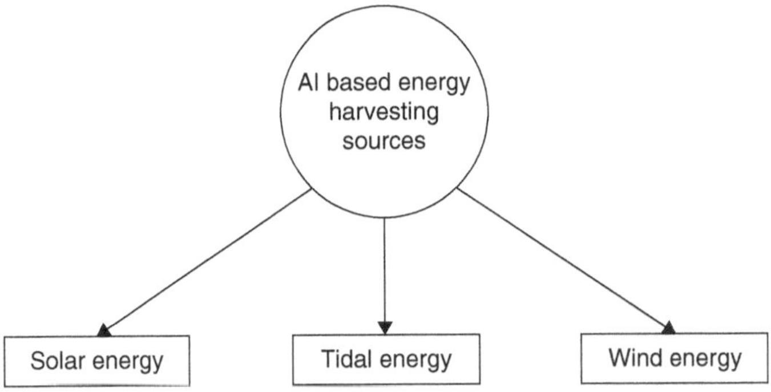

FIGURE 12.2 Energy harvesting sources for ICT based on AI technique.

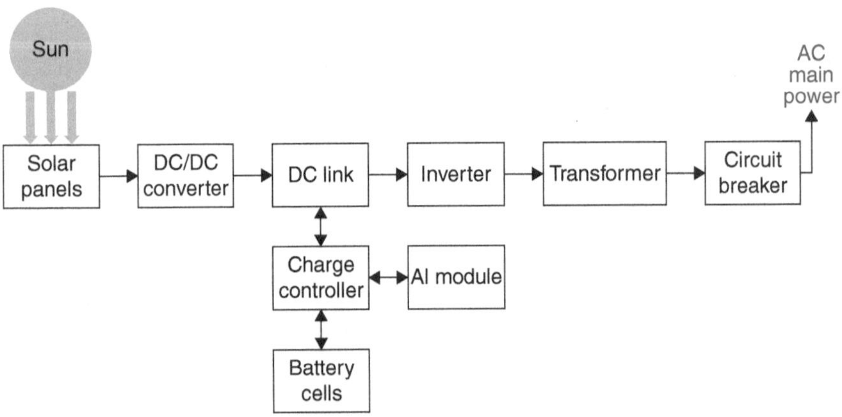

FIGURE 12.3 AI-based solar energy harvesting process.

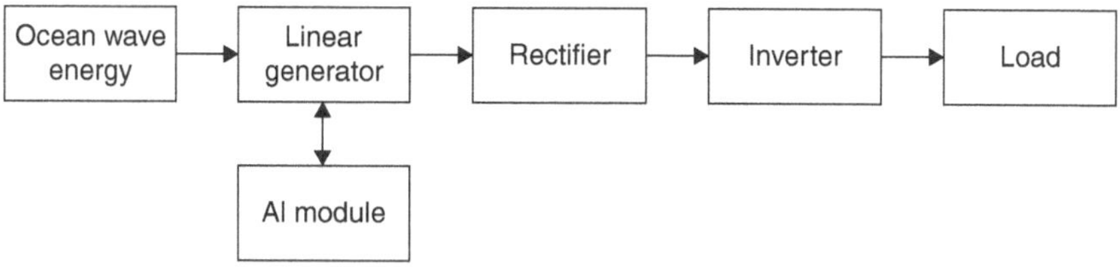

FIGURE 12.4 AI-based tidal energy harvesting process.

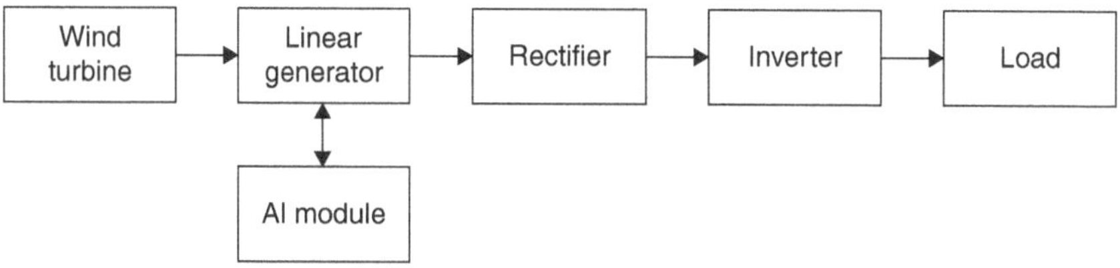

FIGURE 12.5 AI-based wind energy harvesting process.

- In order to harvest tidal energy, accurate wave height prediction is important and various AI techniques are used to order to give correct predictions. AI techniques are also used for the design and also control of wave energy converters. Figure 12.4 shows the AI-based tidal energy harvesting method.
- The performance of wind farms is monitored by supervisory control and data acquisition (SCADA) systems in wind turbines. AI algorithms are used to reconstruct the data lost in supervisory control and data acquisition (SCADA) systems in wind turbines of wind farms. Figure 12.5 represents the AI-based wind energy harvesting process.

Since energy harvesting does not provide a constant source of a fixed amount of power throughout the operation, designing energy harvesting techniques is difficult. The techniques are time-dependent and also depend on energy availability. Since as in the case of wind energy the harvesting period varies, proper resource allocation is required more in energy harvesting techniques. For energy harvesting, the energy consumption should always be less than the stored energy. Energy utilization optimization is time-dependent for energy harvesting communication.

12.4 ADVANTAGES OF AI-BASED ENERGY HARVESTING

AI-based energy harvesting techniques have bloomed because of its numerous advantages. Not only do these have a significant contribution in the reduction of the carbon footprint by simply reducing the primary cell disposal numbers, it ensures that huge electronic systems can run continuously for a large time without the need for battery replacement. AI in energy harvesting has ensured self-sustainable capability of such systems. This also allows easy deployment in areas that are not easily accessible. The advantages of AI-based energy harvesting is represented by Figure 12.6.

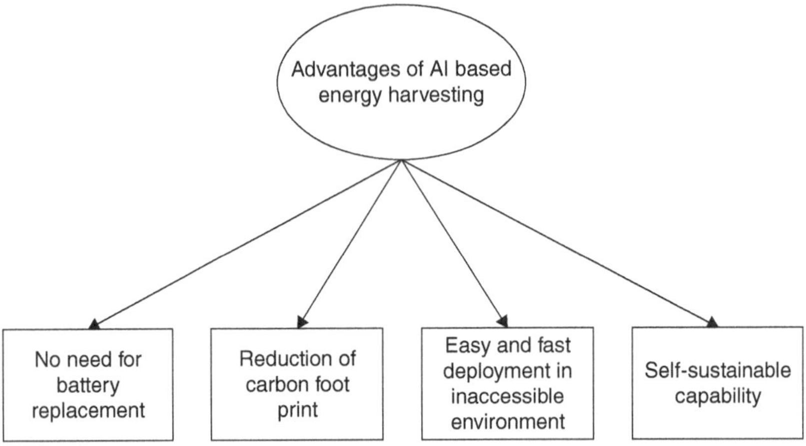

FIGURE 12.6 Advantages of AI-based energy harvesting.

12.5 FEATURES OF GREEN AI-BASED COMMUNICATION SYSTEMS

AI-based green communication systems have gained popularity because of their many perks. This method compared with traditional methods has an upper hand because it solves the problems intelligently.

- AI models can provide network reconfiguration in advance as they can predict channel conditions, mobility patterns etc. Since many base stations also perform the function of storage and computation, AI is used to optimize the computation offloading and content caching policies. This optimization reduces the energy consumption and thus provides the green solution.
- Since energy harvesting constitutes a major component of green communication, several energy harvesting techniques are implemented. And AI techniques are desirable for uncontrollable but predictable scenarios. The relationship between future harvesting power and related factors are also mapped using AI. These results are then used to reconfigure the network beforehand. Thus, green AI-based communication systems have improved efficiency.

12.6 TECHNIQUES OF GREEN AI-BASED COMMUNICATION

There are various techniques of green communication such as the following:

- Green AI-based D2D communication
- Green AI-based mmWave communication
- Green AI-based massive MIMO
- Green AI-based HetNets
- Green AI-based-IoT

12.6.1 Green AI-based D2D Communication

D2D enable direct communication between devices. This radio access technology has the advantage of low power consumption, low latency, increases both energy efficiency and spectral efficiency and reliable link. In this type, the emission of CO_2 is low as the base station here has sleep mode function, thus energy consumed is less. However, for proper implementation of D2D communication, various challenges need to be solved (Ioannou et al. 2020). Over the years, artificial intelligence has been used to solve most of these challenges. These are challenges are as follows:

1. Discovery of device: The device discovery process is done with two devices before they start communicating between each other. This is done in order to know whether these two devices are close enough for D2D communication to take place.
2. Selection of mode: Once device identification is complete next the selection of the mode is performed. This determines whether communication will be done directly or through the conventional network between the candidates. This mode decision is of various categories as shown below:
   a. Inband D2D communication:
      Reuse: Both D2D communication and existing cellular communication use the same resources. This mode selection method ensures high spectral efficiency however interference issues between the D2D and cellular user equipment may exist.
      Dedicated: D2D user equipment uses dedicated resources as channel resources are present in abundance.
      Cellular: Here both the user equipment will use the cellular network for communication as cellular user equipment.
   b. Outband D2D communication
      Controlled: Two interfaces are present in this mode of which one uses an unlicensed spectrum for sharing and the second, is for a licensed spectrum.
      Autonomous: Communication in this mode can happen only in the unlicensed spectrum and do not have access to the base station.
3. Management of Interference: Interference in a network is directly related to mode selection. In order to avoid this interference management is important.
4. Power control: Power control in D2D communication is necessary as though higher transmission power provides a better coverage area, the D2D user equipment battery is drained and this also causes interference.
5. Security concerns: The proper security insurance is important in D2D communication because chances of malicious attacks are high as user data transmission routing of one user is done by other user devices.
6. D2D device handover: Handover of the devices communicating with each other is necessary when they are moving away from each other or away from an access point.

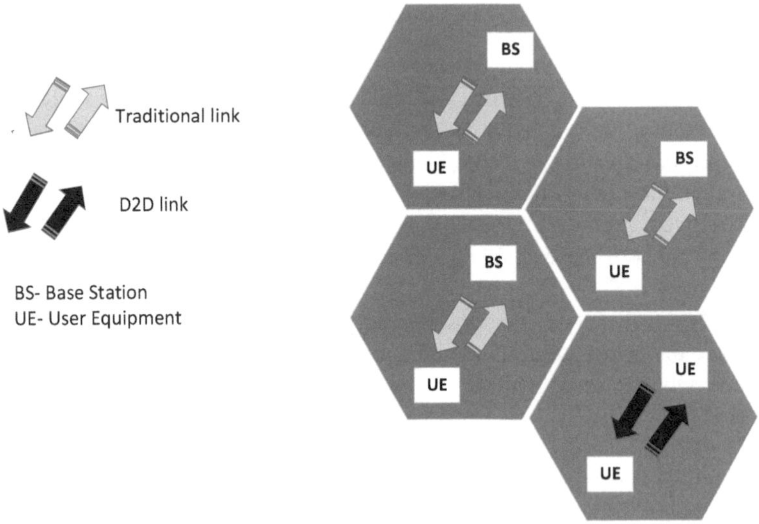

FIGURE 12.7 Green AI-based D2D link.

Figure 12.7 shows the green AI-based D2D link.

12.6.2 Green AI-based mmWave Communication

Millimeter wave communication (mmWave communication) is one of the most suited options to overcome the challenges of 5G and make the communication green by making it energy efficient. However, in order to have communication in the mmWave band antennas need to be efficient for it. One of the major drawbacks of mmWave is its time and power consumption. In order to reduce this overhead and also improve efficiency of beam selection AI techniques are used (Kao et al. 2018). Various AI aided beam selection methods like a machine learning algorithm, which provides offline training and this will reduce on-line processing overhead. Here the pre-training reduces the overhead of the off-line stage. In the following beam management aspects, AI is used (Li et al. 2020). These are the following:

- Training of beam formation: The paths used for communication in mmWave communications are limited therefore training of beamforming by machine learning takes into account the number of dominant beam pairs is much less when compared to the total beam pairs. AI techniques are also used in order to solve radio resource management issues.
- Tracking of beam: The device mobility issue is solved by beam tracking. It is used to compensate for the changes in the orientation of the communication system. Future beam prediction accuracy is increased using AI. Figure 12.8 based on green AI based mmWave communication.

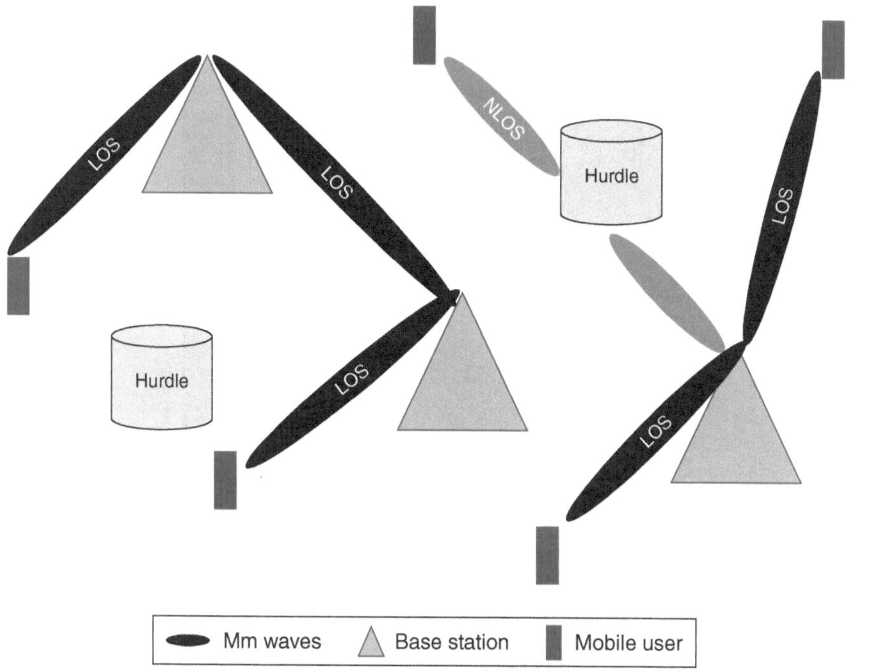

FIGURE 12.8 Green AI-based mmWave communication.

12.6.3 Green AI-based Massive MIMO

Massive multiple-input multiple-output (Chien and Yang, 2015) is gaining importance because they give enhanced throughput, low latency, high-capacity gain and better energy efficiency. In this system, base stations have an array of antennas and multiple users are connected to the base station using a single antenna. Gain is shared by all users as multiplexing is done. The mMIMO has more antennas than the simple MIMO. Massive MIMO is integrated through the following ways:

- Network mMIMO
- Single mMIMO
- Distributed mMIMO

Figure 12.9 represents the green AI-based massive MIMO communication system. AI-based green massive MIMO using the same spectrum band enhances the user experience to many folds. In order to optimize performance and reduce the complexity of massive MIMO AI-based techniques are widely used. These AI-based techniques that green massive MIMO are used for are the following:

- To pre-code and schedule
- To estimate and detect channel
- To manage interference and handover

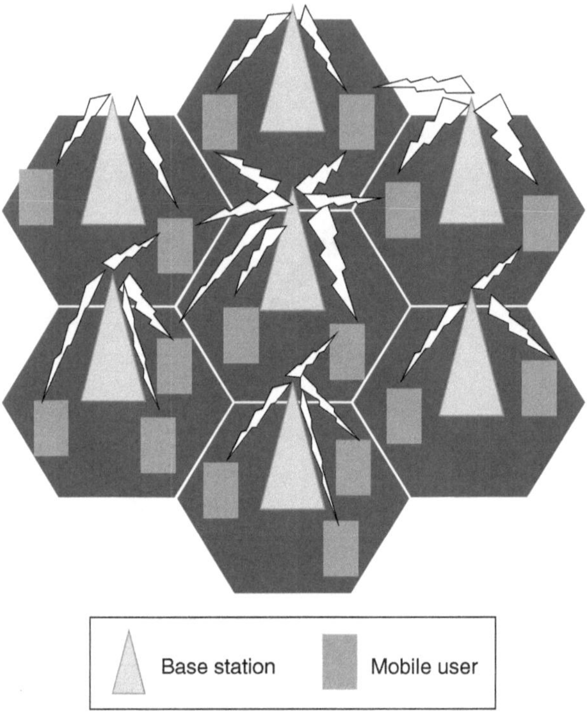

FIGURE 12.9 Green AI-based massive MIMO.

- For intelligent reflective surface
- To sense the spectrum and detect anomalies
- For energy-efficient optimization

12.6.4 Green AI-based HetNets

Heterogeneous networks (HetNets) are a combination of various cells, i.e., having one macrocell and various small cells like microcells, picocells, and femtocells. Power consumption is less in HetNets as fewer macrocells are there. HetNets provide a robust link and allow frequency reuse, thus reducing the bandwidth issue. The smaller cells remain in the rest mode when there is no work, and thus there is better power utilization (Gupta et al. 2020).

Heterogenous networks though provide energy-efficient systems; their proper organization, management, and optimization are the challenges that need to be solved (Wang et al. 2015). AI techniques are implemented in order to solve these. These techniques have the ability to resolve the issue in a more intelligent manner. AI-based techniques are used for the smart infrastructure of HetNets. These include the following:

- Self-configuration: It is considered for the planning of HetNets, for determining the base station's locations and to identify the links between them.

All this needs to be decided before the actual installation of the equipment. Self-configuration includes coverage planning and system configuration. AI techniques can be easily used to implement proper cell planning and system configuration.

- Self-healing: HetNets face the issue of robustness. Self-healing property helps to provide this by actively tracking, measuring the faults, and performing remedial actions. Human intervention is reduced significantly by self-healing. Self-healing can be provided by AI techniques. It includes fault identification and re-configuration.
- Self-optimization: Self-optimization is necessary in order to improve the overall work. However, implementing self-optimization in real-time is a complex process. It includes the following:
  - Load balancing and coverage optimization: Here antenna settings are adjusted and thus radio coverage is shaped. Antenna tilt self-optimization can be done by optimization using a fuzzy neural network (Fan et al. 2014).
  - Optimizing mobility: It is used to remove the unnecessary handover and also provide the right handover timing. Efficient neighborhood maintenance plays an important role in mobility optimization.
  - Device to device communication: This gives the high capacity to Hetnets. It also manages the interference.
  - Link quality estimation: Neural network based estimators (Caleffi et al. 2009) are used in order to provide link quality estimation with high reliability, thus ensuring secure transmissions in HetNets.
  - Multi-hop relay: This ensures more service coverage area and thus improves the sustainability of HetNets.
  - Vertical handoff: For users which cross different cells with technologies of different link-layers, providing seamless mobile services to them is done by vertical handoffs. AI-based techniques are used in order to provide optimized decisions for all parameters in complex conditions (Çalhan et al. 2010, Chai et al. 2011, Ma et al. 2012).

Figure 12.10 shows the HetNet architecture based on green AI technique.

12.6.5 AI-based Green IoT

Green IoT technique is used for green communication. Here energy-efficient strategies are used like applying sleep modes, data being collected slowly from neighboring nodes and quickly from far away nodes. AI-based green IoT ensures better performance with little to no human interference (Parasher et al. 2019b, Kaur et al. 2018) It provides the following advantages:

- Operational efficiency is boosted.
- The scalability of IoT is increased.
- Risk management is improved.
- Unplanned downtimes are reduced.

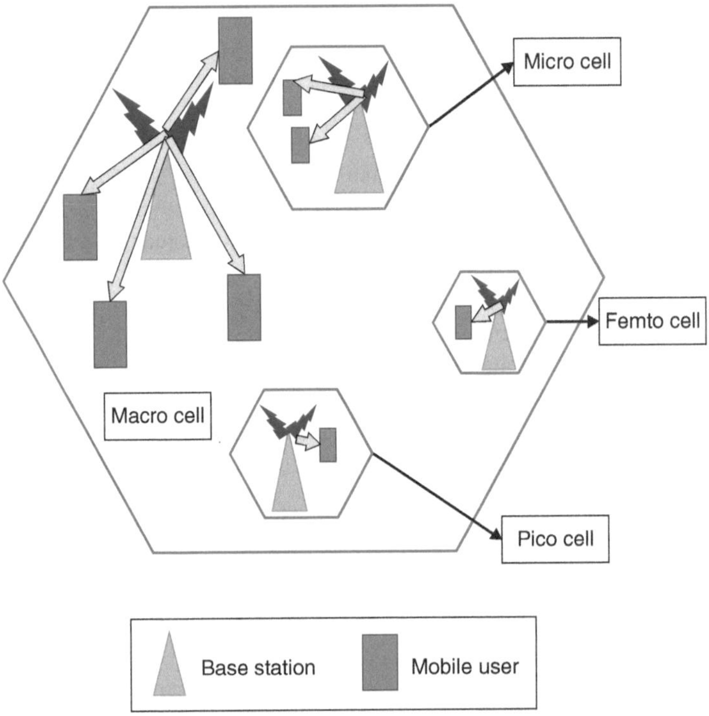

FIGURE 12.10 Green AI-based HetNet architecture.

In IoT systems, several fields are integrated in order to have energy-efficient systems as shown in Figure 12.11.

12.7 GREEN AI-BASED WIRELESS RESOURCE MANAGEMENT SYSTEM FOR MULTIMEDIA STREAMING COMMUNICATION SYSTEM

Green communication systems ensure monitoring and thus reduce energy consumption and improve resource utilization. The main components of the green communication system are the server cluster, the green communication module and the communication sub-net (Srivastava et al. 2021). The server cluster containing various servers connected in parallel distribution are deployed in different regions and these various green communication modules. The channel used in the communication sub-network and the green communication module is a duplex channel. This includes feedback special channels and energy utilization techniques. The user terminal of the user cluster contains various smartphones. The green computing module is connected

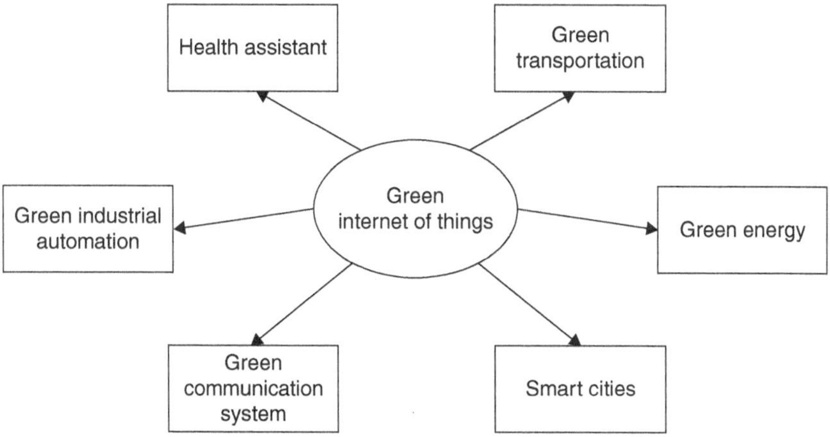

FIGURE 12.11 AI-based green IoT.

to the communication sub-net by the energy report collection, which will then report the energy utilization to the multimedia server cluster. Dedicated computer cluster makes up the green communication module. Figure 12.12 represents the green AI-based wireless resource management system.

12.8 CHALLENGES OF GREEN AI-BASED COMMUNICATION SYSTEMS

Although green AI-based communication systems are one of the most promising solutions it does face various challenges when being implemented (Gandotra and Jha, 2017). These are cost, security issues, and practical deployment, bandwidth, security issues. These are discussed below:

- These communication systems are dependent on the prediction; their reliability is always in question.
- These communication systems though are highly energy-efficient; the changing of entire infrastructure costs a lot. Since hardware is one of the factors on which the amount of energy spent depends on the amount of energy spent.
- As shown by the Shannon's equation transmission rate is legitimately relative to the measure of transmitted power, thus making the system green if transmission capacity is reduced and it will thus reduce the rates.
- Consumed energy for training and running of AI models are often more than required for conventional models, thus having high complexity. If algorithm complexity is reduced it often leads to a decrease in accuracy.

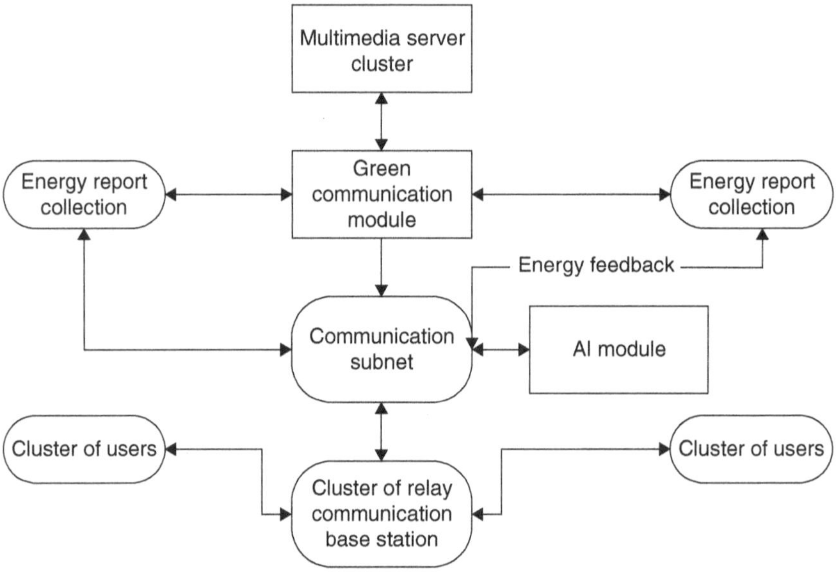

FIGURE 12.12 Green AI-based wireless resource management system for multimedia streaming communication system.

12.9 APPLICATIONS OF GREEN AI-BASED COMMUNICATION SYSTEMS

Being the need of the hour, green AI-based communication systems have a wide range of applications and the scope for using these in the future is increasing every day (Tomar et al. 2019, Parasher et al. 2019a, Singh et al. 2019). Some of these applications are as follows:

- Wireless communication systems, channel models, beamforming are optimized using AI. Optimization helps conserve energy.
- AI and now green AI-based communication has numerous applications for 5G wireless communication. These include large scale antenna array, coordination of transmission/reception of multiple points.
- Handoff strategies, user association, and other such 5G mobile management technologies are done by green AI-based communication systems.
- To ensure energy is not wasted, correct analysis and prediction of network behavior of 5G is done using AI technology. This predicts network collision, network overhead etc.
- Resource management including energy sources, cloud resources, spectrum resources etc. for 5G is also managed by AI thus helping in efficient utilization of energy.

12.10 CONCLUSION

Green AI-based communication systems have been one of the key initiatives to ensure the advancement of communication systems but not at the cost of the environment. These systems ensure operational cost reduction and efficient energy utilization. The key focus is to reduce the carbon footprint and at the same time prove a reliable, fast, and robust communication system. These green AI-based communication systems provide intelligent self-optimization solutions, thus enabling better systems for the user as well as the environment.

REFERENCES

Andreoletti, D., Velichkova, T., Verticale, G., Tornatore, M., and Giordano, S. 2020. A privacy-preserving reinforcement learning algorithm for multi-domain virtual network embedding. IEEE Trans. Netw. Serv. Manage., vol. 17, pp. 2291–2304. doi: 10.1109/TNSM.2020.3022278

Caleffi, M. and Paura, L. Jun. 2009. Bio-inspired link quality estimation for wireless mesh networks. Proc. IEEE Int. Symp. WoWMoM. pp. 1–6

Çalhan A. and Ceken, C. Sep. 2010. An adaptive neuro-fuzzy based vertical handoff decision algorithm for wireless heterogeneous networks. Proc. IEEE 21st Int. Symp. PIMRC. pp. 2271–2276.

Chai, R., Cheng, J., Pu, X., and Chen, Q. Sep. 2011. Neural network based vertical handoff performance enhancement in heterogeneous wireless networks. Proc. WiCOM. pp. 1–4.

Chen, X., Proietti, R., and Yoo, S. J. B. 2019a. Building autonomic elastic optical networks with deep reinforcement learning. IEEE Commun. Magaz., vol. 57, 20–26. doi: 10.1109/MCOM.001.1900151.

Chien, Sf., and Zarakovitis, C.C., Ting, T. and Yang, X.-S. (2015). Bio-Inspired Approaches in Telecommunications. 10.1016/B978-0-12-801538-4.00002-1.

Fan, S., Tian, H., and Sengul, C. Apr. 2014. Self-optimization of coverage and capacity based on a fuzzy neural network with cooperative reinforcement learning. EURASIP J. Wireless Commun. Netw., vol. 2014, no. 1, pp. 57:1–57:14.

Gandotra, P. and Jha, R.K. 2017. A survey on green communication and security challenges in 5G wireless communication networks. Journal of Network and Computer Applications, vol. 96, pp. 39–61. ISSN 1084-8045, https://doi.org/10.1016/j.jnca.2017.07.002.

Gupta, M.S. and Kumar, K. 2020. Application Aware Networks' Resource Selection Decision Making Technique using Group Mobility in Vehicular Cognitive Radio Networks", Vehicular Communications, Elsevier, vol. 26, pp. 100263.

Ioannou, I., Vassiliou, V., Christophorou, C., & Pitsillides, A. 2020. Distributed Artificial Intelligence Solution for D2D Communication in 5G Networks. IEEE Systems Journal, pp. 1–10. doi:10.1109/jsyst.2020.2979044

Kao, W.C., Zhan, S.-Q., and Lee, T.-S. 2018. AI-Aided 3-D Beamforming for Millimeter Wave Communications. 2018 International Symposium on Intelligent Signal Processing and Communication Systems (ISPACS). doi:10.1109/ispacs.2018.8923234

Kaur, G., Tomar, P., and Singh, P. 2018. Design of cloud-based green IoT architecture for smart cities. In Internet of Things and Big Data Analytics Toward Next-Generation Intelligence (pp. 315–333). Springer, Cham.

Kaur, G., Tomar, P., and Tanque, M. 2019. Integrating Artificial Intelligence to Solve Pervasive IoT Issues, Elsevier, Cambridge, MA, USA.

Li, Y.N. R., Gao, B., Zhang, X., and Huang, K. 2020. Beam Management in Millimeter-wave Communications for 5G and beyond. IEEE Access, pp. 1–1. doi:10.1109/access.2019.2963514

Ma, B., & Liao, X. 2012. Speed-adaptive vertical handoff algorithm based on fuzzy logic in vehicular heterogeneous networks. Proc. 9th Conf. FSKD, pp. 371–375.

Patil, A. and Goti, A. 2013. Green Computing in Communication Networks. International Journal of Computer Applications. vol. 68, no. 23, pp. 1–5.

Tomar, P., & Kaur, G. (Eds.). (2019). Green Healthcare for Smart Cities. Green and Smart Technologies for Smart Cities (1st ed.). CRC Press. Pp. 19–130 https://doi.org/10.1201/9780429454837.

Srivastava A., Gupta M. S., and Kaur G. 2021. Impact of Intelligence on Future Green Communication. Artificial Intelligence to Solve Pervasive Internet of Things Issues, Academic Press, ISBN: 978-0-12-818576-6, pp. 1–12.

Srivastava, A. and Kaur, G. 2021. Resource management for traffic imbalance problem in green cognitive radio networks. Physical Communication. Elsevier, vol. 48, pp. 101437.

Srivastava, A., Gupta, M. S., Kaur, G. 2020. Energy efficient transmission trends towards future green cognitive radio networks (5G): Progress, taxonomy and open challenges. Journal of Network and Computer Applications, Elsevier, vol. 168, pp. 102760.

Tomar, P. and Kaur, G. 2019. Artificial Intelligence and IoT-Based Technologies for Sustainable Farming and Smart Agriculture, IGI-Global, International Publisher of Progressive Information Science and Technology Research Books and Journals, USA.

Tomar, P., and Kaur, G. Eds. 2019a. *Green Smart Security System. Green and Smart Technologies for Smart Cities* (1st ed.). CRC Press, pp. 165–184 https://doi.org/10.1201/9780429454837

Tomar, P., and Kaur, G. (Eds.). 2019b. *Green Smart Town Planning. Green and Smart Technologies for Smart Cities* (1st ed.). CRC Press, pp. 19–41 https://doi.org/10.1201/9780429454837

Wang D., Zhang Z., Zhang M., Fu, M., Li, J., Cai, S., et al. 2021. The role of digital twin in optical communication: fault management, hardware configuration, and transmission simulation. IEEE Communications Magazine, vol. 59. doi: 10.1109/MCOM.001.2000727

Wang X., Li, X., and Leung, V. C. M. 2015. Artificial Intelligence-Based Techniques for Emerging Heterogeneous Network: State of the Arts, Opportunities, and Challenges. IEEE Access, 3, 1379–1391. doi:10.1109/access.2015.2467174

13 Green Microwave and Satellite Communication Systems

Divya Sisodiya, Yash Bahuguna, Akanksha Srivastava, and Gurjit Kaur

CONTENTS

13.1 INTRODUCTION

This chapter discusses the importance of microwaves and satellites for communication purposes. The main purpose of this chapter is to describe the use of microwaves

DOI: 10.1201/9781003264477-13

and satellites in a communication system. Microwave and reflex klystron production are discussed.

13.2 MICROWAVES

Microwaves are electromagnetic wavelengths from 1 meter to 1 millimeter or frequencies from 500 MHz to 100 GHz. Microwave communications use a variety of microwave technologies to transmit signals wirelessly. Microwave communication is called the line of sight because communication takes place between towers in a line-of-sight manner and the signal does not interrupt the other signals. Western Union invented the technology that is employed in microwave communications. The first microwave messages were transmitted from base stations in New York and Philadelphia. Microwave communication is carried out in two ways: Analog and digital. Digital microwaves are the best means of communication, but both analog and digital methods provide exceptional quality to users. Compared to digital communications, analog microwave communications are cheaper. Digital microwave communications are based on better and more reliable technologies (Gupta and Kumar, 2020).

Microwaves are commonly utilized in television broadcasts to send signals from a large distance. Different bands are utilized by microwave communications including L, S X, K, Ka bands. This frequency blocks additional UHF frequencies, reduces the absorption of ultrasonic frequencies in the air, and creates a wider bandwidth. C-band satellite TV supports traditional cable and satellite services. X-band and Ku-band communication are the most common modes of military communication. Radar determines the distance, speed, and other attributes of objects in a given region by using microwave power. Most radio astronomy systems use microwaves. The different bands of microwave regions are described in Table 13.1.

- Microwave transmitter: The signal that will be transmitted must be transmitted at uplink frequency. After the signal has been appropriately encoded and modulated, the converter multiplies the signal frequency to the uplink frequency. The frequency is upconverted and then applied to power dividers. The high-power amplifier receives the output of the power divider. Normally, multicavity klystron amplifiers or travelling wave tube amplifiers are utilized. These tubes necessitate a significant amount of cooling. The modulation takes place at a 70 MHz intermediate frequency, which is then upconverted to a 6 GHz uplink frequency. In a power combining amplifier, the output of numerous high-power amplifiers is combined, and the output is then sent via a bandpass filter and circulators. In today's RF and microwave systems, traditional kinds of power amplifiers are utilized. This traditional approach, on the other hand, wastes the majority of its input power as heat dissipation, which is unavoidable due to the nature of the power components used to drive amplification models (Zubir et al., 2017). Because it has a theoretical efficiency of 100 percent, a highly efficient switching mode amplifier like Class E has the potential to give a solution to the problems. When compared to traditional power amplifiers, this would reduce power wastage. The primary goal of this research is to investigate

TABLE 13.1
Different Bands of Microwave Regions and Their Frequency Range

S. No.	Types of bands	Applications	Frequency range
1.	L Band	Military telemetry GPS	1 to 2 GHz
2.	S Band	Weather radar Microwave ovens Microwave devices	2 to 4 GHz
3.	C Band	Long-distance communications	4 to 8 GHz
4.	X Band	Radar Terrestrial broadband Space communications	8 to 12 GHz
5.	Ku Band	Satellite communications	12 to 18 GHz
6.	K Band	Radar Astronomical observations Automotive radar	18 to 25 GHz
7.	Ka Band	Satellite communications	25 to 42 GHz
8.	Q Band	Terrestrial communications	30 to 50 GHz
9.	U Band	Satellite communications	40 to 58 GHz
10.	V Band	Satellite communications	51 to 75 GHz

Source: Kok Yeow You et al., 2018.

elements of switching mode power amplifiers, which are theoretically capable of achieving 100% efficiency. Next, to avoid interferences, frequency stability and power regulation are required. Figure 13.1 shows the block diagram of microwave transmitter.

- Microwave receiver: The front-end converter, which is commonly a double converter, converts the down link frequency signal in the order of GHz into an intermediate frequency of 70 MHz, which is the first stage of the terminal station receiver. To boost signal strength, this intermediate frequency signal is initially sent via a chain of bandpass filters and amplifier combinations. As a result, the IF signal is demodulated and the original baseband signal is obtained. The FM demodulator utilized here is a phase-locked loop (PLL) kind. After passing through the de-emphasis network, the signal is now amplified. To obtain the sound IF, the circuit includes a 5.5 MHz sound trap. Figure 13.2 represents the microwave receiver.

13.2.1 Microwave Generation

Microwaves are made with the klystron, a special vacuum tube invented by the Russell and Sigurn brothers. Klystrons are useful for generating microwave energy. When electrons in a klystron vibrate, then the high-frequency microwave energy is emitted by the klystron that a radar receiver can detect.

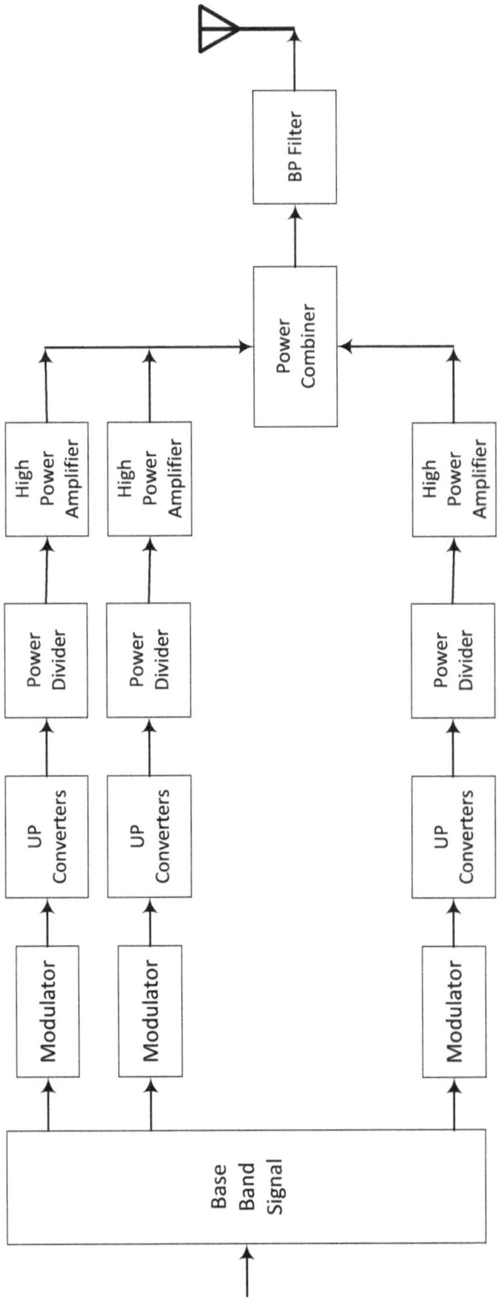

FIGURE 13.1 Microwave transmitter.

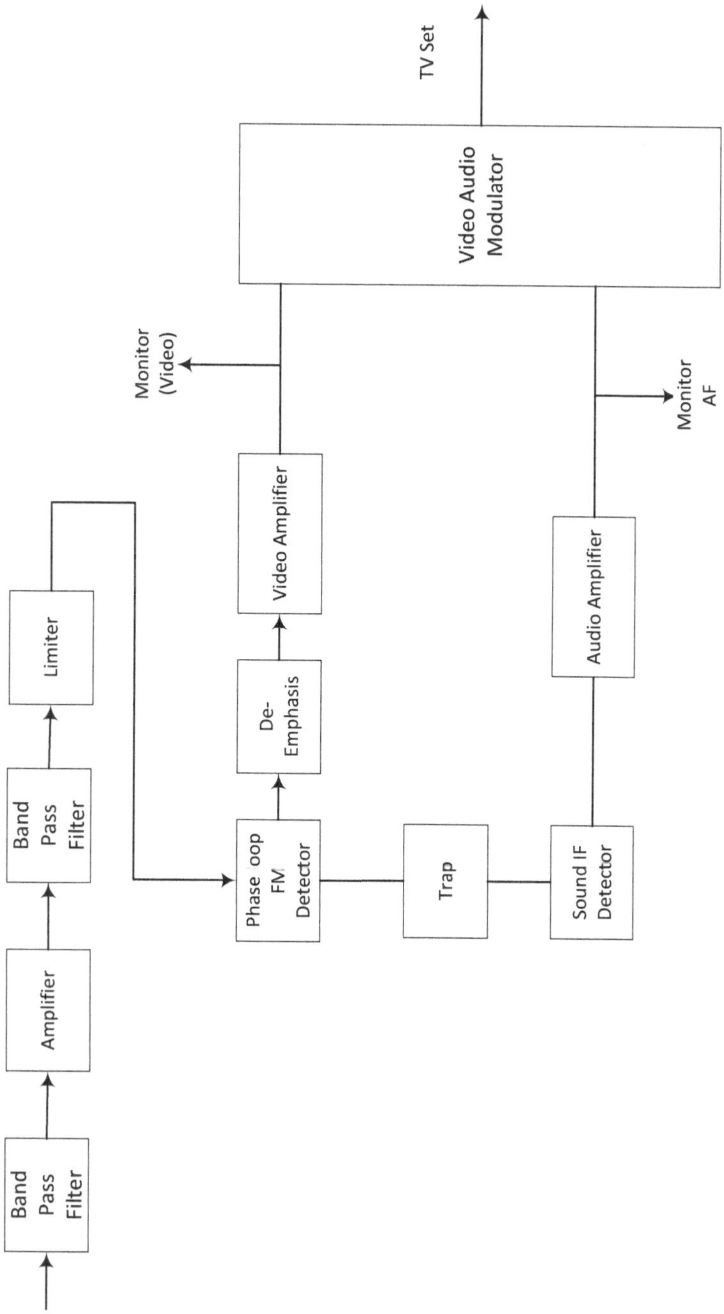

FIGURE 13.2 Microwave receiver.

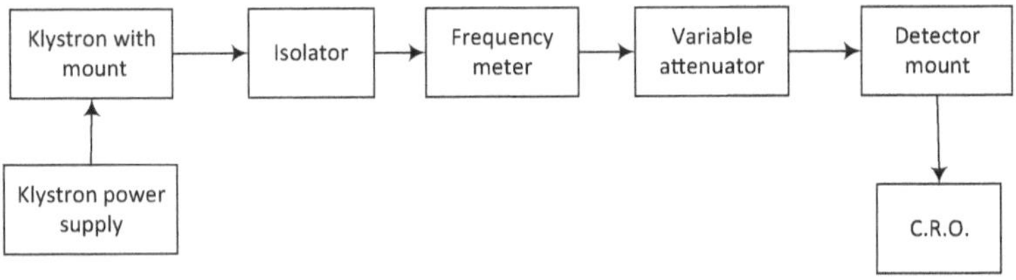

FIGURE 13.3 Block illustration of reflex klystron.

13.2.2 Reflex Klystrons

The reflex klystron is a frequency generator. It possesses low power and low efficiency. The electron beam goes through the resonant cavity, then into the delay field of the reflector, where it is reflected back by the field and re-enters the resonant cavity in the reverse direction. When the space first crosses, a very high-frequency electric field in space regulates the speed of the electrons. Second, in the opposite direction, the electrons reach in the form of groups. The UHF cosmic field slows these stars as they form clusters, converting some of their kinetic energy into UHF vibrational energy. Microwave energy is all that is involved. A reflecting klystron is depicted as a block diagram in Figure 13.3.

The reflex klystron is the most common UHF device. They are designed to operate in the decimeter, centimeter, and millimeter wave bands. Its output power ranges from 5 milliwatts to 5 watts. The efficiency of the reflection klystron varies between 20% and 30% (Parida et al., 2020). The reflection klystron is used as a heterodyne in superheterodyne radio receivers, as a driving oscillator in radio transmitters, as a low-power oscillator in radar, and in radio navigation.

13.2.3 Features of Microwave Communication

Microwave radio transmission is used in point-to-point communication, radar communication, radio navigation systems, sensor systems, radio astronomy, and radio communications. Some common features of microwave communications are given below:

- The microwave waves travel in a straight line due to which transmitter and receiver stations must be in exactly the same direction.
- Multiple receivers can receive serial signals without interference.
- Microwaves cannot enter the building. Therefore, it is not possible to effectively use endogenous receptors.
- Microwaves often penetrate the atmosphere. The reflected rays take longer to arrive at the target than direct sunlight. It leads to a phase change, which is known as multipath fading.
- Due to their short wavelength, microwave connectors are widely used for point-to-point communication and allow routing of small antennas with a fine beam that can be sent directly to receiving antennas. This allows adjacent microwaves to be used at the same time as low frequency radio without interfering with each other.

13.2.4 Green Microwaves

A number of researchers are working to design energy-efficient microwave communication systems. Costanzo et al. (2017) proposed co-design techniques for the energy-efficient ultra-band and ultra-high band frequency range. They proposed that circuit level optimization of a modern radiative RF/microwave system is possible by combining novel methods in nonlinear harmonic-balance-based analysis. For maximum system efficiency, every subsystem should be designed layout wise (Alessandra et al., 2017). By lowering intersystem matching networks, both the goal of minimizing size and the goal of minimizing system losses can be achieved. With the spread of pervasively distributed RF microsystems, considering miniaturized antennas, like wireless sensor networks or RFID, that should be low-power, low-cost, and achieve simultaneous identification, localization, and sense, is now required. Another energy-efficient technique was proposed by (Kawasaki, 2013) for energy harvesting winds based on the RF/microwave power transmission. In Europe, a lot of research based on green technology has been presented for energy harvesting. Additionally, wireless powering to the vehicle by microwave power transmission includes the solar power satellite, and it has been designed in Asia (Gupta et al., 2020). New microwave applications have sparked fast advancements in RF and microwave energy harvesting and transmission technology. The integration of energy harvesting and wireless power transmission (WPT) has recently been termed as "wireless energy transfer and conversion." Yogi et al. (2021) proposed an energy-efficient klystron. They proposed a klystron at saturation. They modulate the cathode high voltage, which was delivered by the modulator. Their proposed system reduces the requirement of power by 25%. Biao Hu et al. (2021) proposed a new magnetron, which was designed by asymmetrical resonant cavity, straps. Their output loop coupling circuit can enhance the interaction of beam wave and can also suppress the mode competition. Further, their system can increase the output coupling with 45% conversion efficiency.

13.3 SATELLITE COMMUNICATION

A satellite is defined as a celestial body that revolves around a larger celestial body on a path called an orbit. Such as the natural satellite of the Earth is the Moon. Similarly, the Earth is a satellite of the Sun. Communication satellites are microwave relay stations for telecommunications, radio, and television signals. There are nearly 750 satellites in space mainly used for communication applications. Satellite communication systems use satellites to wirelessly transmit data between ground stations. The received signal is amplified before being transmitted to Earth (AIAA et al., 2003). This means that the strength of signal of the receiving station is greater than that of the inactive satellite. A typical working channel consists of an active satellite and two or more terrestrial terminals. Radio stations send signals to satellites on frequencies called broadcast frequencies (Yuan et al., 2014). The satellite then amplifies the signal, converts it to a cut-off frequency, and sends it back to Earth. The receiver receives the signal. To cover most of the Earth, at least three satellites are essential.

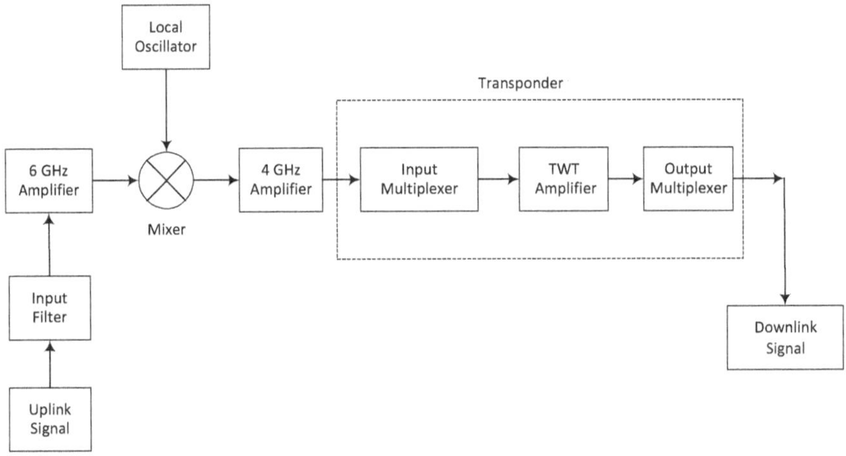

FIGURE 13.4 Block diagram of satellite communication system.

- Block diagram of satellite communication: A microwave repeater node is a communication satellite in space. It is beneficial in mobile communication, television, radio. A repeater circuit improves the strength of the received signal and retransmits the signal. But in this communication repeater circuit functions as a transponder. It means it varies the frequency band of the received signal from the transmitted signal. There are two types of frequency in this communication system: downlink frequency and uplink frequency. The frequency with which the signal is transmitted by the transponder is named downlink frequency, and the frequency with which the signal is transmitted into the space is named uplink frequency. Figure 13.4 shows this satellite communication concept.

Uplink refers to the signal transmission from the first Earth station to the satellite via a channel. Downlink refers to the transfer of a signal from a satellite to a second ground station via a channel. The uplink frequency of first Earth station is the frequency at which it communicates with the satellite. This signal is converted into a different frequency by the satellite transponder and sent down to the second Earth station. Downlink frequency is the name given to this frequency. The same procedure is followed by the second Earth station to communicate with the first. An Earth station is where satellite communication begins. An infrastructure is being built here to send and receive signals from a satellite orbiting the Earth. The information is sent to satellites through high-frequency and high-powered signals from Earth stations.

Figure 13.5 shows the complete scenario of satellite communication system where various devices are connected with the satellite substations. A satellite network is made up of three segments: space, ground, and control. The space section comprises of satellites organized in a constellation, whereas the control segment includes of ground facilities for satellite and traffic command and control. The ground segment includes all network nodes on the planet that serve as hubs or end-user devices for satellite-based service consumption. These stations can range in size from a few

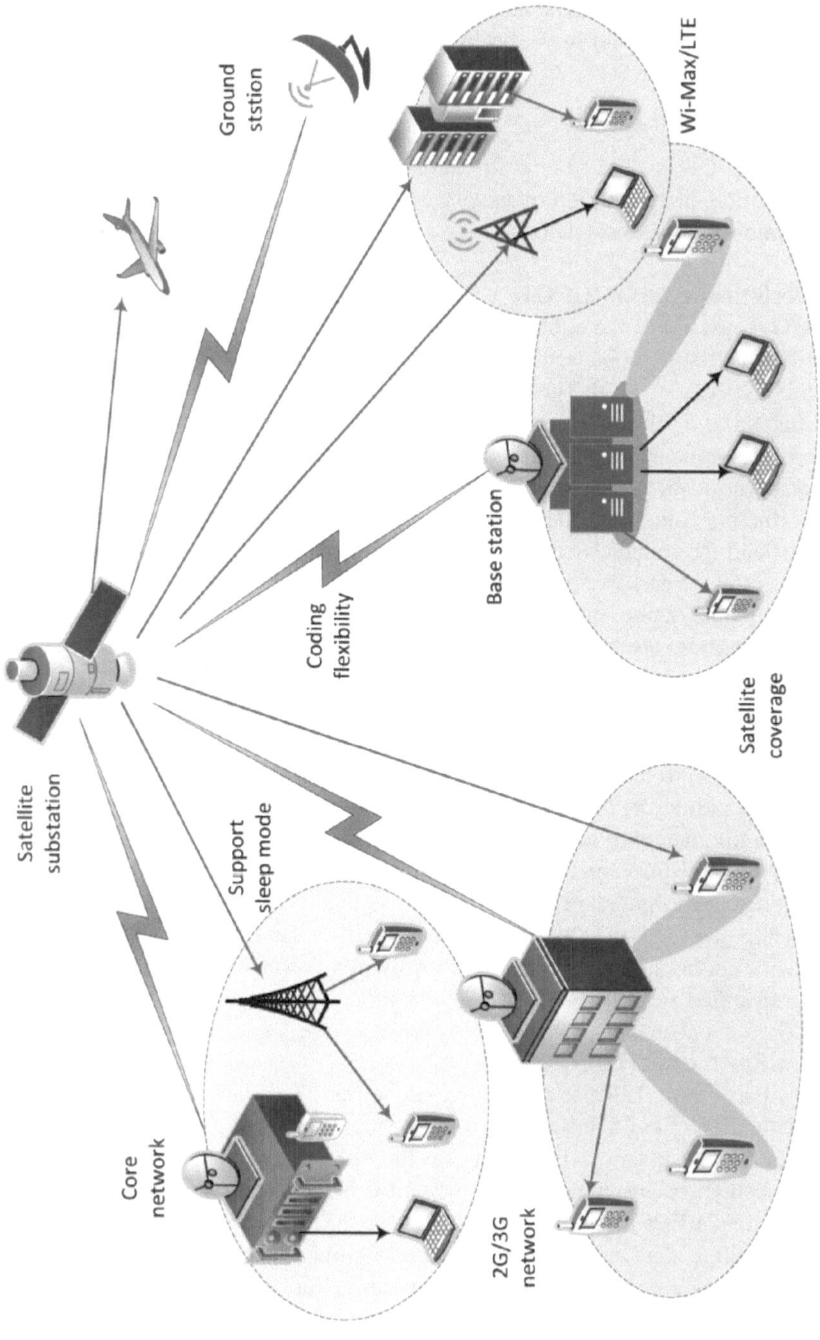

FIGURE 13.5 Satellite network.

centimeters to tens of meters in length. In this circumstance, the end-user device could be directly receiving broadband services through satellite without the need for terrestrial assistance. A terrestrial access point connected to the broadband network through a satellite link could serve the communication endpoint.

13.3.1 Orbits and Related Challenges

Once launched, the satellite or aircraft can usually be placed in one of many orbits around the globe, or sent on interplanetary travel. There are many types of orbits of satellite, which are discussed below:

- **Geostationary orbit (GEO)**
  GEO is just above the equator (0° latitude). This period is equal to the period of Earth's rotation (Ghasemi et al., 2013). This object appears to be associated with an observer on the ground at a specific location in the sky in a specific orbit. So, the relative speed of the Earth and the fixed orbit of the Earth are zero. Communication satellites and weather satellites were placed in a geostationary orbit so that the satellite communication antennas did not have to move in orbit, but could track their celestial position. Because of the constant latitude of 0° and the roundness of the geostationary orbit, the satellites in GEO only have different positions in longitude. Because the GEO satellites cover such a large area, only three equidistant satellites can give nearly worldwide coverage. This is because a satellite can cover a huge region at a time when it is far away from the planet. This is analogous to being able to see many maps from a distance of one meter rather than one centimeter. Geostationary orbits are useful because they make satellites appear stationary relative to a fixed point on the Earth as they rotate, allowing fixed antennas to maintain a link to the satellite. The altitude of the geostationary satellite from the surface of the Earth is 35,786 kilometers or nearly 36,000 kilometers. Satellites that must constantly be in the same place on Earth, such as telecommunications satellites, use GEO. As a result, the antenna can be grounded so that it does not move and remains aimed at the satellite at all times. These satellites can also be used for weather forecasting because they can continuously monitor and record weather trends in a specified area.

- **Low Earth orbit (LEO)**
  Low Earth orbit (LEO) is relatively close to Earth's surface. Altitude is usually less than 1000 km, but there are some that are only 160 km from Earth. It is lower than the other orbits. This means there are many satellite routes available for LEO. This is one of the reasons why LEO is such a great track. For a number of reasons, LEOs are very practical because they are very close to the ground. The satellite that revolves around this satellite can capture high-resolution images. It is also an orbit that makes it easier for sailors to make short trips to the International Space Station (ISS). The satellite orbits at a speed of 7.8 km per second. It takes about 90 minutes for a satellite to orbit the Earth at this speed.

- **Medium Earth orbit (MEO)**
  Medium Earth orbit is in between LEO and GEO. It is similar to LEO because it does not require a specific path around the Earth, and is used by various satellites with many applications. i.e., navigation applications, from exciting aircraft tracking to smartphone guidance.

- **Polar orbit and Sun synchronous orbit (SSO)**
  Polar orbit satellites do not have to traverse the North and South Poles. Polar orbits are defined as deviations of 20 to 30 degrees. Because they are located in lowlands at an altitude of 200 to 1000 km, polar orbits are derived from low Earth orbits. Synchronous solar orbit (SSO) is a sort of polar orbit that is unique. The Sun is synced with satellites transiting the SSW poles. Because they are always in sync, they are in the same "fixed" position in regard to the Sun. Often the SSO satellites are synchronized, so they are always at dawn or dusk. Because if you keep rolling at sunset or sunrise, the Sun will not place the Earth at an angle that blocks you. Satellites in Sun-synchronous orbit usually at altitudes between 600 and 800 km. At 800 km, it reaches 7.5 km per second.

- **Transfer orbits and geostationary transfer orbit (GTO)**
  The transfer orbit is a special type of track used to move from one track to another. When a satellite is launched from the Earth and sent into space with a launch vehicle such as Ariane 5, the satellite does not always go directly into its final orbit. Satellites are usually placed in transfer orbits, in this orbit, satellites or spacecraft can move from one orbit to another using relatively little built-in motor power. This, for example, permits satellites to reach high altitude orbits such as GEO without the requirement for a launch vehicle, which would take more work; it is like taking a shortcut. The geostationary transfer orbit is one of the most typical transfer orbits for reaching GEO in this manner (GTO).

- **Energy efficiency in orbits:** Nanosatellites in low Earth orbit (LEO), such as Cube-Sats, are among the most likely options for this type of communication backup. As a result, they are widely used in testing and other scientific endeavors. However, due to limited energy resources, we must optimize the energy efficiency of the communication connection between the satellite and the terrestrial radio station. We should on the physical layer performance metric, namely, energy efficiency, for designing the green satellite connection design. The spectral efficiency ratio is expressed as the total end-to-end satellite system power consumption divided by the whole end-to-end satellite system power consumption. Energy efficiency for the satellite communication link can be written as

$$\eta = \frac{\log_2(1 + CNR)}{P} \tag{13.1}$$

  where CNR indicates the carrier to noise power ratio at the receiver earth radio station, and P is total end-to-end satellite system power consumption. The CNR depends on various satellite connection parameters, such as antenna gains, effective isotropic radiated power (EIRP), loss factors, equivalent noise temperature.

13.3.2 Transponders

A transponder is an electronically controlled automatic device that transmits a received signal at various receiving, monitoring, amplifying, and retransmitting frequencies. It is mainly used for wireless communication. The word "sender" consists of two words: transmitter and transponder. Satellite communication channels are also called receivers because each channel is a separate transceiver or repeater. The transponder is not a single unit. It is composed of the duplexer, bandpass filter, broadband receiver, power amplifier, De-Mux in and Mux out, as shown in Figure 13.6. The duplexer is used to allow simultaneous transmission and reception. The duplexer is a bidirectional microwave gate that allows to receive the carrier signal from the antenna and transmit the carrier signal to the antenna. A basic bandwidth of 500 MHz is provided at C-band frequencies in the input link frequency range of 5.925 to 6.425 GHz. To reduce noise and interference, these frequencies are routed via a broadband bandpass filter (BPF). After that, it moved to a broadband receiver to give all channels with common frequency down conversion.

A wideband receiver is a set of microwave circuits and filters that separates the 500 MHz bands from the transmitter channel into separate passband channels. After the multiplexer, a power amplifier is provided to isolate the transmit channels, where the power level reaches the level required for downstream retransmission.

13.3.3 Lagrange Points

For a spacecraft orbiting in multiple orbits, even in distant orbits such as GEO, the approach to Earth may be too bad for a mission. The Lagrange point or L point has

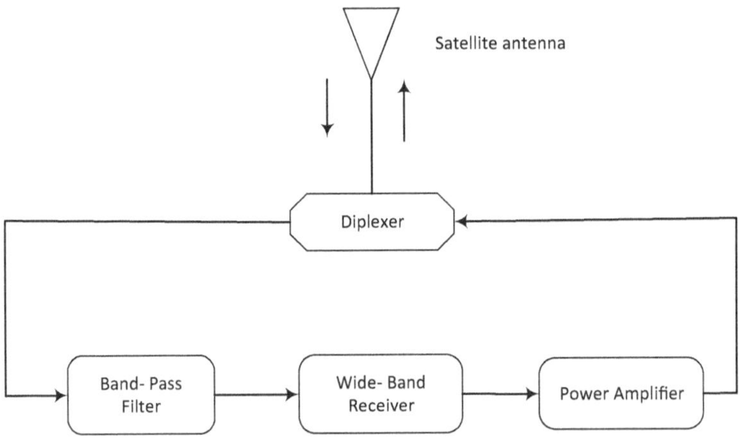

FIGURE 13.6 Block diagram of transponder.

a very long orbit (over 1 million kilometers) and does not rotate directly around the Earth. These are certain distant points in the Universe, and the gravitational fields of the Earth and the Sun combine to keep the orbiting spacecraft stable and "tethered" to the Earth. Of course, when a spacecraft is launched from somewhere in space far from Earth, it orbits the Sun. The spacecraft will soon leave Earth, making communication difficult. In contrast, the spacecraft remaining at this point L will remain stationary and approach the Earth with minimal effort, without going into another orbit. L1 and L2 are the most regularly utilized L points. L1 and L2 are quadruple the distance between the Earth and the Moon (1.5 million kilometers vs. 36,000 kilometres for GEO), yet this is still only around 1% of the distance between the Earth and the Sun.

13.3.4 Features of Satellite Communication

Services provided by satellite communications include voice, video, Internet, fax, television, and radio. Satellite communications enable communication and can operate in any environment or situation without using other forms of communication. Some of the common features of satellite communications are given below:

- The satellites used for satellite communications are usually in Earth's orbit. Some of them are also oval.
- Satellite communication can provide services around the world. This communication can set up not only on land but also in the sea. This will allow us to drive long distances without problems.
- One of the main advantages of satellite communication is its high reliability compared to other forms of communication. No ground infrastructure is required to operate.
- Satellite communication works very well because compatibility and speed differ from other communication formats.
- Satellite communication has higher implementation costs than most types of communication.
- Satellite communication often used for defence because it is not vulnerable to other forms of communication.
- Satellite communications also provide weather information.
- Services rarely fail, useful in the event of a disaster.
- It can transmit large amounts of satellite data.
- Communications satellites include transmitters, receivers, antennas, communications loads, switching and command systems, and system controls.

13.3.5 Green Satellite Networks

Compared to terrestrial networks, satellite communication has many advantages including coverage, content delivery, flexibility, and energy savings. Interference between terrestrial networks and satellite is, therefore, a key feature of 5G. The combination of satellite networks and terrestrial networks is a type of communication that can achieve dissipation at about 90% EE (energy efficiency), with only a 3% loss of SE (spectral efficiency) in high-density networks (Zhang et al., 2016). Hybrid separation of the control network and the user-level (C/U) network is a promising method for future communication systems. The C-plane satellite communication system provides holistic knowledge of base stations and terminals, including contextual information and user behavior, that can be used to implement information-oriented networks (ICNs) with a more flexible and dynamic deployments' strategy and bandwidth allocation scheme. Multichannel connections in a hybrid network can simultaneously receive satellite and small cell traffic, improving satellite control and planning. Hybrid U-plane networks, on the other hand, reduce signaling costs by allowing smaller cells to benefit from higher data throughput and longer idle time (Zhang et al., 2016).

13.4 THE EVOLVING ISSUES OF MICROWAVE AND SATELLITE NETWORKS BASED ON AN ENERGY-EFFICIENT PERSPECTIVE

- **Cognitive radio-based dynamic spectrum access concept:** This concept alters how wireless devices access spectrum and view them as wiser creatures. In wireless nodes, cognitive radios (CRs) can enable improved energy consumption minimization strategies. The first step toward this goal is energy awareness. That is, a node that is capable of modeling or estimating its energy usage can identify possible energy optimization areas. This necessitates data sensing and learning-based computational systems. Furthermore, the effective management of energy, as well as energy measurement in wireless networks, are critical issues for increasing energy efficiency. CRs are guaranteed to execute all of these tasks: the CR understands the power supply source, requirements of other power source, and the energy efficiency measurement of different adaptation methods (Srivastava et al., 2021). The sensor and computational infrastructure required to enable these capabilities is referred to as CRs. Another important characteristic of CR is its capacity to provide wireless communication infrastructure with upgradeable and longer-lasting radio equipment. This is especially critical for commercial end-user devices, because the lifespan of consumer devices is dwindling as the speed of new communication and application capabilities accelerates. From the standpoint of SDR, CR capabilities will be applicable to satellites, as SDR allows for remote upgrades of communication capabilities and possibly mission expansion/upgrades during the lifespan of a satellite (Srivastava et al., 2020).
- **Cross-layer framework**: This framework supports improving the energy efficiency of the satellite network. Because optimizing energy consumption necessitates a thorough examination of the low-level radio hardware, but performance can only be achieved by considering the protocol stack's features, the energy management challenge necessitates a cross-layer solution approach (Gupta et al., 2019). To avoid sacrificing the QoS of essential users, network nodes can manage their energy consumption utilizing several control dimensions such as switching to alternative network interfaces, adaptive sleep cycles, application-specific policies, and class-based policies. These communication

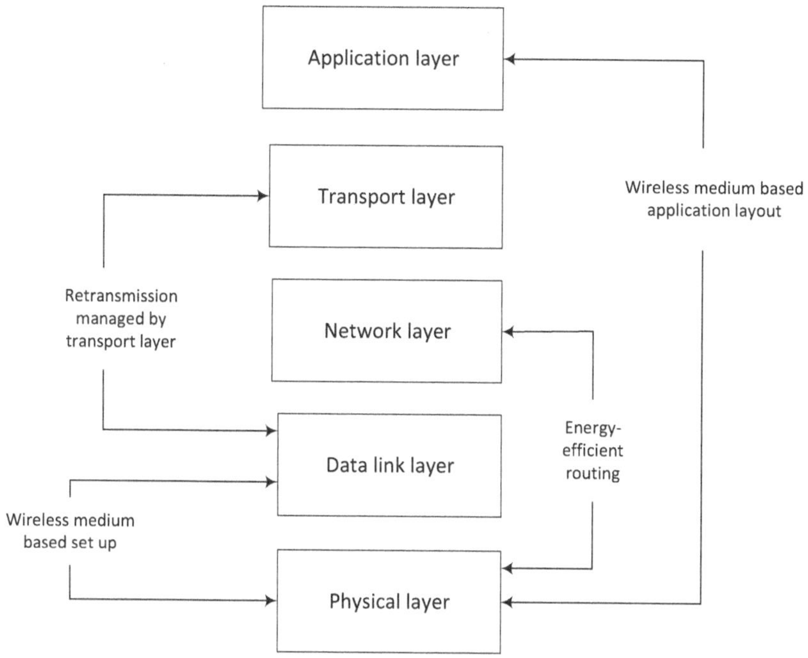

FIGURE 13.7 Cross-layer framework for energy efficiency.

technologies will be used in dynamic contexts where a single energy management solution will not be enough. As a result, cross-layer solutions that are versatile are required. Figure 13.7 depicts the cross-layer framework for energy efficiency.

- **Development of smart grids:** This method provides ample possibilities for satellite networks to become an enabler for improving the overall energy efficiency of infrastructure systems. The smart grid, or next-generation power infrastructure, is projected to address the key flaws in the current grid. The smart grid is envisioned as a sophisticated interconnected system that will provide utility companies with new capabilities such as real-time monitoring and ubiquitous control over their equipment and services. Smart grids require a high level of resilience. They must be self-healing and resistant to system flaws or harsh conditions. Satellites can provide connectivity and crucial services for the widespread implementation of smart grid technology. In addition to supporting substations and power plants, satellite is becoming a more realistic alternative for other smart grid applications, such as the ones listed below.
    - Security and productivity: Satellites can deliver internet services to remote substations for security applications such as video surveillance, as well as phone and data services for productivity. Connectivity for distribution automation, ensuring connectivity across the service territory.

- Remote renewable generation site management, such as solar or wind farm management, is monitored and controlled. Redundant communications at critical substation and distribution sites to backup terrestrial communications.
- **Integration of various networks:** Future mobile networks will likely include systems based on a variety of technologies, including PAN, WiFi, LAN, WiMAX, UMTS, LTE, 5G network, and satellite networks. To address this diverse environment, the International Telecommunication Union (ITU) has developed a hybrid and integrated framework in the context of next-generation networks. The IP based next-generation communication systems offer high-speed data transfer and connectivity to terrestrial data networks and satellite networks (Gupta et al., 2021). The integrated architecture should have a favorable influence on capacity or QoS optimization, resulting in energy savings. In the grid-connected terrestrial base station, there is a demand for minimal energy usage because excessive heat generation is a severe concern. In this integrated setup, energy-efficient multimedia transmission is also achievable. Satellites' intrinsic broadcast capabilities make them a promising candidate platform for delivering point-to-multipoint services. Their close synergy with terrestrial mobile networks has advantages for both mobile and satellite operators. The envisaged benefits of an integrated architecture for satellite operators are investment savings on research and development, reduced total cost of deployment, and the potential to penetrate a broader market than the vertical niche markets that mobile satellite systems typically address.

13.5 CURRENT CHALLENGES FOR SPACE TECHNOLOGIES

13.5.1 Technical Challenges

In the long term, the challenge lies in the development and implementation of technologies, such as air breathing hypersonic rocket engines, which will be used in hybrid launchers to reduce the need for large amounts of oxygen that current vehicles must carry. It should also develop a launch vehicle that can take off and land like an airplane, without the need for a large number of expensive services between missions. Similarly, space propulsion provides opportunities for improvement, especially in the area of electric propulsion systems, which are also hybrid systems that use different operating modes (Aglietti et al., 2020). The performance of the propulsion system is also critical for interplanetary missions because it can provide faster travel and larger payloads when needed.

13.5.2 Protection of Humans

The challenge is to create an artificial environment that fully supports the physical and mental health and well-being of people and protects them from the negative effects of the spatial environment. Whether you are looking for a man-made deep-space vehicle, a large-scale space platform, or a planetary colony for human habitation, a few challenges overlap here. These issues include the need for effective feedback systems to replenish resources and reduce waste (Guglielmo et al., 2020).

Most of the craters created by the largest celestial bodies since the formation of the Earth have been destroyed by geological processes on the planet, but dozens of large craters can still be seen (a good example is the Ballinger meteorite, one kilometer wide). It formed in Arizona about 50,000 years ago when a metal object about 50 meters in diameter collided, a reminder that such a collision could have occurred.

13.5.3 Earth Environment

Globally, satellites provide objective information to develop and validate models that track the environment and enhance forecasting capabilities. However, more effort is needed to provide information on specific needs and concerns. The Earth observation (EO) market, whether institutional or private, requires extremely high coverage, resolution, and analysis time to enable real-time Earth observation.

13.6 APPLICATIONS OF MICROWAVE AND SATELLITE COMMUNICATIONS

Microwaves are used because of their shorter wavelength and high frequency, which allows them to carry a lot of information. Microwaves are most suitable for communicating with geostationary satellites because they have higher frequencies and smaller wavelengths. They are not the best choice for communication with low-orbit satellites. Satellites emit microwaves with a wavelength between 1 and 10 cm. The wavelength-emitting plate has a larger diameter. This does not produce wide diffraction, so it emits a narrow beam that does not diffuse. Some important applications of microwave and satellite communications are listed below:

- **Low latency systems:** Except for line-of-sight propagation delay, cable free low latency versions of microwave lines use low latency technology, with no delay between packets being delivered and received at the other end. Microwave transmission in the air is roughly 40% faster than through fiber optics, resulting in a 40% reduction in latency for clients when compared to fiber optics. Furthermore, because of the reality of building architecture, street ducts, and the requirement to utilize existing telecom equipment, fiber optic set ups are almost never in a straight line. The fiber run can be 100 percent longer than the direct line-of-sight path between two end locations. As a result, cable free low latency microwave solutions are widely used in low-latency applications like high-frequency devices and other applications.
- **Mobile carrier backhaul (MCB):** In MCB, microwave links are a useful tool: traditional PDH 16xE1/T1, modern IP Gigabit ethernet backhaul connectivity, and Greenfield mobile networks, can all benefit from microwave technology. When compared to installing or leasing fiber optic networks, microwave is significantly faster to set up and has a lower total cost for cellular network operators.

- **Industry-based applications**: Microwave products can be used for a variety of industry applications, including disaster recovery, machine to machine connectivity, network redundancy, and temporary connectivity for applications such as video, voice, and data services, various engineering services, medical imaging, and fixed-line carrier bypass, thanks to their scalability and flexibility.
- **Television broadcasting:** Satellites in geostationary orbit are ideal for communication because the microwave antenna can point at the satellite without knowing its location. Broadcasters also use geostationary satellites to distribute services to regional centers. These satellites are also used in cable television networks.
- **Internet application:** Microwaves are used in wireless LAN protocols. If you have used Bluetooth or Wi-Fi before, you have probably tried one of the microwaves. Much of modern life can now be done online using a microwave. Without them, shopping, paying bills, or booking movie tickets would be impossible with a smartphone or laptop. Microwaves are used in satellite communications, but they are not the only ones. Microwave antennas are used for high frequencies, which can transmit a lot of information, and for short wavelengths, where it is easy to aim directly at the receiving antenna.

13.7 SPACE ENVIRONMENT AND RELIABILITY OF SATELLITE COMMUNICATIONS

One of the first discoveries of Earth's first satellites was the accidental large number of charged particles trapped in the Earth's magnetic field. Soviet instruments have detected signs of this, but it was the U.S. Explorer 1 satellite that helped determine that the Earth is surrounded by what is now called the Van Allen radiation belt, which was used by scientists who designed instruments on Explorer 1 and correctly interpreted the reading. These early probes also experienced other characteristics of the space environment, some were expected, some were unexpected. These characteristics include a very hard (i.e., relatively very pure) vacuum, so-called zero gravity, high levels of sunlight, radiation, and micrometeorite hazards. The free fall of satellites in low Earth orbit creates conditions commonly referred to as zero gravity. Technically, the term is inappropriate. The gravity in low Earth orbit is hardly weaker than the gravity on the Earth's surface (Jansky, 2003), it is the motion of the satellite that produces the weightlessness effect. However, the term zero-gravity has begun to be commonly used and will continue to exist. Because slight accelerations occur even on satellites, mainly due to air resistance and satellite motion, the recent term microgravity has also been commonly adopted.

Unfiltered solar radiation can cause the light-emitting parts of the spacecraft to rise to high temperatures. At the same time, the shadows of the spacecraft radiate their heat into space and cool to below the freezing point of common fluids such as water and storable rocket fuel. All these containers and fluid lines are usually equipped with electric heaters, and the overall temperature is adjusted by rotating the spacecraft along an axis perpendicular to the spacecraft-Sun line. This is called passive thermal control. Unmanned spacecraft heading to the inner planets must be equipped with

Sun visors to reflect unwanted solar heat. Those who are sent to the outer solar system or the surface of the moon often use radioisotope heaters during the two-week night.

The satellites in LEO are shielded by the magnetosphere from the Sun's charged particles and most of the cosmic rays from outer space. However, vehicles operating in GEO or interplanetary missions will be fully affected by these radiations (Yuan et al., 2016). As we all know, cosmic rays penetrate the integrated circuits of spacecraft autopilots and change data and commands. During solar storms, the space version of static electricity can accumulate in other spacecraft, generating electric sparks, which can cause serious problems with onboard electronic equipment. The empirical design of such systems reduces the impact of these effects (Ippolito et al., 2008).

13.8 RELIABILITY OF SATELLITE COMMUNICATIONS

The probability that the system will operate normally within a given service life is called reliability. The reliability of a complete satellite communication system depends on the reliability of its two main components: satellite and ground station. In order to have a long-term reliable satellite system, reliability requirements must be part of the inherent design. From a programmatic point of view, this includes the following steps:

- Establish design goals for the life cycle of the task and the likelihood of completion.
- Conduct reliability trade-off studies to select the best design with redundancy.
- Establish design constraints to meet design goals, especially end-of-life parameter reduction and degradation requirements.
- Planning reliability plan; parts and material control plan; subcontractor and supplier control plan.
- Results monitoring.

13.9 RESEARCH OPPORTUNITIES FOR GREEN MICROWAVE AND SATELLITE COMMUNICATION SYSTEMS

There are a number of research opportunities for green microwave and satellite communication systems.

- **Ground station network for next-generation satellites**
  With the continued rapid development of the aerospace industry, companies that have the best technology to meet their new requirements are ready to prosper. The ground station network allows cost efficient, real-time, and easy access to micro/nanosatellite data, and these ground stations make it a suitable market for small satellite operators. Historically, the ground part of a space mission was mainly designed, deployed, and operated by mission operators themselves. In contrast, smaller satellites have less stringent requirements for the amount of data transmitted by each satellite but require flexible operation and low latency.

- **Satellite and autonomous vehicles**
  Driving autonomous vehicles on city roads and highways requires support from mobile phones and ground-based Wi-Fi networks. Satellites play an important role in delivering necessary security software updates and other important information (such as map data) to connected vehicles. Connected autonomous vehicles will forever change the way people and products are transported. The connected environment is also changing the way automakers rely on software to deliver more powerful performance. The savings that come from reducing one-time service visits during a car's lifecycle is a compelling argument in favor of networking cars.

- **Future satellites for oceans cover over 70% of the Earth's surface**
  They provide us with food, regulate our climate, and are the road connecting us all. Satellite and airborne technologies provide enhanced possibilities to monitor your health at a more detailed granularity (Chenhao et al., 2020). Satellites will be the main driving force for future research including tracking plastic waste, shallow water, remote sensing to quantify coral reef crises, monitoring of ocean health, plankton, and acidification.

- **Transferring data from observation satellites**
  Many observation satellites have been developed to continuously monitor the Earth to view the environment and forecast the weather. The satellite is equipped with a radio station capable of recording data and sending it to Earth via an electronic connection.

- **Data transmission using deep space probes**
  The power, quantity, and quality of the equipment mounted on the deep space probe is strictly limited, so the transmission power and the diameter of the antenna are also limited. Ground stations are therefore often equipped with high-power transmitters and multiple receiving antennas to overcome the limitations on board (Jing et al., 2020).

- **Data transmission within the satellite clusters**
  The functions normally performed by large satellites transmitted to the various smaller interconnected and collocated satellites. Each satellite can be allocated different functions, but generally speaking, the group will act as a large satellite. Synchronization of manoeuvring and cooperative control is essential to maintain safety restrictions in the placement tank. The LEO satellite monitoring and control system is controlled through the extended connection of the GEO satellite (Alvarez et al., 2016).

13.10 CONCLUSION

The advantage of microwave communication is that the high frequency of the microwave provides a microwave band and a very large capacity for transmitting information.

The microwave band has a bandwidth 30 times greater than any remaining radio spectrum below. While satellites use microwave frequencies to communicate with each other using inter-satellite links and ground stations or VSAT. It discusses different types of satellites based on applications and their orbits, where we use them in real life and how they are very useful.

REFERENCES

Aglietti. Guglielmo S., "Current Challenges and Opportunities for Space Technologies," Frontiers in Space Technologies, p. 1, 2020.

Alvarez, J. and Walls, B., "Constellations, clusters, and communication technology: Expanding small satellite access to space," 2016 IEEE Aerospace Conference, 2016, pp. 1–11, doi: 10.1109/AERO.2016.7500896.

American Institute of Aeronautics and Astronautics (AIAA) and Takashi, "Progress In Astronautics and Aeronautics: Satellite Communications In the 21st Century: Trends and Technologies, AIAA, 2003, vol. 202."

An, Yuan., Li, Jiancheng, Fang, Weiwei, Wang, Baohua, Guo, Qiwang, Li, Jing, Li, Xiaolong, and Du, Xing ."EESE: Energy-efficient communication between satellite swarms and earth stations," International Conference on Advanced Communication Technology, IEEE, 2014.

Chenhao, Qi, Huajian, Chen, Yansha, Deng, and Arumugam, Nallanathan, "Energy Efficient Multicast Precoding for Multiuser Multibeam Satellite Communications, IEEE Wireless Communications Letters, vol. 9, no. 4, pp. 567–570, 2020.

Costanzo, A., Masotti, D., Fantuzzi, M., and Del Prete, M., "Co-Design Strategies for Energy-Efficient UWB and UHF Wireless Systems," IEEE Transactions On Microwave Theory And Techniques, vol. 65, no. 5, May, 2017.

Ghasemi, A., Abedi, A. and Ghasemi, F. Propagation Engineering in Radio Links Design, Springer Science & Business Media, 2013.

Gupta M. S. and Kumar K., "AFSOS: An auction framework and Stackelberg game oriented optimal network's resource selection technique in cognitive radio networks," IEEE Transactions on Network and Service Management, vol. 19, no. 1, pp. 61–72. 2021.

Gupta M.S. and Kumar K., "Application Aware Networks' Resource Selection Decision Making Technique using Group Mobility in Vehicular Cognitive Radio Networks," Vehicular Communications, Elsevier, vol. 26, pp. 100263, 2020.

Gupta, M. S. and Kumar, K., "Progression on Spectrum Sensing for Cognitive Radio Networks: A Survey, Classification, Challenges and Future Research Issues," Journal of Network and Computer Applications, Elsevier, vol. 143, pp. 47–76, 2019.

Hu B., Cao Z., and Zhou M., "Energy-minimized scheduling of real-time parallel workflows on heterogeneous distributed computing systems," IEEE Transactions on Services Computing, 2021.

Hu, Jing, Li, Guangxia, Bian, Dongming, Shi, Shengchao, Ge, Ruixing, and Gou, Liang. "Energy-Efficient Cooperative Spectrum Sensing in Cognitive Satellite Terrestrial Networks, IEEE Access, vol. 8, pp. 161396–161405, 2020.

Ippolito. "The RF Link," Satellite Communications Systems Engineering Atmospheric Effects Satellite Link Design and System Performance, 08/21/2008.

Jansky D. and Kubik R. D., "Chapter 4 Regulatory and Spectrum Considerations," Springer Science and Business Media LLC, Springer, Boston, MA, 2003.

Kawasaki, S., "The Green Energy Harvesting Winds by the RF/Microwave Power Transmission," 2013 IEEE Wireless Power Transfer (WPT), pp. 111–114, IEEE.

Louis, J., and Ippolito, Jr. "Satellite Communications Systems Engineering," A John Wiley & Sons, Ltd, Publications, 2008.

Parida, Majhi S., Das, S., and Kumar, S., "Wireless Powered Microwave and mmWave based Communication Networks-A Survey," ICICT, IEEE, 2020.

Srivastava A., Gupta M. S., and Kaur G., "Energy efficient transmission trends towards future green cognitive radio networks (5G): Progress, taxonomy and open challenges," Journal of Network and Computer Applications, Elsevier, vol. 168, pp. 102760, 2020.

Srivastava, A. and Kaur, G., "Resource management for traffic imbalance problem in green cognitive radio networks," Physical Communication. Elsevier, vol. 48, pp. 101437, 2021.

Yogi, S. and Vinay, N., "Microwave based Approach for Wireless Power Transmission," Journal of Microwave Engineering & Technologies, vol. 8, pp. 1–5, 2021.

You, K.Y., "Introductory chapter: RF/microwave applications," Emerging Microwave Technologies in Industrial, Agricultural, Medical and Food Processing. IEEE Transactions on Services Computing, vol. 34, no. 12, pp. 3869–3886, 2021.

Yang, Y., Xu, M., Wang, D., and Wang, Y., "Towards Energy-Efficient Routing in Satellite Networks," IEEE Journal on Selected Areas in Communications, vol. 34, no. 12, pp. 3869–3886, 2016.

Zhang, J., Evans, B., Imran, Muhammad Ali, Zhang, X., and Wang, Wenbo, "Green Hybrid Satellite Terrestrial Networks: Fundamental Trade-Off Analysis," 2016 IEEE 83rd Vehicular Technology Conference (VTC Spring), 2016.

Zubir, F. and Radhakrishnan, R. R., "Highly Efficient Power Amplifier for Microwave Transmitter," Asian Simulation Conference, Springer, Singapore pp. 736–745, 2017.

14 Green Energy Efficient Wired and Wireless Charging Techniques for IoT Enabled Healthcare Systems

Arshpreet Singh and Yaman Parashar

CONTENTS

DOI: 10.1201/9781003264477-14

14.1 INTRODUCTION

With rapid technological advancement in the current era, the IoT has turned into a massive growing concept, which holds the potential to influence our lives by interconnecting all daily use electronics devices (Lenaerts and Puers 2007, Chen et al. 2009, Sun et al. 2011, Sun et al. 2012) like the computers, refrigerators, gadgets, sensors etc. together in a comprehensive way. All this simply comprises of smart devices that are associated through a communication channel to the Internet. Where the cloud gives many services to associate distinctive devices upon request. Since the IoT worldview is to associate numerous "things" together, which will empower smarter and more intelligent devices to lift the quality of life to much higher levels. With such massive involvement of these tiny devices all around us, the projection of these smart things is expected to reach the mark of a billion in the next decade (Xie et al. 2006, Carpi et al. 2011). But with such huge involvement of these devices, several other challenges may arise. Their interconnection led them to communicate with each other, thus creating a mesh network that can handle any types of services that one needs to get accustomed with over a different period of time. It simply consists of smart devices (Tomar et al. 2018) that are connected to the internet with a particular communication channel. With such an outburst, the projection of IoT is expected to reach trillions in the near future or so (Wong et al. 2004, Naganuma et al. 2011, Lee et al. 2011). What makes it more significant is its applications in the healthcare sector, through which it is supporting a billion lives on a daily basis. It helps in designing remote health monitoring system for patients who cannot avail medical facilities easily. A typical IoT (Kaur et al. 2018; Lenaerts and Puers 2007) enabled healthcare system usually consist of biomedical sensors or implants, gateways like SBCs, a communication channel, cloud space for data storage and a system for fetching and visualization data at the remote end (Chiu et al. 2010). Though it is harnessing the very crucial need of the patients, its dependency on an implantable battery driven sensor is what makes them inefficient in a number of scenarios. Since the implant has to be placed inside the body of the patient, we cannot rely on running a bunch of wired cables through the body of the subject. It will definitely make the patient much more prone to infections and many other adversities. Thus in order to cope with this problem effective wireless charging solutions needs to be incorporated with IoT that can help patients in avoiding unnecessary surgeries and infectious attack that may intervene if the conventional method is adopted. Apart from this, over dependency on millions of charging devices to support IoT (Tomar et al. 2017) architecture can only put a huge and massive pressure on the environment as a whole by accumulating a large number of e-waste that needs proper disposal and recycling. Therefore, new inventions (Si et al. 2008) and research has been going on in this area to provide effective solutions that can benefit mankind from this fatal issue or problem.

Apart from implantable sensors, a wearable device is another thing that also support the IoT healthcare enabled monitoring system in a great way (Schluter et al. 2005). They need to be smart, multi-functional, portable, along with lots of similar features that makes them distinct in their approach. However, their dependency on batteries and wired cable limit their capability to a certain extent. It is therefore very crucial to find innovative better solutions that can help in extending their life time without having any disruption in the process of monitoring.

14.2 MOTIVATION

The application of transferring wireless power (Sun et al. 2011) is a huge leap forward for current biomedical implants it offers a boundless remote power source and enhances general features of the overall system. Since the power can be remotely transferred to the human body, the utilization and the characteristics of these medical systems can be altered according to the requirement. For example, previously, the battery in the traditional pacemaker always changed in a time of 5–10 years (Ghallab et al. 2005) but due to adaptation of this technology, pacemakers are now able to charge wirelessly, avoiding chances of potential multiple surgeries for battery replacement. The most useable battery in every type of electronic device in today's world is the lithium-ion battery that stood above its peers due to low leakage, high energy density, long lifetime, and low memory effect. However, their high cost, aging effect, and sensitivity to high temperature limit their usage on a much wider scale. Apart from this, another factor that puts it in the downstream is its improvement rate, which is increasing at a much lower rate in comparison to the rate of innovations and improvement in the semiconductor technology. Due to this very reason, batteries need to be much heavier, costlier, and larger in sensors and wearable devices.

At the global level, the wireless power transfer is surging at a stunning rate. Various researchers around the world are working on numerous sorts of ways to improve system stability, conversion efficiency and so on, by developing many kinds of circuit technologies and specialized antennas for the wireless power transfer technology. In the future there is a strong possibility that the deployment of these major changes in the technology will improve the performance of the implants so as to help human recuperation from various vulnerable conditions they are suffering with. In a similar manner, the primary objective of this chapter is to present an introduction to the in-depth technologies related to wireless power transfer for a number of medicinal applications. This chapter therefore presents a brief history of WPT with applications, WPT system designs with possible challenges like efficiency, size, and safety and so on. It also describes the modeling of systems and components, like power converters, antennas, and management circuits etc. and a brief description of standards with prospective future work.

14.3 FUNDAMENTAL OF WIRELESS CHARGING

The principle of charging techniques, currently available applications, and design of charging system with respect to architectures, hardware designs, and implementations are explained in this section. Moreover, we present the wireless power propagation models for non-radiative charging systems. Wireless charging is a technology that allows transmission of power from one device to another through air. But what we usually do not know is its sub classified techniques, standards, detail description about their working components, their limitations and challenges that they might face according to a particular type of application. With the rapid technological advancement in the wireless charging sector, the dependency on portable battery powered device has come to a halt. However, despite a number of advantages, there also exist a number of challenges that this technology is facing, which are in regard to size, power management, and implementation. Therefore, in this chapter we have tried to give our readers a comprehensive overview on the working of this technology as a whole, its importance in IoT enabled healthcare systems, technical advancement in networking applications etc. Apart from all this, we further provide a brief introduction to the challenges that the current biomedical systems are facing while setting up or implementing the wireless charging technologies. At last, we provide a brief to the potential architectures that can be adopted after dealing with the issues, challenges, and limitations that we have discussed in this chapter.

14.4 WIRELESS CHARGING TECHNOLOGIES

In today's world there exist primarily two main charging technologies, one of which is radiative radio frequency (RF) based while the other one is non-radiative coupling-based charging. Amongst both of them, the later one is further sub classified into three techniques, which are generally described as the magnetic resonance, capacitive, and inductive couplings. While on the other hand, the former technology is also sorted into two techniques, namely non-directive RF transfer and directive RF power beam forming. Capacitive coupling is one such technique that finds its use in a number of applications. However, the only limitation that is associated with this technique is its dependency on the area of the system. Due to this reason it often fails to deliver the required power density that is needed. Similar constraints are raised when directive RF beam forming technique is employed due to its LOS directivity constraint that needs to be fulfilled. These are some of the reasons why these two techniques are avoided in many new biomedical applications. To cope with this, a few other techniques like the magnetic resonance coupling, inductive and non-directive RF radiation are often employed (Finkenzeller et al. 2010) on a larger scale. The advantages associated with them is what makes them special for a different set of scenarios. The two techniques in the non-radiative radiation category mainly works on the near field category due to their ability to perform well in a short space between the transmitter, receiver as well as the reflecting nearby bodies. The attenuation in this near field power is generally defined by the cube of reciprocal of the charging distance. On the other hand, the non-directive RF beam forming works quite well in the far field. Here the far field power often gets degraded by the square of the reciprocal of the charging distance. Apart from this, the captivation of radiation does and does not disturb the transmitter in the near field and far field technique, respectively. This mainly occurs due to the appropriate coupling amongst the transmitter and receiver in the near field techniques and failure in the far field techniques.

14.4.1 Inductive Coupling

Inductive coupling is a type of wireless charging technique where the primary coil of a transmitter is responsible for producing significantly varying magnetic fields in the nearby secondary coil present within the receiver (Tesla et al. 2007). This magnetic power is quite capable in inducing current/voltage across the secondary coil, which further has been used to derive the associated circuitry and charge a storage system for full operation. The operational frequency of this technique typically lies in the Kilo Hertz range (Thidé et al. 2004). The given system performs well if the secondary coil is adjusted to the operating frequency but in the absence of the value of quality factors the proper required efficiency is usually maintained at a mere range of 20 cm (Tesla et al. 2006).

14.4.2 Magnetic Resonance Coupling

Another technique, which is named as the magnetic resonance coupling is a type of wireless charging technique that holds the potential to transfer energy from one resonating coil to another by varying the magnetic fields that exist in space between both of them. Due to the proper resonance between the two coils at the operational resonating frequency, minimal leakage of high energy transfer is possible. A model with the same architecture and characteristic were able to achieve a transfer efficiency of 92.6% over a very short distance of 0.3 cm in Barrett et al. (1894). Apart from this, a few other previous works employing this technique have also shown their ability to transfer power to a much larger range, in comparison to the inductive coupling, and much high-power transfer efficiency than the non-radiative RF beam forming approach. Its ability to charge multiple devices at a time is what makes it much more unique among its peers. Due to operational frequency in the megahertz range, the quality factor of the technique usually remains high. This often helps in mitigating the decrement in the overall power efficiency due to gradual increment in the charging distance. In Chiu et al. (2010), a considerably stronger magnetic resonance coupling method was proposed with the name of Witricity, which holds the potential to increase the overall power transfer energy more than 90% with 1 m transmission distance. However, the only limitation that this structure has was its inability to reduce the receiver size due to the large space requirement for the secondary coil to operate fully. This is perhaps the main reason behind the inability of Witricity to penetrate the market of small biomedical devices and sensors for quite a long time since its inception.

14.4.3 RF Radiation

RF radiation is another important wireless technology that utilizes microwave as a medium to propagate in the space at the speed of light. Its typical operational frequency range is from 300 MHz to 300 GHz (Yagi et al 1928, Brown et al. 1961, Anderson and Tesla 2002). Although the given energy type can use other EM waves like X-rays and infrared, but due to the possible threat to humans, they are not employed. This type of technology also holds the potential to radiate in a particular specific direction through beam forming. With the rise of the number of antennas, the sharpness of the energy further improves to a much higher level. Apart from this, the given technology is known for providing several fruitful advantages at the receiver side, like deploying smart antenna (Parasher et al. 2019) technologies, which is developed to provide an apt tradeoff between complexity and performance of the whole biomedical WPT system.

14.5 APPLICATIONS

In regard to application type, there are mainly two types in which the whole WPT charging can be summarized in Figure 14.1. A brief introduction to the same has been provided below in this section.

14.5.1 Near Field Charging

Amongst the two types of charging that we have discussed above for biomedical applications, near field charging is one of them. This type of charging came into existence mainly through two types of couplings, one of which is the magnetic resonance while the other one is inductive coupling, which is used primarily in the current biomedical applications due to their ease of implementation and low-cost factor. Depending upon the power usage the near field charging can be further categorized into different levels. However, most of the medical devices and appliances that we deploy for various applications work under the medium power near field charging category that usually ranges from few watts to ten of watts operational power. A number of biomedical implants and sensor that we used in the IoT enabled healthcare system often adapt inductive coupling in most of their applications. These systems are quite capable of achieving more than 50% charging efficiency within a less than 10 mm air gap between transmitter and receiver .While on the other hand, biomedical system with magnetic resonance coupling portrays much more penetration ability in

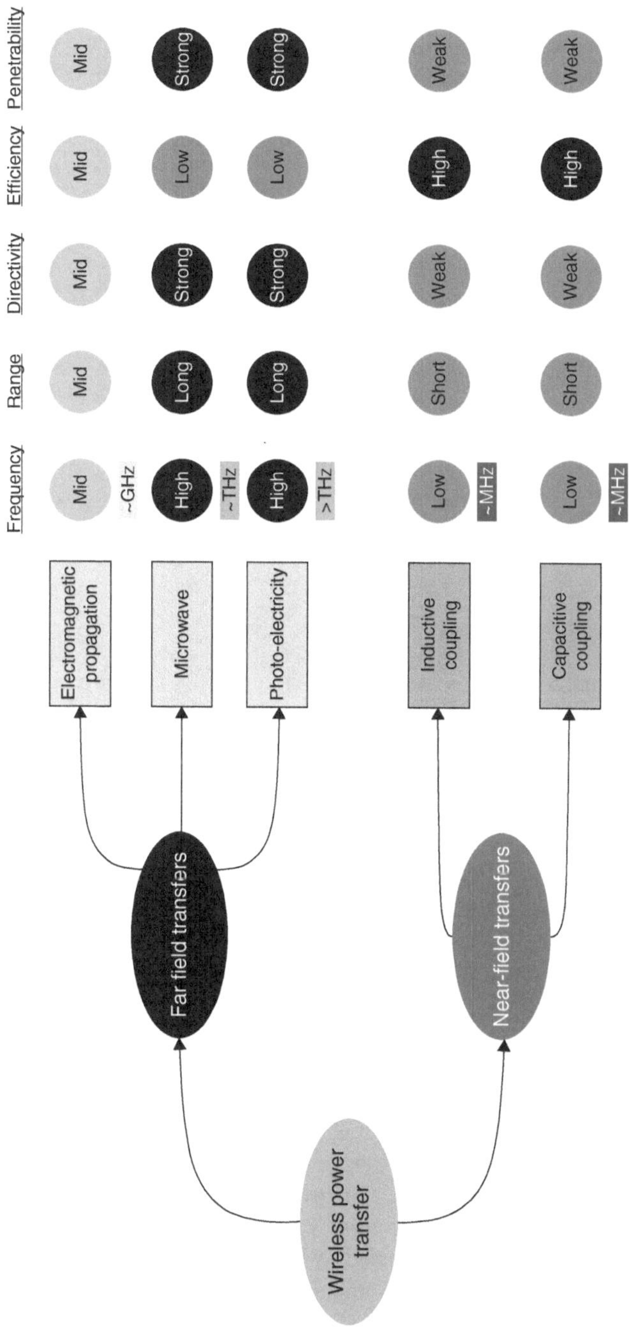

FIGURE 14.1 Wireless power transfer application categories.

comparison to the inductive one and thus proves quite useful in applications, which involve smaller implants and sensors for much larger ranges. Evidence of this has been realized in a system (Kurs et al. 2007) where more than 60% efficiency has been achieved over 20 cm distance with 3 cm and 2 cm wide transmit and receiving coil, respectively. Few improvements in the architecture of these system can further scale it up to 70% efficiency in various scenarios.

14.5.2 Far Field Charging

Amongst the two-charging technique, the second charging method that has been adopted widely across different sorts of biomedical applications is far field charging. These types of charging are mainly adopted through directive RF beam forming or through non-directive RF radiations. The non-directive method is generally less susceptible to relative position and orientation to the antenna at the transmitter end (Yagi et al. 1928, Brown et al. 1961) and holds the capability to conduct charging even without the clear line of sight. While on the other hand directive RF beam forming is widely employed to high power consumption biomedical devices. By bringing technological advancements in these techniques or methods, high power transfer systems have also been realized in today's world to support a myriad of ultra-high power consumption biomedical applications.

14.6 SYSTEM MODELING

A WPT unit in biomedical applications is typically made out of a power transmitter fixed out of the human body and a power receiver inside the body or fixed underneath the human skin. A few sophisticated WPT units often employ a power relay in the middle to deliver power to much larger places. Apart from this there also exist power antennas, converters, power management circuits that also play a significant role in the working of WPT charging technology (Anderson asnd Tesla 2002, Landt et al. 2005). Therefore, the given section tries to describe constraints and design challenges related to each component, which is employed for the wireless power transfer in a number of biomedical applications. The whole model comprises of three components as shown in Figure 14.2, and brief introductions are given down below.

14.6.1 WPT Transmitter

A power transmitter usually performs a lot more function apart from emitting wireless power into space. Some of these functions could be the regulation of the transfer power level to bringing dynamic variations in the operating frequency so that the efficiency of coupling could be optimized. The transmitter is generally placed in a fixed place inside the room from where appropriate charging of the sensor or implant can be done. In Figure 14.2 one could see a perfect WPT charging unit, which is employed in most applications. Talking about the components that fit inside it, there are a number of components like power antenna, DC to DC converter, DC to AC converter, tuning circuit, power management circuit etc. The transmitter usually takes up a DC source

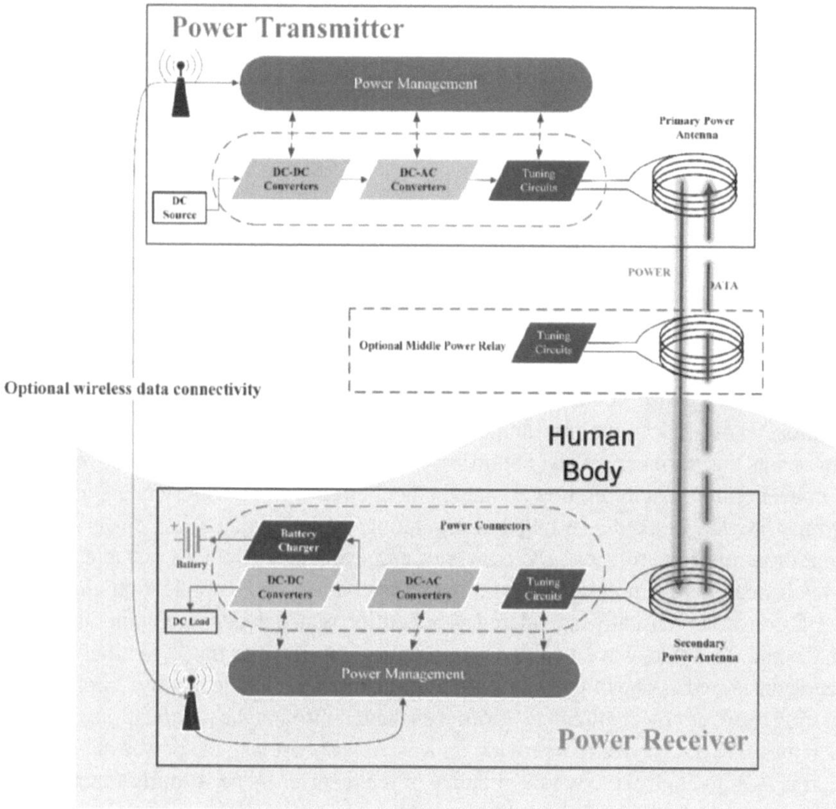

FIGURE 14.2 Schematic of a wireless power transfer charging system.

through a set of batteries or power adapter, where the DC-DC converter converts this DC voltage to another required level that is often needed by the power management circuitry and DC-AC converters for further operation. Amongst the following, the DC-AC converter is usually considered as the key of the WPT transmitter unit. Conversion of DC to AC at a specific operating frequency is done by DC-AC converter. Due to this reason, it is often regarded as a power inverter (Shinohara et al. 2006, Intel 2008). Apart from this, on the other hand, the tuning circuit also plays a vibrant role in the whole functioning of the transmitter. It is responsible to resonate the tiny power antenna present inside the transmitter to a specific desired frequency. The power management circuit on the other hand is responsible for performing a variety of functions that include adjustment in the transferred power level and operating frequency. For this purpose, it is often connected to the receiver on the other end through a wireless link. This connectivity is generally needed to send/receive in commands to/from the receiver for necessary operations or functions like the feedback power information. In a situation where power delivery by the transmitter is not enough, we usually employ an option middle power relay that helps in the delivery of wireless power to farer places by resonating at a particular operation frequency. This type of additional device is usually realized through helical antenna and LC resonators that can be worn on the body of the patient for effective power transfer.

14.6.2 WPT Receiver

The primary fundamental functions of a typical receiver in a WPT charging unit is to provide stable energy into the battery or load to activate further necessary operations linked with it. Apart from this, there are some other functions as well, like smart dynamic adjustment power consumption of load so as to keep up with the power that is recovered on a continuous basis. Also delivery of feedback power to the transmitter for proper regulation in transfer is another vital function that a typical receiver performs. Due to the micro size, the receiver generally comes with the flexibility to get attached in a number of places. These places mainly include the human body or mended on the skin. The components that receiver consist comprises of a power antenna, DC-DC converter, AC-DC converter, tuning circuit, power management circuit, and a battery charger. Starting from the DC-DC converter, it helps in regulating the unregulated DC power from the output of rectifier or from AC-DC converter and eventually provides a stable power to the load. Load here is nothing but a rechargeable battery or an application circuit. The AC-DC converter is usually considered as a rectifier, which depending upon the requirement, could be a half wave to full wave rectifier, or a switch mode rectifier etc. It is indeed considered as one of the most important parts of the whole receiver circuitry. The operation receiver power efficiency is often evaluated through the rectification efficiency, which these AC-DC converters provide on a usual basis. At last, the power management circuit is used to control the whole circuitry of the receiver by performing functions like monitoring power level that has been recovered along with sending or updating the transmitter with the power information through a wireless link.

14.6.3 WPT Antenna

The power antennas that we use in a typical biomedical application plays a significant role in the wireless power transfer charging. A typical WPT unit usually consists of three power antennas within the whole system. One of them is situated at the transmitter end while one of them is on the receiver side. Apart from these two places there is also a third place where such power antennas are often employed. Such places are often called as the optional middle power relay that serves as an agent in linking the transmitter power with the distant receiver within a given biomedical system. The antenna at the transmitter side is responsible for exciting the magnetic field within the space after covering the appropriate electric power delivered from the circuit. While on the other hand, the optional middle power relay is usually employed for enhancing the nearby magnetic flux density for wireless energy transfer to a much farer range. At last, the power antenna at the receiver side is responsible for inducing current and voltages that helps in recovering magnetic power in space back into the electrical power within the circuit. Taking about consideration the antennas, there are multiple factors that largely affects this point. The limitation of size, the quality factor, medium loss constitutes some of these things. That is perhaps the main reason behind the sophistication, which is involved in designing a power antenna for the whole three components that we have described here. The power antenna that we have described here generally consists of loop coils. In today's world most of the biomedical systems that are meant to perform short range applications usually employ these types of power antennas. But with the recent technological advancement in the current structure, a number of improvements in various antenna's structure and techniques have been proposed, which ultimately helps in improving the power coupling efficiency. Depending upon the type of applications, these antennae are often refined to reduce their size, realize their full directionality, and enhance the coupling efficiency to much greater levels. The total power transfer efficiency is calculated by coupling efficiency in space and power conversion efficiency in circuits. Basically the coupling efficiency in the space is much lower than the power conversion efficiency in the circuit. In addition of efficiency, there are other considerations for the power antennas such as power level, size limitation, quality factor, medium loss, and so on. Consequently, the design of power antennas is quite a comprehensive task. Typically, the power antennas are loop coils. Tuning capacitor would be adopted to form a LC resonant circuit. Presently, many biomedical microsystems are employing these power antennas for its decent coupling performance over a short distance (Chen et al. 2009, Tesla et al. 2003, Tesla et al. 1892). To further improve the coupling performance, various kinds of new antenna techniques and transfer structures were proposed in recent years. Some of these antennas accentuate to enhance the power efficiency (Wong et al. 2004, Anderson and Tesla 2002), while some emphasize for the reduction of antenna size (Naganuma et al. 2011), realize full directionality (Landt et al. 2005), and so on.

14.6.4 WPT Converters

Power converters are considered as a vital unit of components that has always been a crucial need at both transmitter and receiver end. The transmitter usually includes a DC-DC and DC-AC converter while the receiver includes AC-DC and DC-DC converter to carry out their related operational tasks. Apart from this, the circuit at the receiver end, which is meant to recharge the battery are also considered as a power converter. These power converters often need auxiliary circuits to support a number of things like reference voltage, oscillator, as well as a bias circuit and so on. Researchers around the world are trying to design a power converter that can provide a high conversion efficiency that can further escalate to perform various functions in the suitable units more effectively. Due to large drop out voltages associated with the conventional power converters, state of the art highly efficient rectifier designs has been proposed for wireless power transfers, some of which are comparator based, ZCP prediction based and so on.

14.6.5 WPT Power Management

For a general simple biomedical application, power management circuits are often not employed. But in the case of sophisticated complex applications, the vital picture of a power management circuit comes into place. The PMC that serves as an essential part of the biomedical systems, is responsible for carrying out a number of functions and are often considered as a central controller to the receiver or the transmitter. A typical PMC connects itself to all power converters within the system, including determination of power level by the transmitter, operational frequency, and adjustment in the power consumption of the battery or load. It also often works to monitor power in the receiver. Apart from this, it is also used to trigger an alarm whenever a current, temperature, or voltage gets out of control to its permissible values.

14.7 CHALLENGES IN SYSTEM DESIGN

As mentioned earlier, wireless power transfer is a budding method to strengthen up microsystems used in biomedical applications. Yet, the transfer is still hindered by some obstacles, which include the unsatisfied power efficiency, the limited transfer distance, the unpredictable reliability, the thermal issues, and various other issues. Consequently, this section inspects in-depth the challenges of designing the wireless power transfer in the field of biomedical microsystems. The overall systematic design challenges are discussed first. Thereafter transmitter and receiver end design challenges are discussed, respectively. There are mainly five system design challenges that arise in the proposed context, which mainly includes the required power levels, transmitter, and receiver size, the transfer medium and distance, and the lateral and angle misalignments.

Challenge 1: Size of Transmitter and Receiver

In a number of IoT enabled wireless charging healthcare applications, the two fundamental power transmitter and receiver components of a typical biomedical implant or sensor are usually placed inside and outside the body or patched of human skin. Due to this, the receiver generally tends to be designed in a much smaller size in comparison to the transmitter, which is obviously present outside the body of the patient. This can be visualized through the hearing aid that usually comes in the size of millimeters along with pacemakers, which have sizes varying between a few centimeters. In the current time, one of the developing trends in the design of technologies for biomedical applications is the miniaturization. But with scaling down the size of these components, there usually are a number of problems. Like for example, with the reduction in the size of the transmitter, the quality factor of the antenna would tend to degrade, which ultimately will decrease the efficiency of coupling in the transmitter and receiver and therefore will force to bring an increment in the energy loss in the tissue due to an increase operational frequency requirement.

Challenge 2: Transfer Distance

Another of the most important features that comes into existence after transmitter and receiver size of the WPT based IoT (Tomar et al. 2019) enable healthcare systems is the transfer distance. It can be evident from the fact that the ratio of the transfer distance and size of transmitter and receiver stood to be a vital key parameter in the concept of system design. Usually when we restrict the size of the antenna, the transfer distance often becomes much larger leading to poor coupling efficiency. That is the main reason why transfer power efficiently becomes a great challenge in the biomedical system design. This could be seen through an example of endoscopic capsule, with a diameter of somewhere between 1 cm, whose needed range of transfer distance lies between 1 to 10 cm. But due to related system challenges as described above, the final output power efficiency comes out to be merely only 0.3% in the worst case.

Challenge 3: Transfer Medium

Another factor that affects the design of biomedical system for efficient wireless power transfer is the transfer medium. A typical biomedical application usually involves three mediums in general, namely human tissue, air, and metals parts of the system and surrounding. Amongst them, human tissue absorbs a great amount of electromagnetic energy in comparison to air. This energy usually results in heating, which often proved to be fatal in most of the cases. Thus, in order to provide safety to an individual, these biomedical implants and sensors are protected through metal shells that help in reducing eddy current effects they may appear inside the system. In this way, most of the accumulated heat energy is wasted in these types of metal shells.

Challenge 4: Angle and Lateral Misalignments

Proper alignment between the two fundamental power units of a biomedical system is considered a vital aspect during the use wireless power transfer. Two types of misalignments, namely angle and lateral often result in the degradation of a certain level of coupling factor amongst the power transmitter and receiver of the biomedical system. This usually results in the decrement of transfer efficiency, which ultimately results in poor system performance. Thus, it very crucial to estimate the appropriate misalignment factor that can cause minimum loss to the system. Special antenna and circuit designs should be adopted for various miniaturized sensors and implants, which usually rotates or is displaced from their position more often. One such example of such a device is the endoscopic capsule, which often rotates in the digestive track of the patient, thus leading to improper misalignment.

Challenge 5: Power Level Requirement

At last, the most important system design challenge of the whole biomedical system is the supply of required power level to the power receiver from the transmitter. This could be visualized through the power requirement of low power implantable microsystems like hearing aid, nerve stimulator, signal recorder that often results in the decrement in the power efficiency associated with them when the power level decreases. While on the other hand, high energy demand systems like artificial hearts often result in thermal issue and heat problems. Therefore, it is very much important to design these biomedical systems by taking into account these types of challenges.

14.7.1 Transmitter End Challenges

In order to build an effective wireless power transfer or charging system in biomedical applications, it is often essential to recognize all the vital challenges that may arise at both the transmitter and receiver end. Here, in this section we firstly carry out an investigation on the design challenges that usually arise at the transmitter side. Though different systems may have different design challenges, but in this we have tried to bring into prospective some five common challenges that every transmitter does face in their entire lifetime.

Challenge 1: Transfer Power Requirement

One of the main challenges, which a power transmitter of a biomedical system usually face, is their ability to satisfy the power requirement of the receiver. It could be explained through an example of endoscopic capsule, which generally need a power of around 10 to 30 Mw to operate upon, thus in order to operate effectively it needs a peak power of as a minimum 10 W. In the worst case, a transmission efficiency of 0.3% is achieved by this transmitter. In most of the cases, the transmission efficiency does not get to the appropriate energy levels, required by the receiver, resulting in failure of associated operations. Therefore, due to this reason it becomes quite necessary to determine the apt peak power to ensure proper working of the whole system along with addressing of associated thermal and safety issues that may arise due to inappropriate energy levels.

Challenge 2: Adjustment in the Power Dynamic Range

Another very vital aspect in the transmitter design is the power dynamic range adjustment. In a typical endoscopic capsule, which we already have discussed later in this chapter has power efficiency that ranges from 5.2% to 0.3% in the minimum achieved case. Thus, with the aim of minimizing harmful electromagnetic radiation effects on the patients accompanied with power wastage, most biomedical often dynamically make adjustment in their transmitter power to cope with the varying power consumption of the receiver along with coupling efficiency. Therefore, a transmitter in such systems usually needs to accommodate or regulate the power level in a certain dynamic range.

Challenge 3: Peak Conversion Efficiency

In a typical WPT transmitter, the DC-AC converters often play a dominant role in deciding the power conversion efficiency of the whole system. These converters are typically belonging to class E or class D type power amplifiers where the later one is used more often than the former due to its ability to remain insensitive to the change of power at the output or operational frequency. Apart from the DC-AC unit, operating frequency also plays an important role in disrupting the power transfer efficiency to a certain extent. It could be indicated by the fact that whenever the operational frequency of conversion is increased, there is a significant decrement in the power transfer efficiency of the system. Therefore, it became quite useful for designers to maintain the tradeoff between the operational frequency and transfer efficiency to operate effectively without any type of disruption. However, there are certain other parameters that lead to the selection of a particular operational frequency. It usually includes the power transfer efficiency, absorption rate, antenna size etc. at the receiver end. Dependency on all these parameters often leads to much more complications in the selection of the operational frequency. Final selection of operational frequency gives the system designer the flexibility to improve the transfer efficiency by bringing out new techniques in the circuitry.

Challenges 4: Weight and Size of Transmitter

The size and weight constraints on the power transmitter are usually not considered as a peculiar requirement for an effective WPT charging system. However, for some applications it does have some requirements. It could be seen from an example where the endoscopy capsule, which is used to diagnose ailments inside the patient's stomach, have their power transmitter entrapped in long wired cables, which often makes the movement of the patient inconvenient. Thus, in order to avoid chaos from such long-wired cables, transmitters running on portable batteries must be employed. This will not only reduce the sophistication of the whole system but will also provide convenience to the patient for a longer period of time.

Challenges 5: Better Safety

In order to ensure better safety across the whole biomedical system, two safety issues must be properly addressed at the transmitter side. Amongst both of them, the first one is the thermal problem, which arises due to high power supply at the transmitter end, leading to the accumulation of unnecessary heating. This heat is generated mainly due to the resistance of conductor in the transmitting antenna and power transistor in the DC-AC converter, which is present inside the transmitter system. Apart from this, operational frequency is also adverse to the thermal problem to a much higher level. To deal with it, heat sinks and fans are employed, which swept away the heat inside the transmitter unit. Another safety issue that lies inside the transmitter end is the electromagnetic radiation, which not only interferes with the working of an implant like a pacemaker but also often leads to various health hazards. Prolonged exposure to electromagnetic radiation above 100 kHz often leads to significant absorption of energy within the human body and can result in the onset of many health-related problems. Though there are no particular standards dedicated to the safety against the EM radiations caused by WPT for medical purposes, but there are still two standards that are put into practice to safeguard the hazardous impacts of these devices on humans. Amongst the two which we have just mentioned, one is the ICNIRP guidelines while the other one is the IEEE C95.1 .To cope with both these problems, there must be reduction in the operational frequency value so that there would be less heat dissipation along with little danger from the radiation as well. While dealing with the following constraints there is a very imp question in what way to design a wireless power transmitter unit that consumes little heat, weight and radiation and high-power transfer efficiency, dynamic power range and peak power all enclosed in a small size.

14.7.2 Receiver End Challenges

Now taking the receiver side into perspective. There are a number of common challenges that a typical receiver in a biomedical system face. All of them are described briefly in the points below.

Challenge 1: Demand of Power

One of the major challenges of the receiver, the power demand, usually differs in the type of applications it has been employed for. A typical biomedical microsystem like endoscopy capsule, nerve stimulators often consume a power in between 10–30 mW, whereas on the other hand big systems like artificial hearts absorb a lot more power that usually ranges up to 15 W. Thus, meeting these demands for a particular type of biomedical application has always been the topmost concern in WPT receiver design.

Challenge 2: Supply Voltage Requirement

The load at the receiver end is nothing but the application circuits and rechargeable battery. They always have a precise requirement of the supply power voltage associated with them. Like for example, whenever the load is a digital circuit, it always requires a voltage of like 1.8 or 3.3 V to perform further associated tasks. Since the AC -DC converters provide unregulated DC voltage, DC-DC converters are generally employed to generate regulated voltages on which the load can perform various associated functions. However, despite all these workout methods, there still exists a problem with the rectified DC voltage from the DC-DC converter. This problem often occurred due to the variation in relative orientation and transmission distance between the transmitter and receiver. Due to this reason, the output of the rectifier often seems to get higher or lower than the required load. Therefore, there is an urgent need to bring special circuits that can resolve this problem.

Challenge 3: Power Transfer Efficiency

Amongst all the challenge that we have discussed with regard to receiver, power transfer efficiency is considered as the most vital point in the designing of the wireless power transfer receiver unit. As per the conventional receiver designs, the power transfer efficiency often reduces due to mainly two reasons. One of the reasons behind this is the decrement in the transfer power while the other reason is the increase operational frequency. Therefore, it is a vast challenge indeed to design a receiver with low power and high transfer power efficiency, which works on a high operational frequency. This could be shown through a work where power transfer efficiency of the rectifier with output power 140 μW and 900 MHz operational frequency comes out to be no more than 65%. In contrast to this high-power transfer efficiency of 93% were achieved when the receiver operates at a much lower operational frequency. Due to this, the system designer varies the operating frequency as the power level is already determined by load.

Challenge 4: Reliability of Power

In comparison of a wired cable, the WPT charging power system is always considered unstable in its operation. A few main reasons behind this are due to the sudden changes in the separate distance and relative orientation between the receiver and transmitter. Due to this, the power reliability of the WPT charging unit often becomes unpredictable. Therefore, it is very much necessary to take into consideration power management practices that can help us to clarify all unstable factors that may arise during the functioning of the given system. A few things that bring instability more often are the sudden change in the orientation of the receiver, error operation triggering etc.

Challenge 5: Size of the Receiver

At last, one of the major problems that we need to take into consideration is the micro size of the implantable device that we deploy for a particular set of applications. Usually, the size of receiver mainly depends upon the size of antenna and circuits that fit within the case of the receiver. The size of these antennas can be decreased significantly by incrementing operating frequency, which ultimately result in the reduction of power conversion efficiency. Now, taking circuit size into perspective. The circuitry size of a typical receiver is based on the number of passive components employed. To be precise, regulators, rectifiers often require decoupling capacitors at the output and input terminals, which often result in the bulkiness in the overall structure. Also, DC-DC converters that are employed to scale down or up a particular set of voltage needs off chip inductors for proper functioning. Thus, it could seem from these examples that the size of these passive components is overlarge to be incorporated into the internal microchip integrated circuits, resulting in large circuitry and the inability to be used for different sets of biomedical applications. Therefore, scaling down the overall size of the receiver has always been a challenge that needs to be tackled down in the years to come. The main challenge that stood in front of us is how to build a WPT power receiver that can offer high reliability, high power conversion efficiency, and wide input voltage with a small size in low power high frequency condition.

14.8 WIRELESS CHARGING STANDARDS

There are various proposed wireless charging standards, which includes Qi and A4WP.

14.8.1 Qi Standard

Charging method: Inductive-resonant coupling
Charging delivery method: Charging pad
Distance of maximum charging: 40 mm

Support for multiple devices: Depends upon implementation
Requirement of contact: Yes
Orientation specification: Depends upon implementation
Qi, pronounced as "Chee", was known to the world since 2008 as a renowned wireless charging standard. From the time of its existence, the standard has been widely used by a number of manufacturers around the world, including HTC, Huawei, Samsung, LG, and Blackberry with diverse variety of pads for charging. The Qi standards generally comes out in two configurations. The first one, which is called the orientation-based platforms employ a single inductive transceiver. While on the other hand, orientation less configuration uses inductive transceiver bundles for its implementation. According to a recent report, Qi intends to take off remote charging stations, called "Qi hotspots", in coffeehouses, eateries, air terminals, and libraries, with the goal that remote charging is accessible in a similar way that Wi-Fi is. The main air terminals to execute Qi hotspots are relied upon to be London Heathrow Airport and New York City's John F. Kennedy International Airport. Qi additionally is creating post-retail remote charging items for auto motives, getting ready for the move to electric vehicles later on.

14.8.2 AirFuel Alliance Standard

Charging method: Inductive, resonant, RF
Charging delivery: Contact and contactless
Maximum charging distance: 4–5 m
Multiple devices' support: Yes
Orientation-dependent: No
AirFuel Alliance came into existence by the merger of the Alliance for Wireless Power (A4WP) and Power Matter Alliance (PMA).With the merging, the AirFuel Alliance standards comes with the integration of capabilities of both PMA and A4WP. The PMA, which is identical to the Qi wireless charging standard uses inductive charging for its short range application type, while on the other hand the A4WP uses a resonant type charging method for different types of applications. This feature in turn provides a low distance slow charging capability to the AirFuel Alliance standard. However, the given standards are working on its way to integrate RF charging, which will enable it to cope with the problem of charging range and transfer efficiency. The AirFuel Alliance is currently used by renowned big consumer electronics tech giants like Samsung, Dell, LG, Motorola etc.

14.8.3 Safety against Electromagnetic Waves

Wireless transfer of power suffers from completely different safety concerns as compared to the conventional cable-based power transfer systems. One major issue is the exposure to electromagnetic waves. There is no significant energy absorption or rise in temperature after exposing the power transferring channel with low frequency electromagnetic radiations, however as soon as the frequency crosses the threshold of 100 kHz, it can cause drastic energy absorption leading to dangerous temperature levels. As a perquisite to the same, analysis and resolution of electromagnetic radiation safety concerns is extremely crucial. In this section, two major international electromagnetic safety standards are discussed followed by numerous clinical examples referring to the concerned system designing parameters.

14.8.4 Safety Standards for WPT

Since there is not any type of specific standards dedicated to wireless power transfer in biomedical systems, one can only adopt two international electromagnetic standards that are devoted to public safety. Amongst them, the first one is the IEEE standards for safety levels of human exposure to electromagnetic fields while the other one describes the ICNIRP guidelines for limiting exposure to time varying magnetic, electric, and electromagnetic fields. Most of the applications of wireless power transfer charging generally operate at a specific frequency range, starting from 100 kHz to 100 MHz. To support this, the IEEE standard and ICNIRP guidelines usually covers a frequency range from 1 Hz to 300 GHz and 3 kHz to 300 GHz respectively, to cover a myriad set of application in this domain. Thus in order to ensure safety of the patient, both of these standards must comply with the considerations that needs to be take into account to charge low power and high power biomedical systems. Like for example, the allowable maximum allowable electric field is specifically 61 V/m for the conventional environment according to the ICNIRP standard. Similarly the maximum magnetic field permissible value is different for specific applications under both the standards.

The IEEE standard covers the frequency range 1 Hz to 300 GHz. 3 kHz to 300 GHz frequency range is used by the ICNIRP guidelines. IEEE and ICNIRP cover the typical operating frequency range in the wireless power transfer, which is from 0.1 to 100 MHz. The maximum magnetic flux density and the maximum electrical field strength is governed by given permissible exposure. Mostly wireless power transfers work at the operating frequency in range from 100 K to 50 MHz.

14.9 PROSPECTIVE FUTURE WORK

The wireless transmission of power is an emerging technology developing promptly. Seeing the present research and development scenario, there are quite a few essential future considerations, which include integrating energy harvesting within wireless power transmitter, transmitter equipped with quick feedback, improved range of power transmission, and minimizing the size of receivers apt for biomedical implants. Though all these considerations are directed towards different features of the framework, yet power transfer efficiency is always the most important factor.

14.9.1 Energy Harvester Integration with Power Transmitter

Several patients are implanted permanently. While wireless transmission of power aids patients in eliminating periodic battery swapping surgical procedures, the patient is still required to recharge the implant externally through a transmitter outside their body, which would be connected to an ordinary power source via electric cables. As the modern-day biomedical implants are very small in size with considerably miniscule energy and current consumption, we consider wearable transmitters integrated with energy harvesters, a very likely future prospective. Several research groups are working to develop a specially designed wearable piece of clothing (jacket, shirt etc.) consisting of several energy harvesting units along with a wireless transmitter, that could harvest any form of energy (electromagnetic, solar, kinetic, thermal etc.) and transmit it wirelessly to the implanted device, hassle free. Additional batteries can also be included in the clothing to counter the instability of harvested energy.

14.9.2 Rapid Feedback Transmission

The difference in response from receiver to transmitter is one of the major factors differentiating wireless and wired transmission of power. The transmitter in wired systems senses the power demand through the feedback from receiver via connecting cord. The cord facilitates monitoring of the receiver's current and voltage by the transmitter, so that it can deliver adequate current and voltage without compromising stability. Feedback techniques ensure sufficient power supply for the required power demand. Moreover, it enhances the stability, efficiency as well as reliability of the power transfer. In the case of a regular wireless power transmitting framework, there is no provision of feedback from receiver to transmitter. With the real-time variations in power demand, relative angle, transfer distance etc., the receiver needs to gather more energy than it actually consumes, ultimately resulting in degradation of transfer efficiency. Development of a wireless power transfer system quipped with rapid feedback will be a reality in the near future. Though it may require additional antennas, circuits, more energy, and add to the complexity of the entire framework, it would still provide considerable upgradation and improvement in the overall power transfer efficiency.

14.9.3 Transmission Range Improvement

One of the most sought out and also the most complex future prospective of wireless power transmission is to increase the practical range of transmission, with reduced size of antenna size and improved power efficiency. Numerous attempts regarding the same were made in the past century. Nikola Tesla attempted to transmit power over the Atlantic, afterwards microwave power beams were successfully deployed for long distance power transmission. However, these methods are radiative in nature, i.e., they possess very potent directivity. Moreover, implementation of such techniques of power transmission to moving receivers (patients) is not practically feasible. In addition to this, radiative methods have adverse reactions on the human body. A system was successful in transferring nearly half of the power over a distance of 2 m, in the year 2007. The technique utilized is termed as "powerful magnetic resonance". Unfortunately, it holds no practical application in biomedical electronic systems due to the large helical antenna adopted (having diameter ~ 0.5 m). However, we believe that improved systems with scalability to power moving micro biomedical implants can be realized in the future, that could potentially transform the powering of the biomedical devices completely

14.9.4 Harvested Energy for Power Transmission

Although biomedical implants are usually deployed for a certain period of time in a patient's body, in some cases, they might be needed for life. Though the WPT charging technique has already solved the issue of such devices being dependent on wired cables and batteries, the new problem has raised another need for some invention. Since the required energy levels by these implants are in a very low range, harvested energy techniques (Figure 14.3) therefore need to be taken into account.

Recently, a work (Tomar and Kaur 2017) from UC Berkeley had developed a micro thermal energy harvester that can power the low power implants from the energy of the body. Apart from the thermal energy, there are other harvesting energies as well that can harness the vitality of the biomedical system. Some of these other harvesting methods could be vibrational energy through piezo, solar energy, electromagnetic waves energy etc. But still there is so much work that needs to be done in this area for different biomedical devices that require different powers to operate upon.

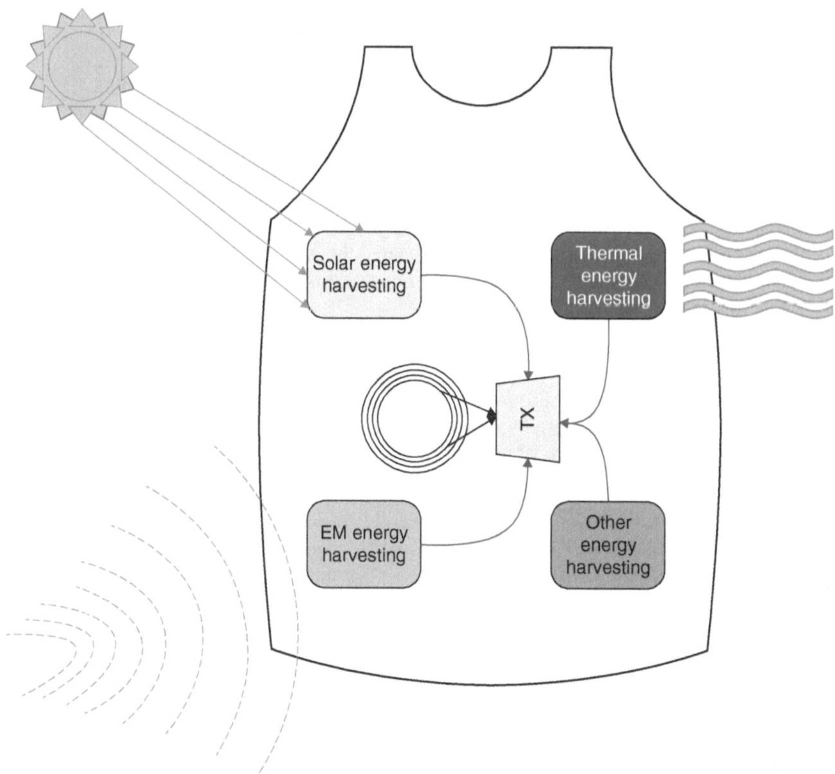

FIGURE 14.3 Integrated energy harvester module for charging.

14.10 CONCLUSION

In this chapter new technology known as the wireless power transfer charging has been discussed. This technology can help in replacing the conventional hazardous methods and provide a promising way to harness the modern IoT based modern biomedical applications with adequate required energy and lesser sophistication and complexity. This chapter consisting of a brief introduction to the technology, its types, applications, system modeling, challenges, safety standards, and prospective future work has been described in a concise and informative manner.

REFERENCES

Anderson, L. I., and Tesla, N. (2002) Nikola Tesla on his work with alternating currents and their application to wireless telegraphy, telephony, and transmission of power: An extended interview: 21st Century Books.

Barrett, J. P. (1894) Electricity at the Columbian exposition, Chicago: R. R. Donnelley, (pp. 168–169).

Brown, W. R. (1961) A survey of the elements of power transmission by microwave beam. IRE, 9(3), 93–105.

Carpi, F., Kastelein, N., Talcott, M., et al. (2011) Magnetically controllable gastrointestinal steering of video capsules. IEEE Transactions on Biomedical Engineering, 58(2), 231–234.

Chen, X., Zhang, X., Zhang, L., et al. (2009) A wireless capsule endoscope system with low power controlling and processing ASIC. IEEE Transactions on Biomedical Circuits and Systems, 3(1), 11–22

Chiu, H.-W., Lin, M.-L., Lin, C.-W., et al. (2010) Pain control on demand based on pulsed radio-frequency stimulation of the Dorsal root Ganglion using a batteryless implantable CMOS SoC. IEEE Transactions on Biomedical Circuits and Systems, 4(6), 350–359.

Finkenzeller, K. (2010) RFID handbook: fundamentals and applications in contactless smart cards, radio frequency identification and near-field communication. New York: Wiley

Ghallab, Y. H., and Badawy, W. (2005) A novel CMOS lab-on-a-chip for biomedical applications, ISCAS (pp. 1346–1349).

Intel, K. (2008) Intel imagines wireless power for your laptop, www.tgdaily.com

Kaur, G. and Tomar, P. (2018) "Handbook of Research on Big Data and the IoT, IGI-Global, International Publisher of Progressive Information Science and Technology Research USA. Total Pages, 568, ISBN13: 9781522574323|ISBN10: 1522574328| DOI: 10.4018/978-1-5225-743, EISBN13: 9781522574330

Kurs, A., Karalis, A., Moffatt, R., et al. (2007) Wireless power transfer via strongly coupled magnetic resonances. Science, 317, 83–86

Landt, J. (2005) The history of RFID. Potentials, IEEE, 24(4), 8–11.

Lee, S.-Y., Su, M. Y., Liang, M.-C., et al. (2011) A programmable implantable microstimulator SoC with wireless telemetry: application in closed-loop endocardial stimulation for cardiac pacemaker. IEEE Transactions on Biomedical Circuits and Systems, 5(6), 511–522.

Lenaerts, B., and Puers, R. (2007) An inductive power link for a wireless endoscopy. Biosensors and Bioelectronics, 22(7), 1390–1395

Naganuma, H., Kiyoyama, K., and Tanaka T. (2011) A 37 9 37 pixels artificial retina chip with edge enhancement function for 3-D stacked fully implantable retinal prosthesis, BioCAS, (pp. 212–215).

Parasher, Y., Singh, P., and Kaur, G. (2019) Green Smart Security System. Green and Smart Technologies for Smart Cities, 165–184.

Schluter, M., Mammitzsch, S., and Lilienhof, H. J. (2005) Detection of a cardiac infarct on a disposable lab on a chip device: 3rd IEEE/EMBS Special Topic Conference on Microtechnology in Medicine and Biology, (pp. 69–71).

Shinohara, N. (2006) Wireless power transmission for solar power satellite. Space Solar Power Workshop, Georgia Institute of Technology.

Shinohara, N. (2006) Wireless power transmission for solar power satellite. Space Solar Power Workshop, Georgia Institute of Technology

Si, P., Hu, A. P., Malpas, S., et al. (2008) A frequency control method for regulating wireless power to implantable devices. IEEE Transactions on Biomedical Circuits and Systems, 2(1), 22–29.

Sun, T., Xie, X. and Li, G., et al. (2011) An omnidirectional wireless power receiving IC with 93.6% efficiency CMOS rectifier and Skipping Booster for implantable bio-microsystems, A-SSCC, (pp. 185–188).

Sun, T., Xie, X., Li, G., et al. (2012) A two-hop wireless power transfer system with an efficiency-enhanced power receiver for motion-free capsule endoscopy inspection. IEEE Transactions on Biomedical Engineering, 59(11), 3247–3254

Tesla, N. (1892) Experiments with alternate currents of high potential and high frequency: IEEE London

Tesla, N. (2003) Nikola Tesla 1856–1943. Belgrade: Nolit

Tesla, N. (2006) On light and other high frequency phenomena. Rockville: Wildside Press

Tesla, N. (2007) Experiments with alternate currents of very high frequency and their application to methods of artificial illumination. New York: Wilder Publications

Thidé, B. (2004) Electromagnetic field theory. Sweden: Upsilon Books.

Tomar, P. and Kaur, G. (2017) Examining Cloud Computing Technologies through the Internet of Things, IGI-Global, International Publisher of Progressive Information Science and Technology Research Books anf Journals, USA p. 311.

Tomar, P. and Kaur, G. (2017). Examining Cloud Computing Technologies through the Internet of Things, IGI-Global, International Publisher of Progressive Information Science and Technology Research Books and Journals, USA, DOI: 10.4018/978-1-5225-3445-7, ISBN13: 9781522534457, Pages 311.

Tomar, P. and Kaur, G. (2018) Green and Smart Technologies for Smart Cities, CRC Press, Taylor & Francis Group, Dec 2019, Boca Raton: CRC Press, Ist Edition, ISBN 9781138318090 – CAT# K391489, https://doi.org/10.1201/9780429454837

Tomar, P. and Kaur, G. (2019) Artificial Intelligence and IoT-Based Technologies for Sustainable Farming and Smart Agriculture, IGI-Global, International Publisher of Progressive Information Science and Technology Research Books and Journals, USA.

Wong, L. S., Hossain, S., Ta, A., et al. (2004) A very low-power CMOS mixed-signal IC for implantable pacemaker applications. IEEE Journal of Solid-State Circuits, 39(12), 2446–2456.

Xie, X., Li, G., Chen, X., Li, X., et al. (2006) A low-power digital IC design inside the wireless endoscopic capsule. IEEE Journal of Solid-State Circuits, 41(11), 2390–2400.

Yagi, H. (1928) Beam transmission of ultra-shortwaves. Proceedings of the IRE, 16, 715–740.

Index

T

U

V

W

X